Pricing Communication Networks

WILEY-INTERSCIENCE SERIES IN SYSTEMS AND OPTIMIZATION

Advisory Editors

Sheldon Ross

Department of Industrial Engineering and Operations Research, University of California, Berkeley, CA 94720, USA

Richard Weber

Statistical Laboratory, Centre for Mathematical Sciences, Cambridge University, Wilberforce Road, Cambridge, CB3 0WB

BATHER–Decision Theory: An Introduction to Dynamic Programming and Sequential Decisions

CHAO/MIYAZAWA/PINEDO–Queueing Networks: Customers, Signals and Product Form Solutions

COURCOUBETIS/WEBER–Pricing Communication Networks: Economics, Technology and Modelling

DEB–Multi-Objective Optimization using Evolutionary Algorithms

GERMAN–Performance Analysis of Communication Systems: Modeling with Non-Markovian Stochastic Petri Nets

KALL/WALLACE–Stochastic Programming

KAMP/HASLER–Recursive Neural Networks for Associative Memory

KIBZUN/KAN–Stochastic Programming Problems with Probability and Quantile Functions

RUSTEM–Algorithms for Nonlinear Programming and Multiple-Objective Decisions

WHITTLE–Optimal Control: Basics and Beyond

WHITTLE–Neural Nets and Chaotic Carriers

The concept of a system as an entity in its own right has emerged with increasing force in the past few decades in, for example, the areas of electrical and control engineering, economics, ecology, urban structures, automation theory, operational research and industry. The more definite concept of a large-scale system is implicit in these applications, but is particularly evident in such fields as the study of communication networks, computer networks, and neural networks. The *Wiley-Interscience Series in Systems and Optimization* has been established to serve the needs and researchers in these rapidly developing fields. It is intended for works concerned with the developments in quantitative systems theory, applications of such theory in areas of interest, or associated methodology.

Pricing Communication Networks

Economics, Technology and Modelling

Costas Courcoubetis
Athens University of Economics and Business, Greece

Richard Weber
University of Cambridge, UK

WILEY

Other Wiley Editorial Offices

John Wiley & Sons Inc., 111 River Street, Hoboken, NJ 07030, USA

Jossey-Bass, 989 Market Street, San Francisco, CA 94103-1741, USA

Wiley-VCH Verlag GmbH, Boschstr. 12, D-69469 Weinheim, Germany

John Wiley & Sons Australia Ltd, 33 Park Road, Milton, Queensland 4064, Australia

John Wiley & Sons (Asia) Pte Ltd, 2 Clementi Loop #02-01, Jin Xing Distripark, Singapore 129809

John Wiley & Sons Canada Ltd, 22 Worcester Road, Etobicoke, Ontario, Canada M9W 1L1

Wiley also publishes its books in a variety of electronic formats. Some content that appears
in print may not be available in electronic books.

Library of Congress Cataloging-in-Publication Data

Courcoubetis, Costas.
   Pricing communication networks : economics, technology, and modelling / Costas
Courcoubetis, Richard Weber.
      p. cm.—(Wiley-Interscience series in systems and optimization)
   Includes bibliographical references and index.
   ISBN 0-470-85130-9 (alk. Paper)
      1. Information technology—Finance. 2. Computer networks—Mathematical models. 3.
Digital communications—Mathematical models. I. Weber, Richard. II. Title. III. Series.

HD30.2 .C68 2003
384′.043—dc21

HD 30.2
.C68
2003   0470 851309
                        2002191081

British Library Cataloguing in Publication Data

A catalogue record for this book is available from the British Library

ISBN 0-470-85130-9

Typeset in 10/12pt Times by Laserwords Private Limited, Chennai, India
Printed and bound in Great Britain by Biddles Ltd, Guildford, Surrey
This book is printed on acid-free paper responsibly manufactured from sustainable forestry
in which at least two trees are planted for each one used for paper production.

We dedicate this book to Dora and Persefoni, the muses of my life (C. Courcoubetis), and to Richard, my father (R. Weber).

Contents

B   Economics   111

C   Pricing   161

Preface

This book is about pricing issues in modern communications networks. Recent technology advances, combined with the deregulation of the communication market and the proliferation of the Internet, have created a new and highly competitive environment for communication service providers. Both technology and economics play a major role in this new environment. As recent events in the marketplace make clear, the success of a communication services business is not guaranteed by new technology alone. An important part of any business plan for selling communications services is pricing and competition issues. These should be taken into account from the start. Traditionally, engineers have devised communication services without reference to how they should be priced. This is because communication services have been provided by large monopolies, with guaranteed incomes. The bundling and pricing aspects of individual services have been secondary. However, services are now sold in competitive markets and an important part of the service definition is how it should be priced. Technology can place severe restrictions on how this can be done. The following are some reasons why the pricing of communications services is now exciting to study:

1. Pricing affects the way services are used, and how resources are consumed. The value that customers obtain from services depends on congestion and on the way services are priced.

2. Communication service contracts provide for substantial flexibility. Pricing plays an important role as an incentive mechanism to control performance and increase stability.

3. Modern networking technology provides new possibilities for producers and the consumers to exchange economic signals on fast time scales. This allows for the creation of new flexible services that customers can control and by which they can better express their needs for quality. This was not possible until a few years ago, since previously services were statically defined and the network operator was in total in control.

4. There is no unique way to price. Issues such as 'flat' versus 'usage-based' charging have important effects on the short and long term network operation and its competitive position. These must be understood by people designing pricing policies.

5. Competition can be greatly influenced by the architecture of a networks and the ability of few players to control bottleneck resources in parts of the network, such as the access. New networks should be designed so that they provide an open competition environment in all parts of the supply chain for services. Competition and regulation issues are important in today's communication market.

6. Communication services are economic goods and must be priced accordingly. There are generic service models that capture aspects such as quality and performance and can be used to derive optimal prices in a services market. They can be used to propose tariffs with the desired incentive properties by pricing the appropriate service contract parameters.

We began this book after five years of research focused in pricing the rich family of ATM services and the newly emerging Internet. We believe there is a need for a book that can explain the provision of new services, the relation of pricing and resource allocation in networks, and the proliferation of the Internet and the debate on how to price it. We have had in mind as readers graduate students and faculty in departments of Electrical Engineering, Computer Science, Economics and Operation Research, telecoms engineers, researchers and engineers who work in research and industrial laboratories, and marketing staff in telecoms companies who need to understand better the technology issues and their relation to pricing. Our experience is that most of these people have only part of the background needed to follow such important subjects. Readers with engineering and OR background usually lack the economics background. Economists usually know little about communications technology and usually underestimate its importance. We have sought to write in a way that all readers will find stimulating. The book should interest anyone with some technology and mathematics background who wishes to understand the close relation of communication networks and economics. Of course, economists may skip the chapters on basic economics.

When we started this book, ATM technology was already declining in importance as an alternative to the Internet. However, there continues to be a practical demand for services such as ATM and Frame Relay. These can be put into the same generic model as the provision of WAN connectivity services. Similar concepts will apply in future extensions of Internet services that provide quality guarantees, such as differentiated services and integrated services. Consequently, we not only deal with the Internet, but also with effective bandwidths and statistical multiplexing.

The scope of this book is broad. It covers most of the concepts that are needed to understand the relation of economics and communications. We do not claim to provide a complete unifying framework, but explain many concepts that are generic to the problem of pricing. This is not a 'how to price' recipe book. Rather, it explores relevant subjects. It provides the basic models and terminology needed for a non-specialist reader to understand subtle topics where technology, information and economics meet. It explains the architecture of the communications market and provides a simple and intuitive introduction to network services at all levels, from the infrastructure to transport. We have tried to make the book technology independent, emphasizing generic service aspects and concepts.

The reader does not have to be an expert in communications or read several books on networking technology numbering hundreds of pages in order to understand these basic concepts. This may be of great benefit to a reader with an economics or operation research background. The same holds for readers with no economics background. We explain relevant microeconomic concepts in enough detail that the reader can follow many issues in network economics, without having to study advanced economic textbooks. However, we are not economists and do not claim to cover all topics in network economics. We hope that we do provide the reader with a useful summary of many key issues and definitions in basic economics. Those who wish to study these ideas in more depth can turn to economics textbooks. For instance, our section on game theory should remind those readers who have previously studied it of those concepts from the subject that we use in other parts of the book. Readers who have not studied game theory before should find that the section provides a readable and concise overview of key concepts, but they will need to look elsewhere for details, proofs and further examples.

There is no one unifying model for network services. We provide models for several services and leave others of them out. These models allow network services to be priced similarly to traditional economic goods. These models can be used by network engineers as a framework to derive prices for complex transport services such as ATM, Frame Relay, IP VPNs, etc. We model the Internet and its transport services and discuss certain issues of fairness and resource allocation based on pricing for congestion. This provides a deeper understanding of the feedback aspects of the Internet technology, and of the recent proposals to provide for a richer set of bandwidth sharing mechanisms. We also provide the theoretical framework to price contracts in which parameters can be dynamically renegotiated by the users and the network. Finally, we give the reader a simple but thorough introduction to some current active research topics, such as pricing multicasting services, incentive issues in interconnection agreements between providers, and the theory of price regulation. For completeness, we also provide a simple introduction to auction mechanisms which are currently used to allocate scarce resources such as spectrum.

We hope to introduce non-specialists to concepts and problems that have only been accessible to specialists. These can provide both a practical guideline for pricing communication services and a stimulation for theoretical research. We do not review in extreme detail the existing literature, although we provide basic pointers. A guide to references appears at the end of each chapter. We seek to unify and simplify the existing state-of-the-art by focusing on the key concepts. We use mathematics to make the ideas rigorous, but we hope without being unnecessary detailed. About 80% of the results in the book have been published elsewhere and 20% are new. The level of the mathematics is at that of first year university student's knowledge of calculus and probability, and should be accessible to students and engineers in the field. Appendix A covers some important ideas of solving constrained optimization problems using Lagrange multipliers. The book has parts which are more technology specific and other parts that are more theoretical. Readers can take their pick.

We have found it convenient to divide the book in four parts. An overview of their contents can be found at the end of Chapter 1. Possible course that could be taught using this book are as follows:

1. An introductory course on pricing: Sections 1.4, 2.1, 3.2–3.3, 4.1–4.5, 4.10, 5.2–5.4.3, 5.4.7, 6.1–6.3, 7.3, 7.5, 8.1–8.4, 9.1–9.4, and Chapter 10.

2. An advanced course on mathematical modelling and pricing: Section 1.4, Chapter 2, Sections 3.1–3.3 and 3.5, Chapter 4, Sections 5.1–5.4, 5.6, 6.1–6.3, Chapters 8, 9 and 10.

3. A course on telecoms policy issues and regulation: Chapter 1, Sections 2.1, 3.2–3.6, Chapters 5 and 6, Sections 7.1–7.1.2, 7.3–7.5, Chapters 12 and 13, Sections 14.1–14.1.3, 14.2 and 14.3.

4. A course on game-theoretic aspects of pricing: Sections 5.1–5.4, 6.1, 6.4, 7.1–7.2, Chapters 9, 10, 11, Sections 12.4–12.5, 13.1, and Chapter 14.

5. An introductory network services and technology course: Sections 1.1–1.2, 2.1 and Chapter 3.

Acknowledgment

There are many people with whom we have enjoyed stimulating discussions while working on this book. These include especially Frank Kelly and Pravin Varaiya, who have done so much to inspire research work on pricing communications. They include also our partners in the Ca$hman and M3i projects, and Panos Antoniadis, Gareth Birdsall, Bob Briscoe, John Crowcroft, Manos Dramitinos, Ioanna Constantiou, Richard Gibbens, Sandra Koen, Robin Mason, Georges Polyzos, Stelios Sartzetakis, Vassilis Siris, Georges Stamoulis and Jean Walrand.

List of Acronyms

ATM	Asynchronous Transfer Mode
ABR	Available Bit Rate
BGP	Border Gate Protocol
BSP	Backbone Service Provider
CAC	Connection Acceptance Control
CBR	Constant Bit Rate
CDVT	Cell Delay Variation Tolerance
CLP	Cell Loss Probability
CPNP	Calling Party Network Pays
CS	Consumer Surplus
DS	Differentiated Services
DWDM	Dense Wavelength Division Multiplexing
ECPR	Efficient Component Pricing Rule
ERP	Enterprise Resource Planning
FCFS	First Come First Serve
FDC	Fully Distributed Cost
IBP	Internet Backbone Provider
IGMP	Internet Group Management Protocol
IETF	Internet Engineering Task Force
ILEC	Incumbent Local Exchange Carrier
IS	Integrated Services
ISDN	Integrated Services Digital Network
ISP	Internet Service Provider
LAN	Local Area Network
LMDS	Local Multipoint Distribution Service
LRIC	Long Run Incremental Cost
MAN	Metropolitan Area Network
MC	Marginal Cost
MPEG	Moving Picture Experts Group
MPLS	MultiProtocol Label Switching
M-ECPR	Market determined Efficient Component Pricing Rule
PCR	Peak Cell Rate
NAP	Network Access Provider
PHB	Per Hop Behaviour
POP	Point of Presence
QoS	Quality of Service
RBOC	Regional Bell Operating Company

RFC	Request for Comments
RSVP	Resource Reservation Protocol
SCR	Sustainable Cell Rate
SDH	Synchronous Digital Hierarchy
SLA	Service Level Agreement
SMG	Statistical Multiplexing Gain
SONET	Synchronous Optical NETwork
SW	Social Welfare
TELRIC	Total Element LRIC
TCP/IP	Transmission Control Protocol/Internet Protocol
TCA	Traffic Conditioning Agreement
UNE	Unbundled Network Element
UBR	Unspecified Bit Rate
UDP	User Datagram Protocol
VBR	Variable Bit Rate
VC	Variable Cost
VC	Virtual Circuit
VPN	Virtual Private Network
WAN	Wide Area Network
WWW	World Wide Web
XSP	Access Service Provider

Part A

Networks

1

Pricing and Communications Networks

This chapter describes current trends in the communications industry. It looks at factors that influence pricing decisions in this industry, and some differing and conflicting approaches to pricing. Section 1.1 is about the market for communications services. Section 1.2 is about present developments in the marketplace. Section 1.3 is about issues that pricing must address. Section 1.4 presents some introductory modelling.

1.1 The market for communications services

1.1.1 The Communications Revolution

We are in the midst of a revolution in communications services. Phenomenal advances in fibre optics and other network technology, enhanced by the flexible and imaginative software glue of the World Wide Web have given network users a technology platform that supports many useful and exciting new services. The usefulness of these services is magnified because of *network externality*. This is the notion that a network's value to its users increases with its size, since each of its users has access to more and more other users and services. This is one of the facts that spurs the drive towards worldwide network connectivity and today's Internet revolution — a revolution which is changing the way we engage in politics, social life and business. It is said that the electronic-economy, based as it is upon communications networks that provide businesses with new ways to access their customers, is destined to be much more than a simple sector of the economy. It will someday be *the economy*.

In a world that is so thoroughly changing because of the impact of communications services, the pricing of these services must play an important role. Of course a price must be charged for something if service providers are to recover their costs and remain in business. But this is only one of the many important reasons for pricing. To understand pricing's other roles we must consider what type of product are communications services and the characteristics of the industry in which they are sold.

1.1.2 Communications Services

The number of connections that can be made between n users of a network is $\frac{1}{2}n(n-1)$. This gives us *Metcalf's Law* (named after the inventor of Ethernet), which says that the value of

Pricing Communication Networks   C. Courcoubetis and R. Weber
© 2003 John Wiley & Sons, Ltd   ISBN 0-470-85130-9 (HB)

a network increases as the square of the number of users. It relates to the idea of network externality and the fact that a larger network has a competitive advantage over a smaller one, because each of the larger network's users can communicate with a greater number of other users. It makes the growth of a large customer base especially important. With this in mind, a network operator must price services attractively. In this respect, communications services are like any economic good and fundamental ideas of the marketplace apply. One of these is that deceasing price increases demand. Indeed, it is common for providers to give away network access and simple versions of network goods for free, so as to stimulate demand for other goods, build their customer base and further magnify network externality effects.

The above remarks apply both to modern networks for data communication services and to the traditional telecommunications networks for voice services, in which the former have their roots. Throughout this book we use the term 'telecommunications' when referring specifically to telephony companies, services, etc., and use the broader and encompassing term 'communications' when referring both to telephony, data and Internet. It is interesting to compare the markets for these networks. For many years the telecommunications market has been supplied by large regulated and protected monopolies, who have provided users with the benefits of economy of scale, provision of universal service, consistency and compatibility of technology, stable service provision and guaranteed availability. Services have developed slowly; demand has been predictable and networks have been relatively easy to dimension. Prices have usually been based upon potential, rather than actual, competition. In comparison, the market for modern communications services is very competitive and is developing quite differently. However, the markets are alike in some ways. We have already mentioned that both types of network are sensitive to network externality effects.

The markets are also alike is that in that network topology restricts the population of customers to whom the operator can sell and network capacity limits the types and quantities of services he can offer. Both topology and capacity must be part of the operator's competitive strategy. It is helpful to think of a communications network as a factory which can produce various combinations of network services, subject to technological constraints on the quantities of these services that can be supported simultaneously. Severe congestion can take place if demand is uncontrolled. A central theme of this book is the role of pricing as a mechanism to regulate access to network resources and restrict congestion to an acceptable level.

Traditional telecoms and modern data communications are also alike in that, once a network of either type is built, the construction cost is largely a *fixed cost*, and the variable operating costs can be extremely small. If there is no congestion, the *marginal cost* of providing a unit of communications service can be almost zero. It is a rule of the marketplace that competition drives prices towards marginal cost. Thus, a danger for the communications industry is that the prices at which it can sell communications services may be driven close to zero.

In summary, we have above made three elementary points about pricing: lowering price increases demand; pricing can be used to control congestion; competition can drive prices to marginal cost.

1.1.3 Information Goods

It is interesting to compare communications services with *information goods*, such as CDs, videos or software. These share with communications services the characteristic of being costly to produce but cheap to reproduce. The first copy of a software product bears all the production cost. It is a sunk cost, mainly of labour. Many further copies can be produced at almost no marginal cost, and if the software can be distributed on the Internet then its potential market is the whole Internet and its distribution cost is practically zero. Similarly, once a network is built, it costs little to provide a network service, at least while there is no congestion. This also shows that information goods and network services can sometimes be viewed as public goods, like highways. Assuming that the installed network capacity is very large (which is nearly true given today's fibre overprovisioning), the same information good or network service can be consumed by an arbitrary number of customers, increasing its value to its users (due to externalities) and the value to society. This is in contrast to traditional goods like oranges and power; a given orange or kilowatt-hour can be consumed by a single customer and there is a cost for producing each such additional unit.

The similarity cannot be pushed too far. We must not forget that a network has a continuing running cost that is additional to the one-time cost of installation. This includes network management operations, amongst which accounting and billing are particularly costly. The cost of selling a single copy of a piece of software is small compared to the cost of maintaining, monitoring and billing a network service. It is not surprising that cost, among many other economic factors, influences the evolution of networking technology. One reason for the acceptance of Internet technology and the Internet Protocol (IP) is that there it is less costly to manage a network that is based on a single unifying technology, than one that uses layers of many different technologies.

There are some lessons to be learned from the fact that information goods can sell at both low and high prices. Consider, for example, the fact that there are hundreds of newspaper web sites, where entertaining or useful information can be read for free. It seems that publishers cannot easily charge readers, because there are many nearly equivalent sites. We say that the product is 'commoditized'. They may find it more profitable to concentrate on differentiating their sites by quality of readership and use this in selling advertising. In contrast, a copy of a specialist software package like AutoCad can sell for thousand of dollars. The difference is that its customer base is committed and would have difficulty changing to a competing product because the learning curve for this type of software is very steep. Similarly, Microsoft Word commands a good price because of a network externality effect: the number of people who can exchange documents in Word increases as the square of the number who use it. These examples demonstrate another important rule of the marketplace: if a good is not a commodity, and especially if it has committed customers, then it can sell at a price that reflects its value to customers rather than its production cost.

We have noted that both traditional telecoms and modern communications services are sensitive to network topology and congestion. This is not so for an information good. The performance of a piece of software running on a personal computer is not decreased simply because it is installed on other computers; indeed, as the example of Microsoft Word shows, there may be added value if many computers install the same software.

1.1.4 Special Features of the Communications Market

One special feature of the market for communications services, that has no analogy in the market for information goods (and only a little in the market for telecommunications), is that in their most basic form all data transport services are simply means of transporting data bits at a given quality level. That quality level can be expressed such terms as the probability of faithful transmission, delay and jitter. A user can buy a service that the operator intended for one purpose and then use it for another purpose, provided the quality level is adequate. Or a user can buy a service, create from it two services, and thereby pay less than he would if he purchased them separately. We say more about the impact of such *substitutability*, *arbitrage* and *splitting* upon the relative pricing of services in Section 8.3.5.

Another thing that makes communication transport services special is their reliance on *statistical multiplexing*. This allows an operator to take advantage of the fact that data traffic is often bursty and sporadic, and so that he can indulge in some amount of overbooking. He need not reserve for each customer a bandwidth equal to that customer's maximum sending rate. Statistical multiplexing produces economy of scale effects: the larger the size of the network, the more overbooking that can take place, and thus the size of the customer base that can be supported increases more than proportionally to the raw quantity of network resources. It is intuitive that a network service that is easier to multiplex should incur a lesser charge than one which is more difficult to multiplex. There are many multiplexing technologies and each is optimized for a particular type of data traffic. For instance, SONET (Synchronous Optical NETwork) is a multiplexing technology that is optimized for voice traffic (which is predictable and smooth), whereas the Internet technology is optimized for data traffic (which is stochastic and bursty).

Simple economic goods are often specified by a single parameter, such as number of copies, weight, or length of a lease. In contrast, contracts for data communications services are specified by many parameters, such as peak rate, maximum throughput and information loss rate. Contracts for services that support multimedia applications are specified by additional parameters, such as ability to sustain bursty activity, and ability and responsibility to react to changing network conditions. Since service contracts can be specified in terms of so many parameters, their potential number is huge. This complicates pricing. How are we to price services in a consistent and economically rational way? Moreover, contracts are more than simple pricing agreements. For example, a contract might give a user the incentive to smooth his traffic. Customers also benefit because the quality of the service can be better and lower priced. This poses questions of how we can reasonably quantify a customer's network usage and price contracts in a way that makes pricing a mechanism for controlling usage.

1.2   Developments in the marketplace

In the next two sections, we look at some important factors that affect the present market for communications services. We make some further arguments in favour of the importance of pricing. We describe the context in which pricing decisions occur, their complexity and consequences. Some of these issues are subject to debate, and will make most sense to readers who are familiar with present trends in the Internet. Some readers may wish to skip the present section on first reading.

There have been two major developments in the marketplace for telecoms services: the development of cost-effective optical network technologies, allowing many light beams to be packed in a single fibre; and the widespread acceptance of the Internet protocols as the common technology for transporting any kind of digitized information. Simultaneously, the Internet bubble of late 1990s has seen an overestimation of future demand for bandwidth and overinvestment in fibre infrastructure. Together, these factors have created a new technology of such very low cost that it threatens to disrupt completely the market of the traditional telephone network operators, whose transport technologies are optimized for voice rather than data. It has also commoditized the market for transport services to such an extent that companies in that business may not be able to recover costs and effectively compete.

One reason for this is that the Internet is a 'stupid' network, which is optimized for the simple task of moving bits at a single quality level, irrespective of the application or service that generates them. This makes the network simple and cheap. Indeed, the Internet is optimized to be as efficient as possible and to obey the 'end-to-end principle'. To understand this principle, consider the function: 'recovery from information loss'. This means something different for file transfer and Internet radio. The end-to-end principle says that if such a function is invoked rarely, and is not common to all data traffic, then it is better to install it at the edge of the network, rather than in each link of the network separately. Complexity and service differentiation is pushed to the edges of the network. The reduction in redundancy results in a simpler network core. Customer devices at the edges of the network must provide whatever extra functionality is needed to support the quality requirements of a given application.

The fact that the Internet is stupid is one of the major reasons for its success. However, it also means that a provider of Internet backbone services (the 'long-haul' part of the network, national and international) is in a weak bargaining position if he tries to claim any substantial share of what a customer is prepared to pay for an end-to-end transport service, of which the long-haul service is only a part. That service has been commoditized, and so in a competitive market will be offered at near cost. However, as noted previously, the cost of building the network is a sunk cost. There is only a very small variable cost to offering services over an existing network infrastructure. The market prices for network services will be almost zero, thus making it very difficult for the companies that have invested in the new technologies to recover their investments and pay their debts. As some have said, *the best network is the hardest one to make money running* (Isenberg and Weinberger, 2001).

This 'paradox of the best network' does not surprise economists. As we have already noted, there is little profit to be made in selling a commodity. The telephone network is quite different. Customers use only simple edge devices (telephones). All value-added services are provided by the network. Network services are constructed within the network, rather than at the edges, and so operators can make money by being in control. Similarly, video and television distribution use service-specific networks and make good profits. Telephone networks are optimized for voice and not for data. Voice streams are predictable in their rates, while data is inherently bursty. Due to the overspecified requirements (for reliability and voice quality), the technologies for voice networks (SONET and SDH) are an order of magnitude more expensive than the technology for providing simple bit moving services of comparable bandwidth, as provided by the Internet using the new optical transmission technologies. The extra quality per bit offered by telephone network infrastructures does not justify their substantially greater costs. Moreover, the large network capacity available may let the quality of the bits provided by the new Internet technology networks approach that provided by the telephone network. Unfortunately, these voice-centred technologies are not so old as to be easily written-off. Existing operators invested heavily in them during the late 1980s and mid 1990s, encouraged by regulators who allowed them a 'return on assets', that is, a profit proportional to the assets under their control. This makes it hard for operators to abandon their voice-centred infrastructures and build new networks from scratch.

The above arguments suggest that network operators deploying the new Internet over fibre technologies should be able to carry voice at substantially less cost than traditional network operators, and so drive them out of business. They will also be able to offer a rich set of high bandwidth data services, which are again cheaper for them to provide.

However, things are not entirely rosy for these new network operators. They have their own problem: namely, a bandwidth glut. During the Internet bubble of the late

1990s investors overestimated the growth in the demand for data services. They believed there would be an unlimited demand for bandwidth. Many companies invested heavily in building new fibre infrastructures, at both the metropolitan and backbone level. DWDM (Dense Wavelength Division Multiplexing) made it possible to transport and sell up to 80 multiple light waves (using present technology) on a single strand of fibre. Gigabit Ethernet technologies combined with the Internet protocols allowed connectivity services to be provided very inexpensively over these fibre infrastructures. Using present technologies each light wave can carry up to 10 Gbps of information, so that a single fibre can carry 800 Gbps. Although DWDM is presently uneconomic in the metropolitan area, it makes sense in the long-haul part of the network. It has been estimated that there are now over a million route-miles of fibre installed worldwide, of which only about 5% is lit, and that to only about 8% of the capacity of the attached DWDM equipment. Thus there is potential for vastly more bandwidth than is needed. Some experts believe that fibre is overprovisioned by a factor of ten in the long-haul part of the networks. Further bad news is that demand for data traffic appears to be increasing by only 50% per year, rather than doubling as some had expected.

The result is that the long-haul bandwidth market has become a commodity market, in which demand is an order of magnitude less than expected. A possible reason is miscalculation of the importance of complementary services. High-capacity backbones have been built without thinking of how such 'bandwidth freeways' will be filled. The business plans of the operators did not include the 'bandwidth ramps' needed, i.e. the high-bandwidth access part that connects customers to the networks. The absence of such low priced high-bandwidth network access services kept backbone traffic from growing as predicted. Besides that, transport services have improved to such an extent that technology innovation is no longer enough of a differentiating factor to provide competitive advantage. Prices for bandwidth are so low that it is now very hard for new network operators to be profitable, to repay the money borrowed for installing the expensive fibre infrastructure, or to buy expensive spectrum licenses.

Existing operators of voice-optimized networks are also affected. Their income from highly priced voice calls has reduced, as voice customers have migrated to the Internet technology of voice-over-IP networks, while the demand for voice remains essentially constant. They have not seen a compensating increase in demand for data services, which in any case are priced extremely low because of competition in that commoditized market. Some local service providers are even selling data services at below cost because of their expensive legacy network technology, while simultaneously installing the new IP over fibre technology in parts of their networks to reduce their costs. Of course infrastructure is not the only cost of providing traditional access and voice services. A larger part of the cost is for orders, repairs, customer service and support. This cost will always be reflected in customers' bills. Thus local operators, who have traditionally been in a monopoly position, do live in a somewhat protected environment because they have a steady income from their large and loyal base of telephone customers. Competition is fiercest in the long-haul part of the network, where new technologies can be easily deployed, economies of scale are great, and many operators compete.

It may seem paradoxical to have such severe sustainability problems in a growth industry such as telecommunications. Although the pie is growing, the business models seem to have some serious flaws. This is due to miscalculations, and because companies have tried to become simultaneously both retail and wholesale service providers, with the result that they have been competing with their own customers. Some experts envisage extreme scenarios. In one such scenario, the regulator acquires and controls the complete fibre infrastructure in the US, and leaves telecoms operators to compete in providing 'edge' services, which are better differentiated by innovation and service customization, and hence more profitable. Others believe that the industry will self-regulate. Cash-rich companies will buy the ailing telecoms companies at low prices and enter the telecom market. As profit margins are small, companies offering infrastructure and connectivity services will consolidate so as to gain economies of scale. This suggests that horizontal integration may be more sensible than vertical integration. Other telecoms companies may benefit from increased complexity at the edges of the 'stupid' network, and manage this complexity on behalf of their customers. This outsourcing of the management of the communication assets of large companies may be a substantial source of income and a new business model in the telecom industry. In this new service-centred industry, network (service) management software will play an increasingly important role. However, we should caution that it is very hard to predict the evolution of a complex industry such as telecommunications. Predictions are very sensitive to time assumptions: no one knows how long it will take for new technologies to dethrone old ones. Well-established services do not disappear overnight, even if less expensive substitutes are available. Brand name plays an important role, as do factors such as global presence, and the ability to provide one-stop shopping for bundles of services.

1.3   The role of economics

We believe that economics has much to teach networking engineers about the design of networks. First, it has much to say about decentralized control mechanisms. Secondly, we feel that the design and management of networks should adopt a 'holistic' view. We consider these two points in turn.

First, let us note that economics is traditionally used to study national economies. These can be viewed as large decentralized systems, which are almost completely governed by incentives, rather than by strict hardwired rules. On a smaller scale, economic incentives also manage the flow of vehicle traffic in a congested part of town during rush hours. Each driver estimates the repercussions of his actions and so chooses them in a way that he expects to be best for his self-interest.

Things are similar in a large network, such as the Internet, in the sense that central control tends to be relaxed and many decisions must be taken at the edges of the network, both by users, and by providers who have different profiles and incentives. This similarity makes economics very relevant. Just as economic theory explains what can be achieved in the national economy by the incentives of wages, taxes and prices, so economic theory is useful in explaining how distributed control mechanisms, based on incentives such as price and congestion level can be used to ensure that a complex system like the Internet will perform adequately. As in a national economy, agents are to take decisions at points where the information required to take them is actually available, rather than on the basis of some central 'full information' about the system state (which would be impossible to obtain in practice). Theorems of economics can guarantee that such distributed control dynamically moves the system to an equilibrium point where resources are used efficiently, and performance is the same as if the solution had been obtained using full information.

Now we turn to the second reason that economics is relevant to networks. Engineers are used to designing mechanisms that achieve optimum system performance. This 'performance' is usually measured in terms of packet delay, call blocking, and so on. We suggest that it is better to think in terms of 'economic performance', which includes the above measures, but also wider-ranging measures, such as flexibility in the use of the network, and the ability to adapt and customize the service to the particular needs of the customers. This economic perspective looks at the network and its customers as a whole and defines system performance to include the value that customers obtain from using the network services. In this 'holistic' approach, the customers and network cannot be seen as separate entities. Network mechanisms must take account of their interactions. Flexibility suggests the use of incentive mechanisms where economic agents (users, autonomous infrastructure and service providers) are provided sufficient information to take decisions, each acting rationally, in his best interest. Prices are mainly used in such mechanisms to convey information about resource scarcity and congestion cost.

We next discuss several issues for networks that are essentially economic ones. We begin by looking at the use of pricing by a network operator who wants to control congestion and smooth bursty customer demand. We argue that even if there is a fibre glut for the near future, and new light waves can be provided at a small marginal cost, there remains the possibility of congestion, and thus a need for pricing (and an understanding of its economic theory).

Given all the above, including the commoditization of the market, what role remains for pricing? In the next section we argue that even if there is a fibre glut for the near future, and new light waves can be provided at a small marginal cost, the possibility of congestion always remains present. Hence pricing remains useful to a network operator who wants to control congestion and smooth bursty customer demand.

1.3.1 Overprovision or Control?

As we have seen, there is much uncertainty about growth in demand for communications services. Just as it was once overestimated, it may now be underestimated. It is hard for any operator to predict demand, how technology will evolve, to tell where the future bottlenecks in service provisioning will be, or to predict the price and quality of interconnection with other networks.

What we do see is that lower networking costs have spurred the creation of demanding new applications: such as the automatic downloading of complete web sites, Internet radio, outsourcing of back-office applications for ERP (Enterprise Resource Planning), video streaming and new peer-to-peer computing paradigms like the Grid (a technology that lets users tap processing power off the Internet as easily as electrical power can be drawn from the electric grid), and Storage Area Networks (SANs). An important characteristic of these applications is that they are run by software on machines rather than by humans. We expect that the vast majority of future Internet traffic will be generated by programs and devices connected to the Internet. Since these can ultimately greatly outnumber humans, network traffic has the potential to grow extremely rapidly. It is an open question as to which will grow more rapidly: capacity or demand. The answer greatly affects the extent to which congestion remains a dominating factor, the role of pricing and the evolution of network management mechanisms.

Let us examine this idea a bit more. It is reasonable to assume that as network services play an increasingly key role in the future economy, businesses will want services of high quality, with attributes such as low latency and information loss. How can the network meet the demand for high quality services without becoming overcongested? There are two possibilities. Either the network is extremely simple, but there is so much capacity that it is never congested. Or there is less capacity, but sophisticated control mechanisms are used to provide high quality services to applications that need it. A good analogy can be made with freeways. In the absence of any special controls a freeway can provide only a 'best-effort' service. To provide a better quality of service there are two strategies. Either one can overdesign the freeway, by building enough lanes so that all customers receive the better quality of service. Or one can build a smaller freeway, but implement a priority service; perhaps a number of lanes are reserved for customers who are prepared to pay an extra fee. Both strategies are costly, but in different ways. Quality differentiation allows for price differentiation. The cost and complexity in the second strategy is in ensuring that customers are charged differentially and that only those who have paid for the service can use the priority lanes.

Some commentators believe that future networks will be overdesigned. We see this in today's local area networks and personal computers. Experience shows that people so value high responsiveness that they are willing to overdimension their private networks and their computing platforms by taking advantage of the low cost of the new technologies. It may be that simple overprovisioning can solve the problem of congestion and can be justified by the rapidly decreasing cost of bandwidth. But can the whole network be overdesigned? Although overprovisioning may be reasonable in the backbone of the network, which consists of a fairly small number of links, it may not be reasonable in the metropolitan part of the network, and even less so in the access part. In the present Internet, a large amount of fibre capacity connects major cities in the US and around the world, but there is substantially less fibre installed at the access network part that connects customers to the backbone. The core network infrastructure is shared by all customers, but that part of the infrastructure that lies in the metropolitan and the access network is used by much fewer customers. This is where the largest cost of the network lies. Indeed, some experts believe that it would take twenty to thirty times as much time and expense to overprovision the fibre in the local part of the network as it has taken to install the present fibre infrastructure in the backbone. For these reasons it may be very costly to overprovision all of the network.

If the above arguments are correct then congestion and overload are always dangers. Controls will always be needed to safeguard network operation. In implementing such controls the network must monitor new connections, implement rules for deciding which connections to block, and then effectively block them.

An alternative to overprovisioning is the second strategy: equip the network with some form of control that operates at all times, even when no overload occurs. This control can be of variable complexity, and essentially can provide a controlled access to the network resources by various customer types, allowing for service (quality) differentiation. By optimizing the operation of the network, less capacity is needed to meet a given demand than is required by simple overprovisioning. However, it may be extremely costly to deploy a new control mechanism in an existing network if the mechanism was not put place when the network was originally designed. For example, it would difficult to win universal acceptance for adding a new control mechanism to the existing Internet protocols. Moreover, if any control is to be effective, it must be combined with appropriate tariffs so as to attract the right customers. It is awkward for the network itself to differentiate and assign priorities amongst customer traffic without taking into account the actual value of the service to the customers that will be affected.

This last observation is extremely crucial and will be further explored in Chapter 5. As we see, the social value of a system is increased when users are given incentives to choose the levels of service most appropriate to them. Prices can produce just the right incentives, and so help to ensure that customers do not waste important resources that they do not value. Indeed, pricing can be viewed as a control mechanism for *shaping* demand. This is better than blocking demand in an ad hoc way. Note that simple usage pricing may be a sufficient control mechanism. Consider a city suffering from congestion in the provision of parking spaces, but presently not charging for parking space. Many of the spaces may be used by people who would be willing to use public transport rather than pay a parking fee. The city could impose rules that reserve certain places for specific people. However, a better way to reduce congestion would be to introduce a simple per-hour parking charge. Of course, there is a cost to installing parking meters and policing their use.

Perhaps the future lies in all networks being just on the borderline of being overprovisioned, even in the access part. Since there is a nonzero variable cost for providing new capacity (at least equal to the marginal cost of lighting the existing fibre), the only strategy that is economically defensible is to provide just enough capacity so that the marginal cost of extra capacity equals the marginal benefit to customers of reduced congestion. In other words, customers should pay for the extra value they obtain by increasing service quality. This is the only rational business model for network operators who operate in a competitive environment. It is now interesting to think about models for capacity expansion. Capacity should expand at the rate needed to guarantee some fixed congestion levels at all times, given the dynamics of the demand. Since capacity cannot be provided in arbitrarily small increments, congestion will eventually appear and signal the need for capacity expansion. The facts that demand is not predictable and capacity expansion cannot be provided in 'real-time' in response to every increase in demand, suggest that pricing can serve an important role during these transient phases by increasing stability and reducing quality fluctuations. As the network transport service market will be constantly in such a transient phase, we believe that pricing will always play an important role in safeguarding network performance.

1.3.2 Using Pricing for Control and Signalling

We continue with the theme of pricing as a means of control, and describe the role it has in signalling. By increasing prices an operator can reduce demand, reduce congestion, and ensure that services are provided to the users that benefit most and are most willing to pay. On a short timescale, pricing can provide a type of flexible policing mechanism. On a long timescale, it can be used as part of a feedback loop to stabilize the network through a sort of flow control. Viewed as a control mechanism, charging has the advantage that it scales easily with the size of the network.

Pricing can also be viewed as a mechanism by which the network operator communicates with his users and gives them incentives to use the network efficiently. By this means, he can improve the value of services to users and provide stability and robustness. Conversely, the way that users respond to charges can tell the operator something about user preferences and their intended network use. For instance, a customer's choice amongst a menu of mobile phone charging plans can signal whether he plans to phone mostly during the working week or at the weekend. The mobile network operator can use this information to dimension network capacity and allocate resources where they are most probably needed. A tariff design is said to be *incentive compatible* if it induces customers to choose tariffs that accurately reflect their actual usage plans, and while doing so increases the aggregate utility of all users. There is nothing for a user to gain by disguising how he plans to use the service. A user who plans mainly to call friends at the weekend will have no advantage in choosing the tariff designed for the working week.

Clearly, it is important that charges should provide the right signals: both with regard to incentives to users and information that can be used in network control. Properly designed tariffs accurately convey information between the network and its customers. Charges should be simple, but not simplistic; they should be understandable, implementable and competitive.

Price information that is signalled to the edges of the network can play a significant role in providing rational end-users and applications with the appropriate incentives to control their flows. This is almost what happens in the Internet. As it is presently engineered, the decision as to when a user should increase or to decrease his traffic flow is not made by the Internet itself, but by the TCP protocol running on the user's computer. A major task of the Internet is to send congestion signals to its users. The congestion signals are generated by the user's packet losses. When TCP receives a congestion signal from the network, it reduces the sending rate; otherwise it increases it. Interestingly, all users of the Internet cooperate by implementing TCP identically; but no one forces them to do so. Although it would not be trivial to implement, in principle, a user might cheat by rewriting his software to disobey the TCP protocol and send at a greater rate than TCP says he should. This would not be an issue if the congestion signals were to actually impose a monetary charge. All a user could do would be to observe the rate at which he is being charged for his lost packets and choose the rate at which he wishes to submit packets. His choice would depend on how much he is prepared to pay to run the application he is running. There is incentive compatibility, in that a user has no reason to pretend he values bandwidth differently than he really does.

Pricing can also solve the congestion problem. When there is congestion along a route the users of that route can be made to see an increasing price. This price increases until the users reduce the rate at which they send packets and congestion is reduced. Interestingly, the Internet as it is presently designed, can be interpreted as indirectly implementing a charging mechanism that treats all users equally and that assumes every flow has equal value. The network provides congestion signalling and the users respond. Although no actual charging takes place, the TCP protocols act as if the rate of congestion signals had the interpretation of a rate of charges, and hence a greater rate of congestion signals provides the incentive to reduce the flow. A current challenge is to extend this mechanism to models in which different users have different utilities for different network services.

1.3.3 Who Should Pay the Bill?

In the previous section we saw that there could be advantage in charging end-users of the Internet, as both a function of their sending rate and the congestion level of the network. This is controversial. The history of the telephone network has shown that charging according to usage reduces network use since users are reluctant to incur charges. Many studies suggest that users of telecommunications services appreciate the simplicity and predictability of flat rate charging. Yet flat rate charging is not fair to all customers and can lead to a waste of resources. For example, in an all-you-can-eat restaurant, customers pay a flat fee, but there is an incentive to overeat.

Economic theory suggests that efficiency is greater when the charge takes account of actual usage. Waste is reduced and resources are reserved for the customers that value them the most. Furthermore, to optimize economic efficiency even further, prices could be changing dynamically to more accurately reflect demand. Such pricing schemes are far more complex than simple flat fee schemes and hence raise questions of feasibility. Users

facing such complex schemes may be deterred from using the network services and slow
the expansion of the Internet. Should we sacrifice short-term network efficiency to increase
the extremely valuable long-term demand for new network services and applications?

There are two important technological facts that play a role in answering this question.
First, users (individuals and end-customers of large organizations) can today make use of
intelligent edge devices which can absorb the decision complexity. Software running at the
user machine can make decisions about network usage and absorb the complexity of the
network tariffs and the fluctuating prices. Such an 'intelligent agent' can simply follow
policy rules set by the user and optimize decisions at the user-network interface.

Secondly, today's technology allows charging to be done in complex ways, basically
by programming. It is possible to implement charging structures in which sophisticated
charges are attributed to potentially many stakeholders in the value chain of the service
provisioning, and arrange that end-users face only simple tariffing structures, which could
be flat rate.

What happens is that the end-user purchases a high-level service, such a contract for
a web browsing and email service of a given quality, an Internet telephony connection,
or viewing of some multimedia content. This generates a demand for a transport service
of some quality. As far as the end-user is concerned the transport service and high-level
service are bundled and priced as a single service. It is the provider of the high-level service
who must find and buy an adequate quality transport connection from a transport service
provider. It is he who has to deal with the complexity of the transport tariffs and perhaps
dynamically fluctuating prices and quality. Perhaps he will use some sort of insurance
contract to protect him from excessive price fluctuations. One could even imagine that
he uses financial instruments, such as futures and options, to manage the risk involved in
buying and selling network services.

1.3.4 Interconnection and Regulation

It is to users' advantage that the networks of different operators interconnect. Creating
larger networks from smaller ones is key to unleashing the power of network externalities.
Interconnection is a service provided among networks to extend their services to larger
customer bases. Consider, for example, three networks, A, B and C, covering different
geographical locations, with B located between A and C. Network B can provide
interconnection service to network A by carrying A's traffic that is destined for C, or by
terminating the traffic that originates from A's customers and is destined for B's customers.
In the first case, the customers of A and C benefit; in the second case the customers of A
and B benefit. In a broader sense, interconnection allows users that can be reached through
one network to become customers of services provided by another network. If networks
are 'perfectly' interconnected, then services are offered in a truly competitive environment
in which a customer is free to choose the best service on offer. Otherwise, the network that
'physically' owns the customer is in a position to restrict this choice to services offered
only by that network and its allies. Competitive markets improve service quality and result
in lower prices, to the advantage of the consumer. Further details of interconnection are
pursued in Chapter 12.

The previous discussion suggests that it is not always to a network provider's advantage to
offer interconnection services. By refusing or asking unaffordable prices for interconnection,
a large network may reduce the value of smaller networks and eventually force them out of
business. In our previous example, if A is small compared to B then, after interconnection

with B, his customers enjoy the same benefits as the customers of B, while the operating costs of A may be significantly lower (since he has no need to maintain a national backbone). A typical historical example of using interconnection as a strategic tool for dominance is the case of the Bell System in the US. In the early 1900s, the Bell System controlled about half of the phones in the US and was the only company offering long-distance service. As the value to customers of long-distance service increased, the Bell System refused to offer interconnection services to independent local telephone companies. This made customers switch to the Bell System which eventually became the dominant local and long-distance carrier under the corporate name of AT&T and remained so until its breakup in 1984.

Although such large natural monopolies can be very beneficial to consumers, by deploying nationwide expensive infrastructures and creating de facto interoperability standards for network and consumer devices, eventually they lose momentum and become superseded. Opening the competition in these monopoly markets requires careful intervention by the regulator, who must set new goals that clearly take account of new developing technologies, the state of the market and its desired evolution, the market power of certain players and, most importantly, convey a new vision. The regulator is the public authority responsible for the overall health of the telecommunications market. He must intervene where competition is reduced and network operators use their market power in a way that is not socially optimal. He also uses pricing as a control. His aim is to 'open' networks to competitors (make components of services sold by a network to its own customers available for a price to competitors), and exert control over such prices so as to induce operators to compete fairly. True competition results in network resources being used efficiently and for greatest benefit of the industry and users of communications services overall. We return to the subject of regulation in Chapter 13.

A key to success is motivating (we use a softer term than 'obliging') networks to interconnect in order to achieve truly competitive markets for communications and value-added services. If successful, with no artificial barriers, an enormous number of players will be free to unleash their creative and inspired product and service ideas in the competitive information services marketplace. However, problems of interconnection can be difficult. It can be difficult to manage interconnection agreements, e.g. to offer a service with a quality of service guarantee that is respected across networks. It is also difficult to share fairly amongst networks the charges that users pay. The economic models that have been proposed for interconnection are complex, and it is not obvious how to provide the right incentives for interconnection. If interconnection prices are unpredictable, this can deter investment and competition. It can be difficult to introduce new network technology, as this requires agreement and implementation effort by all network providers. For example, now that IP is the incumbent protocol for the Internet, operators are reluctant to change that technology or add new features.

If interconnection problems prove too difficult, then network operators may prefer to grow their networks vertically and so reduce the risks associated with interconnecting with others and pricing bottleneck services. Perhaps the Internet will not evolve to become a single network that provides high quality end-to-end service between any two access points. Instead, a small number of vertically integrated private Internets may evolve, each guarding its customer base by providing proprietary services that encompass the whole range from broadband access to content. To protect its customer base such a network might artificially degrade the services that customers of other networks receive. This can be done by degrading the quality of interconnection services to other networks. Of course, such a scheme will be stable only if customers mostly use the Internet for consuming content rather than for interacting and communicating with other customers. If it is mainly communication and interactivity that is sought, then market demand will push for high quality interconnection and thus for a true Internet.

But how far should the regulator reach? Experts believe that although the IP protocol has allowed the creation of open, interconnected networks, in reality the networks can only be as open as the various conduits used to reach them. Should there be more competition in this 'first mile' (the part of the network that reaches individual customers)? What is the best way to ensure this competition and for what type of infrastructures? Do the incumbent local telephone operators that own the copper local loop infrastructure face enough competition from new technologies such as the unlicensed multihop wireless networks, the mobile service networks, the low cost IP over fibre networking technologies deployed by the new competitors, and the broadband capabilities of the cable modems, or should they be treated as still having monopoly power? Should the regulator expand his reach beyond the local and long-haul wireline network to include wireless, cable and fibre facilities as well as facilities in which traffic is multiplexed and demultiplexed? Would such moves deter companies from taking risks and investing in infrastructure? How should such risks be compensated? Given that the current fibre glut and infrastructure over provisioning makes it hard for companies to recover their sunk costs and pay their debts, should the fibre assets of the telecommunications companies be nationalized, and these companies then made to focus on using the infrastructure to provide advanced value-added services? Or should the regulator favour horizontal consolidation of the infrastructure companies, so as to create a sustainable market of a few players? What incentives will facilitate the rapid introduction of those truly broadband services that can only be provided over fibre? These are few of the difficult questions faced by the regulator.

Some trends in modern technology challenge traditional regulation concepts. New access technologies such as wireless Ethernet, which consume public spectrum without license from a central authority, are essentially self-regulating. Such new decentralized and self-managed networks evolve dynamically in an ad hoc fashion and pose new questions for regulators accustomed to making decisions for systems that evolve on longer timescales.

1.4   Preliminary modelling

As we have seen, network services are economic goods, which a network provides through use of its resources of links, switches, hardware, software and management systems. This section introduces some of the basic economic concepts that are useful in reasoning about markets and in making pricing decisions. We look at some examples, compare the merits of flat rate and usage-based charging, and identify some important structural properties of good tariffs. We see how a price can be used to share a congested resource. These ideas are pursued much further in the economics tutorials of Chapters 5 and 6.

1.4.1   Definitions of Charge, Price and Tariff

Our consistent terminology in this book is that the *charge* is the amount that is billed for a service. By *price* we mean an amount of money associated with a unit of service; this is used to compute the charge. The *tariff* refers to the general structure of prices and charges. A example of a tariff is $a + pT$, where a is a price for setting up, p is a price per second for using the service, and T is the duration of the connection in seconds.

A tariff is that part of the contract between two parties that specifies the way the charge will be computed for the service. Its structure can affect the parties' behaviour. Consider, for instance, the tariff used to compute a taxi fare. It is common for such a tariff to be of the form $a + bT + cX$, where a is the amount paid at the start, and T and X are the duration and the distance of the ride respectively. A feature of some taxi meters is that the metering of T and X are mutually exclusive: if the speed of the taxi is less than a certain amount then time is metered; otherwise distance is metered. What incentives does this tariff give to the taxi driver? Observe that if b is very large this gives the driver an incentive to prolong the duration of the ride, rather than to complete the trip quickly. However, if b is very small there is an incentive to avoid congested areas, and no taxi may be available in parts of the city. The driving pattern is also affected since, when driving between traffic lights, the driver has the incentive is to drive as fast as possible between the lights and then spend as much time as possible waiting for red lights to turn green; thus stop/start driving is encouraged. A similar encouragement of 'bursty' behaviour is also encountered (but for other reasons) in the case of traffic contracts in communication networks. Interestingly, the demand for taxis also influences the way their drivers will drive. If there is little demand for taxis, as during the nights, and total X is known from previous experience, then drivers have the incentive to maximize duration of trips. During the day when demand for taxis is high, the fixed charge a, gives the right sort of incentive; drivers are encouraged to use minimum distance routes and minimize the length of rides. Apart from the stop/start driving between lights, this suits the customer well.

1.4.2   Flat Rate versus Usage Charging

An 'all-you-can-eat' restaurant provides an example of how a flat-fee tariff can give the wrong incentives. Since customers pay one flat fee to enter the restaurant and are then free to eat as much as they wish, they tend to over eat. This wastes food (which is analogous to wasting network resources). Interestingly, the health of customers also suffers because they are encouraged to over eat. The flat fee must cover the cost of the average customer if the restaurant is to recover its cost. Light eaters will feel cheated if they have to pay for more than they consume; the customer base will decrease and the restaurant will make less profit. Note that many Internet tariffs are presently of a flat fee type.

How can one provide incentives that avoid the overeating problem? A simple remedy is to charge a customer for what he actually consumes; this happens when a restaurant has an á-la-carte menu. Now each customer chooses the meal that provides him with the greatest satisfaction and value-for-money. The customer has complete control over his choice of meal, can see its price on the menu and predict his charge. Unfortunately, the charge is not as predictable when usage-based charging is used for network services. A network user cannot usually predict accurately the traffic volume that will result from his interaction with the network and so predict his charge (though he might be able to do so if he were using a specific application, such as constant bit rate video).

Is a simple usage charge enough? If an á-la-carte restaurant charges only for the food consumed then there is danger that some customers might occupy their table simply to socialize and not order any food. A tariff that has the right incentives should take account of the fact that resource reservation is costly in itself, independently of the cost of the actual resources consumed. This is why restaurants make sometimes make a 'cover charge'.

The telephone network and the present Internet are alike in that they transport bits at a single quality. By some measures the telephone network provides better bit quality, but it is also more expensive to build. Extensions of the Internet protocols and technologies such as ATM allow data bits to be transported at different levels of quality. The relation between quality of service and price is a major theme of this book. As we will see, it makes more economic sense for customers to choose bit qualities that are matched to their needs, than for the network to allocate all users the same bit quality.

1.4.3 Dynamic Pricing in an Internet Cafe

An interesting approach for pricing Internet access is used by a popular chain of Internet cafes in Europe (easyInternetCafe). The price per unit time that is charged for using a computer terminal is not fixed throughout the day but varies dynamically to reflect demand. A user pays a fixed price for a ticket, say $3, and then gets more or less Internet access time, depending on the time of day and the number of terminals that are busy at the time he buys the ticket. The day is divided into three periods: the 'peak' period (11am–3pm), the 'off' period (1am–9am), and the 'normal' period (all other times). In the off period a ticket buys 150 minutes. In other periods, the price depends upon the number of busy terminals, n, where $0 \leq n \leq 450$. During normal time, the user receives 150, 120 or 90 minutes as n lies in the range 0–150, 151–300 or 301–450, respectively. During peak time, the minutes are reduced to 90, 60 or 30 minutes, respectively for the same ranges of n. (This is not exactly the same charging scheme as used in the stores but illustrates the same ideas.) If no terminals are available a customer has to wait for one to become available.

Observe that, although the price for a ticket is flat, the amount of usage varies. To obtain more time a user can buy more tickets. What are the merits of such a pricing scheme? Customers value (in addition to good coffee) small waiting time and convenience (of accessing the Internet when they need it, rather than postponing it to a different time).

- Setting lower prices for off-peak times reduces demand during peak times since customers that do not value convenience can choose a cheaper time.
- Use of dynamic prices makes it less probable that a customer must wait for a terminal. This is because when demand is high (i.e. there are few free terminals), customers will spend less time on-line due to the greater price per minute. They use their time more efficiently by wasting less time in being connected to the Internet when it is of no economic value, and so more customers can use the system.
- When there is no 'congestion' indications, i.e. n is small, the time is not unnecessarily reduced, offering the best possible value to the customers. This nice self-regulating effect is not achieved by a flat time ticket.
- If the cost incurred by waiting is very high, one may simply create one more usage zone, say for $400 \leq n \leq 450$, and reduce the ticket time even further. Such simple corrective actions are straightforward to implement and require no sophisticated analysis. Similarly, if the usage of the terminals is observed to be rather low during a particular time period, one may increase the ticket times. Such a system works very much like a thermostat which turns the burner on and off using feedback from temperature measurements. Note that it is easier to build a thermostat than to solve the differential heat equations to compute the exact activity patterns of the burner (the optimal average price independent of n).

The bottom line is that dynamic pricing, which uses feedback from the system, can better control demand for resources. The overall value that customers obtain is greater, leading them to prefer this cafe over others. Indeed, the charging scheme may be used to shape demand and resource usage and to maximize the value of the service to the customers, rather than simply maximizing revenue. The cafe owner can capture some of the extra value creates for his customers by raising the price of coffee. Economic theory suggests that such a strategy may generate greater profit than simply setting prices to maximize revenue from Internet access alone.

1.4.4   A Model for Pricing a Single Link

Suppose a network operator owns a link between Athens and London of capacity C bits per second and that the only service he sells is constant bit rate transport. Suppose that there are N customers who would like to use some of this transport capacity. How might C be divided amongst these users? In other words, given that user i is allocated x_i bits per second, how should the operator choose $x_1, \ldots, x_N$, subject to the constraint that they sum to no more than C?

To make the problem more interesting and realistic let us require that it is the technology of the network that must decide how the bandwidth is shared, rather than the network operator directly. Suppose each customer has an individual access pipe of capacity C to the Athens–London link. If the total bandwidth that the customers would like to use is less than C, then there is no difficulty in providing each customer with his full request. However, since each customer could completely fill the link with his own traffic, the network must implement some sharing policy or mechanism to decide how to share the capacity of the link among the competing customers when their total demand exceeds C. This policy could try to share capacity 'fairly', as defined in some technologically dependent way.

Suppose the network operator can completely control the way capacity is allocated. One of many possible policies is to simply allocate an equal share of the bandwidth to each user, so that $x_i = C/N$. A more sophisticated method, which takes account of customers' requests, is to use the so-called *fair shares algorithm*. At the first step of the algorithm each customer is allocated his requested bandwidth or C/N, whichever is smaller. After these allocations are made, any remaining bandwidth is shared in a similar way amongst the customers whose requests were not fully satisfied at the first step; this is done by redefining the parameters N as the number of remaining customers with unsatisfied requests and redefining C as the bandwidth not yet allocated. The algorithm repeats similarly until all bandwidth is allocated.

However, these methods of allocating bandwidth ignore the fact that customers do not value bandwidth equally. An allocation of x_i might be worth $u_i(x_i)$ to user i. Here u_i is called the *utility function* of user i. If the network is provided by a public authority then a reasonable goal might be to maximize the overall value that customers obtain by their use of the network. To do this, the network operator needs its customers to make truthful declarations of their utilities. In practice, it is usually impossible to gain direct knowledge of utility functions. Let P denote the problem of maximizing the total user benefit. This is

$$P: \quad \underset{x_1,\ldots,x_N}{\text{maximize}} \sum_{i=1}^{N} u_i(x_i), \quad \text{subject to} \sum_{i=1}^{N} x_i \leq C$$

An important starting point for engineering a solution is that the fact that if each u_i is a concave increasing function, then there exists a price $\bar{p}$ such that P can be solved by the simple method of setting this price, and then allowing each user i to choose his x_i to solve the problem

$$\underset{x_i}{\text{maximize}} \left[u_i(x_i) - \bar{p}x_i \right] \tag{1.1}$$

The fact that there exists a $\bar{p}$ to make this possible follows from the fact that $\bar{p}$ is the Lagrangian multiplier with which we can solve the constrained optimization problem P (see Appendix A).

Let $x_i(\bar{p})$ be the maximizing value of x_i in (1.1), expressed as a function of the price $\bar{p}$. We call $x_i(p)$ user i's *demand function*. It is the amount of bandwidth he would wish to purchase if the price per unit bandwidth were p. Under our assumptions on u_i, $x_i(p)$ decreases as p increases. Let us suppose that at a price of 0 the customers would in aggregate wish to purchase more than C, and when p is sufficiently large they would wish to purchase less that C. It follows that, as p increases from 0, the total amount of bandwidth that the customers wish to purchase, namely $\sum_i x_i(p)$, decreases from a value exceeding C towards 0, and at some value, say $p = \bar{p}$, we have $\sum_i x_i(\bar{p}) = C$. By setting the price at $\bar{p}$ the operator ensures that the total bandwidth purchased exactly exhausts the supply and that it is allocated amongst users in a way that maximizes the total benefit to the society of customers taken as a whole.

This solution to problem P has a number of desirable properties. First, the network need not know the utility functions of the users. Secondly, the decisions are taken in a decentralized way, each user rationally choosing the best possible amount of bandwidth to buy. Thirdly, since the price is chosen so that demand equals capacity, the network technology's sharing policy does not intervene. Users decide the sharing amongst themselves, with price serving as a catalyst. Hence, price works as a kind of flow control mechanism to shape the demand.

The operator may or may not be happy with this solution. He has obtained a total revenue equal to $\bar{p}C$, which of course equals $\bar{p}\sum_i x_i(\bar{p})$. Customer i is left with a 'user surplus' of $u_i(x_i(\bar{p})) - \bar{p}x_i(\bar{p})$. The total value to society of the Athens–London link has been maximized and then divided amongst the operator and customers. However, it is has not been equally divided amongst the customers, nor in a way that specially favours the operator or takes account of his costs.

If our operator is not subject to competition or regulation he might like to capture all the benefit for himself; he can do this if he can present each customer with a customized offer. He simply says to customer i, 'you may buy $x_i(\bar{p})$ units of bandwidth for a penny less than $u_i(x_i(\bar{p}))$ — take it or leave it'. User i is better off by a penny if he accepts this offer, so he will do so, but the operator gains all the value of the link, minus N pennies. If the operator cannot make each customer such a take-it-or-leave-it offer, he still might say, 'you may can have any amount of bandwidth you like, but at a price of p_i per unit'. That is, he quotes different prices to different customers. As we see in Section 6.2.1 the operator maximizes his revenue by quoting higher prices to customers who are less price sensitive.

In practice, the operator does not usually know much about his customers, and it is very unlikely that he knows their utility functions. Moreover, he cannot usually tailor prices to individual customers. Nonetheless, we will find that some charging schemes are better than others. Some schemes give customers a greater incentive to act in ways that maximize welfare. Other schemes enable the operator to obtain a greater payment, thereby obtaining a greater part of the link's value for himself.

Let us take the second of these first. There are various ways in which the operator can extract a greater payment. He may present users with nonlinear prices. For example, he can make a subscription charge, or vary the price per unit bandwidth according to the quantity a user purchases. He may offer different prices to different groups of customers (e.g. home and business customers). Or he may define versions of the transport service, such as day and night service, and offer these at different prices.

If the operator is constrained to sell the bandwidth at a single price his objective function is $p \sum_i x_i(p)$, which may be maximized for a p for which not all of the bandwidth is sold. Of course he must always have an eye on the competition, on his desire to grow his customer base, and to fund the costs of building, maintaining and expanding his network.

Thus far, we have taken a very simple view of both the service and the network. Many modern services are not best provided for by simply allocating them a constant bit rate pipe. A customer's service requirement is better-visualized as his need to transport a stream of packets, whose rate fluctuates over time. The customer may be able to tolerate loss of a proportion of the packets, or some delay in their delivery; he may be able to assist the network by guaranteeing that the rate at which he sends packets never exceeds a specified maximum.

Suppose that a customer has utility for a transport service that can be characterized in terms of some set of parameters, such as acceptable mean packet delay, acceptable peak rate, mean rate, and so on. Chapters 2 and 3 describe ways that such services can be provided. Suppose there are J such services types and we label them $1, \ldots, J$. As we show in Chapter 4, it can be a good approximation to suppose that the supplier's link can simultaneously carry $n_1, \ldots, n_J$ connections of each of these services, at guaranteed qualities of bit transfer, provided $\sum_j n_j \alpha_j \leq C$, where $\alpha_1, \ldots, \alpha_J$ and C are numbers that depend on the burstiness of the sources, the link's resources and the extent to which statistical multiplexing takes place. The supplier's problem is to decide how to charge for these J different services. Note that problem P has a new dimension, since the constraint now involves $\{\alpha_1, \ldots, \alpha_J\}$.

We continue discussion of this problem in Chapter 8. One must be cautious in applying economic models. Pricing is an art. No single theory can weigh up all the important factors that might affect pricing decisions in practice. No single prescription can suffice in all circumstances. There are many technology aspects that must be taken into account, such as quality of service, multi-dimensional contracts, network mechanisms for conveying price information, the capabilities to support dynamic prices, and the power and responsibility of edge devices. It is particularly difficult to price a good for which customers have preferences over attributes that are difficult to measure, such as brand name, service reliability, accessibility, customer care, and type of billing. Marketing strategies that take account of such attributes can lead to prices that seem rather ad hoc. This is particularly true in the market for communications services.

1.5 A guide to subsequent chapters

In Chapters 2–4 of Part A, we expound the fundamental framework and concepts that we use to think about network services. We explain the important concepts of service contract and network control. As examples, we describe the services provided by ATM and the Internet. We introduce the idea of effective bandwidths, which are the key to addressing questions of pricing services that have quality of service guarantees.

In Chapters 5 and 6 of Part B, we present some key economic concepts that are relevant to pricing. The material in these chapters will be familiar to readers with a background in economics and a useful tutorial for others.

Part C is on various approaches to pricing and charging for service contracts. No one approach can be applied automatically in all circumstances. The designer of a charging scheme needs to consider the type of service contract that is being priced, and whether the aim of pricing is fairness, cost recovery, congestion control or economic efficiency.

Chapter 7 describes cost-based pricing methods and discusses how such methods are used in practice in the telecommunications industry. Chapter 8 is concerned with charging for guaranteed contracts (those with certain agreed contract parameters, such as the packet loss probability). Chapter 9 discusses congestion pricing. Chapter 10 is concerned with charging for flexible contracts (those in which certain contract parameters, such as peak rate, are allowed to change during the life of the contract).

Part D concludes with discussions of the special topics of multicasting, interconnection, regulation and auctions (Chapters 11–14). Auctions are of interest because they are often used to sell important resources to the telecoms industry. Also, auction mechanisms have been proposed for allocating network resources to users in real time.

1.6   Further reading

There are many excellent books on the digital economy and on the impact of the new technologies, especially the Internet. Shapiro and Varian (1998) give an economist's perspective on the rules that govern markets for information goods. Kelly (1999) gives a wonderful introduction to the Internet economy and the new concepts that apply to it. Another well-written book is that of Downes, Mui and Negroponte (2000), which explains the interaction between the laws of Metcalf and Moore. These laws are, respectively, that 'the value of a network increases as about the square of the number of users', and 'the number of transistors in computer chips doubles every eighteen months'. The web pages of Economides (2002) and Varian (2002) contain references to many papers on issues of network economics, and pointers to other relevant sites.

A great source of articles on the evolution of the Internet and related economic issues is the home page of Odlyzko (2002), and a good source for information on many issues of the Internet telecoms industry is The Cook Report on Internet, Cook (2002). Isenberg and Weinberger (2001) describe *the paradox of the best network*: namely, 'the best network is the hardest one to make money running'.

2

Network Services and Contracts

It is useful to distinguish between 'higher-level' and 'lower-level' services. Higher-level services are those that interface directly with customers. Lower-level services are those that customers use indirectly and which are invisible to them. Consider, for example, the Internet as it is used by students and staff of a university. One higher-level service is email; another is web browsing. Web browsing uses the lower-level service of Internet data transport to exchange data between users' terminals and the servers where web pages reside. The quality of the higher-level web browsing service depends on the quality of the lower-level transport service. That is, the speed at which web pages will be delivered to users partly depends on the quality of the network's data transport service. This will be specified in a contract between the university and the network.

A transport service can be defined in many ways. It can be defined in terms of a guarantee to transport some amount of information, but without any guarantee about how long this will take. It can fully specify the performance that is to be provided, and do this at the start of the service. Alternatively, it can respond to changing network load conditions, and continuously renegotiate some qualities of the information transfer with the data source. We investigate these possibilities in this chapter.

Finally, we note that the provision of a service involves not only a flow of information, but also a flow of value. Flow of information concerns data transport, whereas flow of value concerns the benefit that is obtained. One or both parties can benefit from the flow of value. However, if one party enjoys most of the value it is reasonable that he should pay for the service. For example, if an information server sends a customer advertisements then the information flow is from server to customer, but the value flow is from customer to server, since it is the advertiser who profits. This suggests that the server should pay. If, instead, the customer requests data from the server, then value flows to the customer and so the customer should pay. Note that it is neither the initiator of the transport service, nor the one who sends information that should necessarily pay.

This chapter is about various characteristics of services, independently of charging issues. In Section 2.1 we discuss a classification of the network services according to different characteristics. We also provide a primer to present technology, in which we explain the characteristics of the most common network service technologies. Please note that the figures that we quote for various parameters, such as SONET's maximum line speed of 10 Gbps, are continually changing. The concepts that we present do not depend on such parameter values. In Section 2.2, we discuss generic issues

---

Pricing Communication Networks   C. Courcoubetis and R. Weber
© 2003 John Wiley & Sons, Ltd   ISBN 0-470-85130-9 (HB)

related to contracts for network services, focusing on issues of quality of service and performance.

2.1 A classification of network services

At its most basic level a network provides services for transporting data between points in the network. The transport service may carry data between just two points, in which case we have a *unicast service*. Or it may carry data from one point to many points, in which case we have a *multicast service*.

The points between which data is carried can be inside the network or at its periphery. When a web server connects with a user's browser then both points are at the periphery. When an access service connects a customer's terminal equipment to the network of a different service provider then the customer's point is at the periphery and the point connecting to the different service provider's network is inside. When a network interconnects with two other networks then both points are inside. Thus network operators can buy or sell transport services amongst themselves and collaborate to provide transport services to end-points residing on different networks. We see all these things in the Internet. For simplicity, we often refer to a large collection of cooperating networks that provide a given transport service as 'the network'.

2.1.1 Layering

Service layering is common in communication networks. A higher layer service consumes lower layer services and adds functionality that is not available at the lower layers. Services of various layers can be sold independently, and by different service providers. An example of a higher layer service is an end-to-end transport service that connects customer equipment at two periphery points of the network. This service uses lower layer services, some of which are strictly internal to the network; these lower layer services provide connectivity between internal nodes of the network and the *access service* that connects the users' equipment to the network. The end-to-end service may perhaps add the functionality of retransmitting information lost by the lower-level services.

A simple analogy can be made by considering a network of three conveyor belts. One connects node A to node B. Two others connect node B to nodes C and D. Suppose that each conveyor belt is slotted and provided with fixed size bins that move with the belt. Parcels are inserted into the bins so that they do not fall off the belts while travelling. In order to provide an end-to-end service from A to C and D, some additional functionality is needed. For instance, bins travelling between A and B might be coloured red and blue. Parcels arriving in a red bin at node B are assigned by a clerk to continue their journey on the conveyor belt from B to C, whereas parcels in the blue bins continue on the conveyor belt from B to D. Clerks are needed to read the destination addresses, fill the different colour bins on the conveyor belt, and empty the bins that arrive at nodes C and D. Of course there are other ways to build the same end-to-end service, for instance, we could use bins of just one colour on the belt from A to B, but have a clerk at node B check the destination address of each arriving parcel to decide whether it should next be placed on belt BC or BD. A key feature of this setup is the layering of services: one or more companies may provide the basic conveyor services of conveyor belts AB, BC and BD, while another company provides and manages the bins on top of the conveyor belts. Yet another company may provide the service of filling and emptying the bins (especially if bins are a single colour and the destination address of the parcels must be checked at point B). Thus, our setup has three layers of service. The first layer is the conveyor service AB. In Internet terms it is analogous to an access service, which connects the equipment of customer A to the network B by, say, a dial-up connection. Typically, an Internet service provider provides the other two layers of service (of running the conveyors internal to the network, and managing and filling the bins on the conveyors, including the access part). Sometimes, a third party provides all three layers of service in the access part.

Let us illustrate these concepts in more depth by briefly describing transport service layering in an actual example from the current Internet. We view the Internet as a single network using layers of different technologies. Further treatment of these services is provided in Section 3.3.

2.1.2 A Simple Technology Primer

The basic Internet transport service carries information packets between end-points of the Internet in much the same way as the post office delivers letters. Letters that are going to the same city are sorted into large mail bags, which are loaded onto airplanes, and then delivered to a central point in the destination city. The letters are then regrouped into the smaller mail bags that postmen can carry on their routes.

Just as the post office uses airplanes, vans and foot, and different size containers and mail bags, so Internet transport service uses many different transport technologies. These include Ethernet, Asynchronous Transfer Mode (ATM), Synchronous Digital Hierarchy (SDH), Synchronous Optical Network (SONET), and Dense Wavelength Division Multiplexing (DWDM). These technologies are described in Sections 3.3.2–3.3.5. We introduce the basic technologies in an informal way that motivates their particular use.

For the moment, we emphasize the fact that each of the above technologies provides a well-defined transport service and packages information in different size packets. The packets of one service may act as containers for packets of another service. Suppose, for simplicity, that the post office transports fixed size packets between customers. A transport company provides a container service between local post offices at A and B by running small vans of fixed capacity at regular intervals between A and B. Prior to the departure of a van from A, the local post office fills the van with the packets that are waiting to be delivered to the post office at B. Such a service is a paradigm of a *synchronous container service*, since it operates at regular intervals and hence offers a fixed transport capability between point A and B. The SONET or SDH services are examples of synchronous services in communications networks.

If each van can hold at most k packets then the unit of information transfer between points A and B is a container of size k. If a van departs every t seconds, then the capacity of the container service is k/t packets per second. (For data, we measure capacity in bits per second, or kilobits, Megabits or Gigabits per second.) Observe that containers may not be filled completely, in which case the extra space is wasted. We can extend this type of synchronous container service by supposing that the transport company uses larger vans, of container size $10k$, again leaving every t seconds. These containers can be filled by smaller 'subcontainers' of sizes that are multiples of k, and customers can rent such space in them (provided that the sum of the sizes of the subcontainers does not exceed $10k$). The post office could obtain the same service as before by renting a subcontainer service of size k. Similarly, an operator running a 622 Mbps SONET service between points A and B can sell four distinct 155 Mbps SONET connections between these points (after reserving two of the 622 Mbps to control the connection).

What happens if customers cannot effectively fill the smallest size subcontainers? Say the post office traffic between points A and B has a maximum rate of $0.5k/t$ packets per second, and so can justify using containers of size at most $0.5k$, but there are other potential customers who can also use fractions of k. Then there is a business opportunity for another operator, who buys the k container size service from the original operator and reserves space in each such container for his customers. This is a 'value-added' service, in the sense that he may reserve a different maximum amount of space for each customer, fill the unused space of one customer with excess traffic of another customer, and is able to distinguish packets belonging to different customers when the container is unpacked. The equivalent of this 'smart container packing' service is an ATM *virtual path* service. A simple case of container packing is to reserve a fixed portion of the space to each customer. For instance, an ATM service provider using the 155 Mbps SONET service between points A and B, can provide two independent ATM virtual path connections of sizes 55 and 100 Mbps that may be sold to different customers. Basically, he can flexibly construct any number of such fixed bandwidth bit pipes based on the actual demand. Again notice that a customer such as the post office which buys the above fixed bandwidth service may not fill the capacity of the service at all times. There are more interesting ways that ATM can pack the containers to avoid unused space. In these cases, the virtual paths do not have a fixed static size but can dynamically inflate or deflate according to the actual number of packets that are being shipped.

In the above, the post office plays an analogous role to IP. Since the local post office at B may not be the final destination of a packet, but only an intermediary, the post officer at B must look at each packet in turn and decide whether to deliver it locally or forward it to another post office location. This is the functionality of the IP protocol: to distinguish packets belonging to different customers and deliver them or route them effectively through the other 'IP post offices'. A customer delivering packets at random irregular intervals to the IP post office (destined for some other customers) views the IP service as building a flexible 'packet pipe' through the network that does not reserve some predetermined amount of bandwidth. Note that such connections may have highly variable durations, and their end-points may be unpredictable as far as the IP service is concerned.

In its turn, the IP service provider can sell a number of such packet connections between points A and B (or the capability for activating such connections), by making certain that there is only a small probability of completely filling the fixed bandwidth service that he purchases from the ATM service provider between A and B. Now statistics come into play. Since most of the time only a small number of the IP connections will be sending packets simultaneously, say a fraction p of the total number n, he needs only enough bandwidth between A and B to accommodate pn sources, assuming that these send continuously. Note the large saving in bandwidth compared to what he would need if he were to reserve the maximum bandwidth needed by each source, that is, enough bandwidth for n such sources instead for pn. This controlled overbooking is an effect of *statistical multiplexing* discussed in Section 4.2. It is important to observe that fixed bandwidth services can be used for achieving the reverse effect of *flow isolation*. For instance, if the IP service needs to assign dedicated bandwidth for a packet connection between A and B, then rather than mixing these packets with IP packets from other connections in the same containers, it can purchase a dedicated container service, solely for carrying the packets it wishes to isolate. Such flow isolation may be used to guarantee good performance, since shared containers have fixed

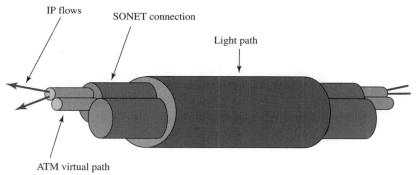

Figure 2.1   A transport service layering hierarchy. Light paths and SONET (SDH) provide large synchronous bit pipes. ATM further divides these pipes, and allows connections to use capacity that is temporarily unused by other connections. IP is used to establish connections between arbitrary network end-points, of unpredicted duration and intensity.

size and packets may have to queue at the IP stations to find free space in containers. This congestion effect is reduced by offering such an exclusive treatment, but comes at an extra cost. We are ready now to proceed with the Internet analogy.

In the late 1990s, many parts of the Internet were implemented as IP over ATM. ATM can run over SDH (or SONET), which in turn can run over an optical network. This transport service layering is shown in Figure 2.1.

More specifically, an optical network technology provides a point-to-point synchronous 'container' service, such as SONET operating at a maximum steady rate of 10 Gbps. In turn, SONET provides subcontainer transport services with rates that are multiples of 155 Mbps. ATM is used to provide flexible partitioning of such large SONET containers for services that require fractions of this bandwidth. IP is responsible for packing and unpacking the fixed size bandwidth services provided by ATM into information streams consisting of variable size objects (the IP packets produced by user applications), whose resulting bit rates are much smaller and bursty. IP is a multiplexing technology that 'buys' such fixed size bandwidth services and makes a business of efficiently filling them with information streams that are variable in both the rate and size of packets. Thus, IP and ATM can be viewed as 'retailers' of 'wholesale' services such as SONET.

Different parts of the overall network may be connected with different container technologies. The idea is to choose a technology for each link whose container size minimizes wasted space in partially packed containers. In the interior of the network many traffic streams follow common routes and so it makes sense to use large containers for links on these routes. However, at the periphery of the network it makes sense to use small containers to transport traffic from individual sources. Thus the business of a network operator is to provide connectivity services by choosing appropriately sized containers for the routes in his network, and then to efficiently pack and unpack the containers. The Internet transport service efficiently fills the large fixed size containers of the lower-level services and connects two end-points by providing a type of connecting 'glue'.

Example 2.1 (IP over ATM over SONET)   A concrete example of transport service layering is shown in Figure 2.2. In this figure Provider 1 aims to fill completely his 622 Mbps container service between points K and L. He may be buying a light path service from a provider who owns the fibre infrastructure between the above points, in which the container service could run up to 10 Gbps. He fills his containers by selling

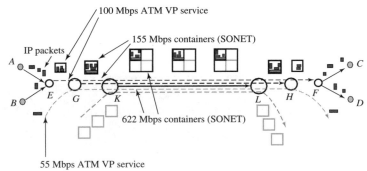

Figure 2.2 An example of transport service layering. Transport service Provider 1 operates a 622 Mbps SONET service between points K and L and sells 155 Mbps SONET services to customers. Provider 2 runs an ATM over SONET network with nodes G, H, and sells a 100 Mbps ATM service between points E and F to Provider 3; to do this he buys a 155 Mbps SONET service for connecting G and H from Provider 1. Provider 3 sells IP connectivity service to customers A, B, C and D by connecting his routers E and F using the 100 Mbps ATM service bought from Provider 2.

smaller container services, in sizes that are multiples of 155 Mbps, such as that which connects nodes G and H of Provider 2. Provider 3 sells Internet services to his customers and runs a two node IP network between routers at E and F. In doing this, he must connect these nodes so that they can exchange Internet data. This data is packaged in variable size IP packets and is sporadic, with a total rate not exceeding 100 Mbps. To connect E to F he buys a 100 Mbps ATM Virtual Path (i.e. a 100 Mbps bit pipe) from Provider 2. Provider 2 uses the ATM technology to subdivide the 155 Mbps SONET container service between G and H, and so sell finer granularity bandwidth services. For instance, he fills the rest of the 155 Mbps containers traversing the F to G link by selling a 55 Mbps ATM connection to some other customer. Note that if Provider 3 has enough Internet traffic to fill 155 Mbps containers, he can buy a pure SONET service between points E and F, if available. This is what happens in IP over SONET. If he has even more traffic, then he can buy a light path service to connect the same points, which is IP over λ. Such a service may provide for 10 Gbps of transport capability for IP packets.

Note that bitrate is not the only differentiating factor among transport services. The IP network E–G allows any pair of customers amongst A, B, C and D to connect for arbitrarily short times and exchange data without the network having to configure any such connections in advance. By contrast, SONET (and ATM) are used for specific point-to-point connections that have a much longer lives.

Finally, each service that is sold to a customer has initial and final parts that give access to the provider's network. For instance, in order to run the ATM service between E and F one must connect E to G and H to F. This access service may be provided by Provider 2 himself or bought from some third provider. Similarly, IP customer A must use some access service to connect to the IP network of Provider 3.

2.1.3   Value-added Services and Bundling

Some services provide much more than simply a data transport service. Consider a web service. It provides a data transport service, but also a data processing service and a data presentation service. The latter two services add value and belong to a layer above that of the transport service. Thus, the web browsing is what we call a *value-added service*, which is complementary to the network transport service. Similarly, an Internet telephony service is a *bundle of services*, which includes a directory service, a signalling service, a data transport service and a billing service. In Section 3.6, we discuss a possible model for Internet services and explain the structure of the value chain in Internet service provisioning.

It is important to distinguish between transport and value-added services. Think of a bookstore which provides the value-added service of retailing books by mail order. A customer chooses his books and says whether he wishes delivery to be overnight, in two business days, or by ordinary post. He pays for the books and their delivery as a bundle, and the bookstore contracts with a delivery service for the delivery. The bundled service has components of attractiveness and timeliness of book offerings, speed of delivery and price. The demand for books drives the demand for the delivery service. Similarly, the demand for information services drives the demand for data transport services. How a customer values the particular content or functionality of a communications service determines the charge he is prepared to pay. Of course, this charge will contain a component that reflects the value of the data transport service, since transport service is what a communications network provides. In Figure 2.3 the user enjoys a video on demand value-added service. Although the customer may make a single payment for the service (to download the software required, run the application and watch the movie at a given quality level), this payment may be further split by the valued-added service provider to compensate the transport service provider for his part of the service.

It is useful to familiarize oneself with some of the formal definitions that regulators use to classify network services. The Federal Communications Commission (FCC) uses the term *information services* for value-added services, and *telecommunications services* for lower-level transport services. The Telecommunications Act of 1996 defines *telecommunications* as "the transmission, between or among points specified by the user, of information of the user's choosing, without change in the form or content of the information as sent and received", and a *telecommunications service* as "the offering of telecommunications for a fee directly to the public, or to such classes of users as to be effectively available to the public, regardless of facilities used". An *information service* is defined as "the offering of a capability for generating, acquiring, storing, transforming, processing, retrieving, utilizing, or making available information via telecommunications". According to these definitions, an

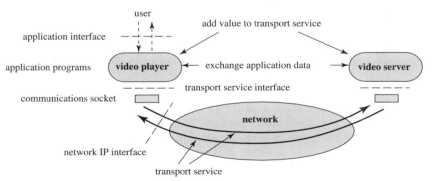

Figure 2.3 Transport and value-added services. The user enjoys a value-added service (such as watching a movie) which combines the transport of data from the video server with the content itself, and probably some additional functionality from the video server (such as back-track, fast-forward and pause). Such an application may require some minimum bitrate in order to operate effectively.

entity provides telecommunications only when it both provides a transparent transmission path and it does not manipulate the form or content of the information. If this offering is made directly to the public for a fee, it is called a 'telecommunications service'. An entity may sell an information service as a bundle of telecommunications (the lower-level data transport services) with content specific applications such as email and web browsing (the valued-added services according to our previous definitions), or sell telecommunications separately as independent services. According to this definition, telecommunications refers to the lower end of the network transport services, where the network offers transparent bit pipes. When, as with TCP/IP, data is processed either inside the network at the routers, or at its edges, the resulting service is closer to an information service according to this definition. In practice, information services are more usually viewed as being associated with content and value-added applications that run at the edges of the network. In the Internet, such applications manipulate the data part of the IP packets according to the particular application logic. Network transport, such as the routing of IP packets, is not considered a valued-added service, as it is offered as a commodity, using open standards. In this book, we deal with network transport services that complement these higher-level, value-added applications. By the FCC definitions they are 'telecommunications services' at lower layers and 'information services' at higher layers.

2.1.4 Connection-oriented and Connectionless Services

We may also classify services by the way data is transmitted. In a *connection-oriented service*, data flows between two nodes of the network along a 'virtual' pipe (or a tree of virtual pipes when multicasting, with duplication of data at the branching points). Data travels along a fixed route, with a specified rate, delay and error rate. In a *connectionless service*, the data does not follow a fixed routing. Instead, the data is transmitted in packets, or datagrams. Successive datagrams, travelling between a source and destination, can take different routes through the network, and can suffer loss.

Connection-oriented and connectionless services may not be substitutable. It is more difficult, or impossible, for a connectionless service to deliver datagrams in a regular way. Take, for example, the postal service, which is a datagram service. It ensures that parcels can be sent to a destination from time to time, with acceptable delay. However, it cannot guarantee delivery of a stream of parcels to a destination at a constant rate, say one per hour. That would require a connection-oriented approach in which a flow of packets is treated as a separate entity. A schematic of connection-oriented and connectionless services is shown in Figure 2.4.

A connection-oriented service can be used to provide a type of deterministic performance guarantee. Consider a connection and a link of the network that it uses. Suppose we reserve periodically reoccurring slots of time on this link for transmission of the connection's packets. It is as if the link were a conveyor belt, and a fixed portion of the belt were reserved for carrying the connection's packets. Each time that portion comes around one of the connection's packets can be sent. We assume that slots are large enough to carry an integral number of packets. In practice, packets may be fragmented into small fixed size pieces (called *cells*), where a slot of the synchronous link (the belt) is large enough to hold a cell. Slots reoccur, being part of larger constructs called *frames*. For instance, a particular connection might be assigned the first two slots in a frame consisting of hundred slots, such that every hundred slots the connection gets the first two slots. Packets are reconstructed at the end from the corresponding cells. If the connection sends a stream of packets at a

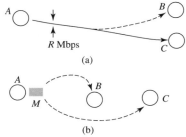

(a)

(b)

Figure 2.4   Connection and connectionless services. Connection-oriented services have the semantics of a directed virtual bit pipe (or perhaps a tree). Connectionless services have the semantics of a datagram service (perhaps to multiple destinations). In (a) a connection-oriented service connects A to B and C with a bit pipe of R Mbps, maximum delay T and bit error rate r. In (b) a connectionless service delivers a message of size M to B and C with maximum delay T and loss probability p.

constant rate, and sufficient time slots are reserved on all the links that it uses, then its packets will arrive at the destination at that same constant rate. Such a transport service is called a *synchronous service*. An important characteristic of synchronous service is that the relative timing of packets at the entrance is preserved at the exit.

This is in contrast to an *asynchronous service*, which makes no such static allocation of slots to connections. Slots are allocated on demand, only when cells are present to be carried. A key characteristic of asynchronous service is that the relative timing of packets at entrance and exit is not preserved. However, we can have reservation of resources (at possibly less than maximum rate) even for asynchronous services (e.g. ABR with MCR, VBR, etc.). For instance, we may specify that every 100 slots, the connection should be able to get at least one slot (if it has cells waiting to be transferred). Note that no particular slot is reserved solely for use of the connection.

Continuing the conveyor belt analogy, when a packet is to be placed on a belt there must be an empty slot, but specific slots on the belt are not pre-allocated to connections. If there is contention for slots, then the connections must wait for the network to assign them free slots. The network does this using some 'contention resolution policy'. We define *synchronous networks* as those that support only synchronous services. They use technology that is optimized for this purpose, breaking information into packets of the size that can be transmitted in a slot and then sending them as streams of slots while reserving specific slots for each connection on the links that the connection uses. In contrast, *packet switching* (or *cell switching*) technology is used for asynchronous services. Information is broken into variable or fixed-sized packets, called cells. These are transported in a store-and-forward manner, without preallocating any slots. Examples of synchronous networks are ISDN (Integrated Services Digital Network), SDH and SONET. Examples of asynchronous networks are the Internet, Frame Relay and ATM. Note that a synchronous service can be provided by an asynchronous network (such as a CBR service in ATM) by performing smart scheduling of the slots. This is used for running telephony over ATM.

Clearly, if customers send data sporadically then a synchronous service may be inefficient, since preallocated slots can go unused. Asynchronous services are better. Note that because of the sporadic nature of asynchronous services it may be sensible for the network to engage in some sort of 'overbooking' when assigning resources to slots. For instance, one may assign a number of slots that is less than what would be required to support the peak rate of the connection.

2.1.5   Guaranteed and Best-effort Services

There is an important distinction between services that do and do not come with guarantees, and which correspondingly do and do not require some reservation of resources. On the one hand, *guaranteed services* come with quality of services guarantees that are expressed in terms of certain parameters of the service's performance. Some reservation of resources is usually required if the guarantees are to be fulfilled. For example, a service that guarantees a minimum transmission rate may need to reserve capacity on a set of links. On the other hand, a service may make no guarantees and reserve no resources; in this case, the performance of the service depends on the quantity of resources it is allocated, and this allocation depends on the network's policy and the set of other services that compete for resources. Since the network usually tries to provide the best quality it can to each of its customers, these services are called *best-effort services*.

Service guarantees may allow some flexibility. For example, it might be guaranteed that no data will be lost if the user's sending rate never exceeds h, but subject to the network being allowed to vary the posted value of h. For more details see Section 2.2.1. This type of flexibility can help the network to improve efficiency by making better use of resources.

The request for a network service originates at an application, and so it is the application's needs that determine the type of connection required to exchange information. For example, a video server needs a minimum bandwidth to send real-time video and so needs a guaranteed service. Other audio and video applications can tolerate performance degradation and can adapt their encoding and frame rates to the available bandwidth. They are examples of *elastic applications*. For these, flexible guarantees may be acceptable.

Note that an elastic application must know the bandwidth that is available at any given time and be able to adapt its rate, rather than risk sending information into the network that may be lost. Thus application elasticity goes hand-in-hand with the network's ability to signal resource availability. Elastic services require this signalling ability. Best-effort services usually do not provide signalling and so elastic applications must implement this signalling functionality themselves (at the application layer). Thus guaranteed services, which provide flexible guarantees, such as in the example above, may be better suited to some adaptive applications.

Example 2.2 (Traditional Internet transport services) The Internet Protocol (IP) is the basic protocol by which packet transport services are provided in the Internet. It operates as a simple packet delivery service. When the IP 'representative' (a piece of software) at the source machine is handed a packet of data and the address of a destination machine (an IP address), it forwards this packet tagged with the IP destination address to 'colleagues' (IP software) running on Internet computers (the routers). These continue to forward and route the packet until it reaches the IP representative at the destination machine. If the network is congested, then packets may be lost before reaching their destination. This happens when a packet arrives at a router and overflows the available storage. This classic IP service is a best-effort service, because it provides no performance guarantees. Today's router implementations permit certain IP packets to receive priority service. However, no explicit guarantees are provided to the flows of such packets.

TCP (Transport Control Protocol) and UDP (User Datagram Protocol) are two transport services that run on top of the IP service, and so are denoted by TCP/IP and UDP/IP. The TCP/IP protocol provides a data transport service with certain performance guarantees. It guarantees zero packet loss to the user by retransmitting packets that are lost because of congestion inside the network. The basic idea is that the TCP software 'listens' to congestion indication signals transmitted by the network and intelligently adjusts its sending rate to the minimum capacity available in the links along the path. Any lost packets are resent. The protocol aims to minimize such retransmissions and to achieve a high link utilization, but it guarantees no minimum or average sending rate. Essentially, the rate at which a connection is allowed to send is dictated by the network. TCP/IP has the interesting property that when it is used by all competing connections then the bandwidth of the bottleneck links is fairly shared. (In practice, connections with longer round-trip delays actually receive smaller bandwidth shares since they are slower to grab any extra bandwidth. This can have severe repercussions).

A connection using the UDP/IP protocols has no constraints, but also has no guarantees. UDP adds little functionality to IP. Like TCP it allows the receiver to detect transmission errors in the data part of the packet. It sends at a maximum rate, irrespective of congestion conditions, and does not resend lost data. UDP is appropriate when one wants to send a small burst of data, but because of its short life, it is not worthwhile to set up a complete TCP/IP connection. UDP is a typical example of a best-effort service with no guarantees. Further details of these protocols are provided in Section 3.3.7.

2.2 Service contracts for transport services

A transport service is provided within the context of a *service contract* between network and user. A part of the contract is the tariff that determines the charge. Beyond this, the contractual commitments of the network and user are as follows. *The network commits to deliver a service with given quality and performance characteristics, and the user commits to interact with the network in a given way.*

If the user violates his side of the contract, then the contract might specify what the network should do. The network might not be bound to any quality of service commitment, or it might restrict the service quality given to the user. Service quality characteristics include geographic coverage, billing services, reliability, up-time, response to failures, help-desk and call-centre support. Service providers can differentiate their service offerings by these characteristics, and so influence the customers' choices of provider. Not surprisingly, it is often hard to quantify the costs of providing these characteristics. No standards exist to constrain the definition of a service contract. It may be arbitrarily complicated and may include clauses specific to the customer.

The part of the service contract that deals with the valued-added part of the service can be complicated, since it can concern issues that are specific to the particular value-added application, such as the copyright of the content provided. Throughout this book, we mostly choose to focus on that part of the service contract that concerns the quality of the transport service. We call this the *traffic contract* part of the service contract. For simplicity, we speak of 'service', rather than 'transport service', when the context allows.

2.2.1 The Structure of a Service Contract

Let us focus on the traffic contract part of the service contract, i.e. the part that deals with aspects of the transport service. In practice, this part of the contract can be described independently of the particular network technology.

As a first example, suppose the network agrees to transport cells between two given points, at a rate no less than m, and dropping no more than a proportion of cells, p. The user agrees to send cells at no greater than a rate, say h, and to access the network for no longer than a maximum time f. If the user sends a total of V bits he will be charged $a + bV$. If the user exceeds his contract limits, the network will provide the excess of the user's traffic with a free and simple best-effort service, but this has no guarantees. The parameters of the traffic contract are m, p, h, f, a and b.

Now let us be more formal. Both the user's traffic and the service quality provided by the network have properties that can be quantified in terms of *measurable variables*. They include, for example, the present rate at which data is sent, the average rate so far, the rate at which cells are lost, and the average delay per cell. (In practice, many contract variables are not measured, as it is too costly to provide the necessary infrastructure. This 'information asymmetry' can provide incentives for contract violation, see Chapter 12.) Let P and Q be predicates over the variables, where P denotes a constraint on the user's traffic and Q denotes a constraint on the performance provided by the network. Then the service contract can be represented as $P \implies Q$. The parameters of the contract are constants used in the construction of P and Q. For instance, the contract clause 'if the sending rate is less than 2 Mbps, then maximum delay per packet is guaranteed to be less than 10 ms' can be specified as '$x_1 \leq 2 \implies x_2 \leq 10$', where x_1, x_2 are variables and 2, 10 are constants. Conceptually, one can imagine that the current values of the variables and constants are continuously displayed at the service interface between user and network.

A service contract is sometimes called a *Service Level Agreement* (SLA). This terminology is more often used for traffic contracts between large customers and network operators, and includes parameters specifying help desk and customer support. In practice, SLAs are specified by long textual descriptions and may contain many ambiguities.

The network's contractual obligations can take the form of either *deterministic* or *statistical* guarantees. An example of a deterministic guarantee is a strict upper bound on delay. Another example is 100% service availability. Statistical guarantees can be framed in terms of cell loss rate, the probability of obtaining access to a modem bank, or the probability that a web server will refuse a connection due to overload. The idea that the network should make some service level guarantees represents a departure from the basic best-effort model of the Internet. In Example 2.3 we describe such guarantees in the context of a particular technology.

Example 2.3 (Quality of Service)   The *quality of service* (QoS) provided by a transport service is defined in terms of the way a traffic stream is affected when it is transported through the network. This is typically in terms of the probability of cell loss, delay, and cell delay variation (or jitter). If access to a resource is required, it may also include the probability that service is refused because the resource is not available. In the case of ATM services (see Section 3.3.5), the QoS measures are

Cell Loss Ratio (CLR): the proportion of cells lost by the network.

Cell Delay Variation (CDV): the maximum difference in the delays experienced by two different cells in their end-to-end transit of the network. CDV is also known as 'jitter'.

Maximum Cell Transfer Delay (max CTD): the maximum end-to-end cell delay.

Mean Cell Transfer Delay (mean CTD): the average end-to-end cell delay.

Minimum Cell Rate (MCR): the minimum rate at which the network transports cells.

A popular way to specify a user's contractual obligation concerning the traffic he sends into a network is by a *leaky bucket* constraint. This is described in Example 2.4. The constraint is on the source's peak rate, average rate and burstiness.

Example 2.4 (Leaky buckets) The *leaky bucket traffic descriptor* can be used to bound the density of a traffic stream at a reference point in the network. It constrains the traffic stream's peak rate, average rate and burstiness (i.e. the short range deviations from the mean rate). Suppose the unit of traffic is a cell. A leaky bucket descriptor is defined by a *leak rate*, r, and *bucket size*, b. Let $X[t, t']$ denote the number of cells of the traffic stream which pass the reference point during the interval $[t, t')$. The leaky bucket imposes the constraint that a *conforming traffic stream* must satisfy

$$X[t, t'] \leq (t' - t)r + b, \quad \text{for all } t < t' \tag{2.1}$$

We call a flow that conforms to (2.1) a (b, r)-flow. Another way to understand (2.1) is to rewrite it as

$$\frac{X[t, t']}{t' - t} \leq r + \frac{b}{t' - t}, \quad \text{for all } t < t'$$

Note that as the window $[t, t')$ increases in width, the average rate permitted during this window becomes bounded above by the leak rate r, but that for a window width of $t' - t$ the stream is allowed to produce a 'burst' of size b above its greatest allowed average amount of $(t' - t)r$. Since this amount b can be produced within an arbitrarily small time window, the leaky bucket descriptor does permit an arbitrarily large peak rate. Note that large values of b allow for large bursts. However, $b = 0$ places a simple bound on the peak rate; at no point can it exceed r.

The simplest way to police a source is by the speed of the access line to the network. This simple mechanism is equivalent to a leaky bucket with $b = 0$ and r equal to the line rate. Figure 2.5 depicts a leaky bucket.

More complicated contractual obligations can be specified from logical combinations of simpler ones. In Example 2.5 the contractual obligation is specified as the conjunction of two simple leaky bucket constraints, each of which addresses a different aspect of the traffic. Such constraints upon the user's traffic are also called *traffic descriptors*.

Example 2.5 (Multiple leaky bucket traffic descriptor) For traffic that is bursty (i.e. which has phases of high and low activity), it is customary to use two leaky buckets to specify conforming traffic. The first leaky bucket constrains the peak rate and the second leaky bucket constrains the time for which the source can send a burst at the peak rate. An

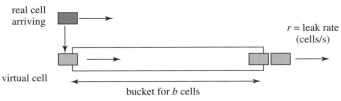

Figure 2.5 A leaky bucket policer. A real cell is conforming if and only if when it arrives there is space in the token buffer to add a virtual cell. The token buffer has space for b virtual cells and is depleted at constant rate of r cells/s.

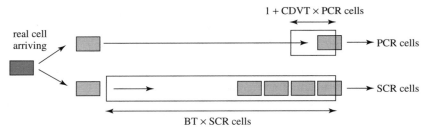

Figure 2.6   A VBR traffic descriptor for bursty traffic defined in terms of two leaky buckets. The top leaky bucket constrains the peak rate. The bottom bucket constrains the burstiness. PCR = peak cell rate. SCR = sustainable cell rate. CDVT = dell delay variation tolerance. BT = burst tolerance.

example of this is the VBR traffic descriptor shown in Figure 2.6. It is used to characterize bursty ATM traffic (described in Chapter 3).

An arriving cell is conforming when it is conforming to both leaky buckets. The leaky bucket that constrains the peak rate is defined in terms of the PCR (peak cell rate) and CDVT (cell delay variation tolerance). When CDVT = 0, then the minimum time that is allowed between two cell arrivals is 1/PCR, whereas if CDVT×PCR = 0.1 (a typical case), then this interarrival time may be temporarily 0.9/PCR. This allows some small fluctuation in cell interarrival times, but the average rate at which cells arrive cannot exceed PCR. Similarly, the leaky bucket that constrains burstiness is defined in terms of SCR (sustainable cell rate) and BT (burst tolerance). Usually the value of BT×SCR is a large integer, allowing for a burst of cells to arrive at greater rate than SCR. Note that the allowed duration of the burst increases with BT and depends on the rate at which cells arrive. Since the average rate of conforming cells cannot exceed SCR, this leaky bucket also constrains the mean rate of the source.

The Integrated Services architecture for the Internet uses a similar approach in its traffic specification (Tspec). This is defined in terms of a dual leaky bucket (similar to the above and with CDVT = 0), a bound on maximum packet size (since in the Internet packet sizes are not fixed as they are in ATM), and a 'minimum policed unit' m (specifying that packets smaller than m bytes should be padded to size m when entering the leaky bucket). There is also a similar use of leaky buckets in the Differentiated Services architecture for the Internet. For more details of these architectures, see Section 3.3.7.

2.2.2   Policing Service Contracts

The network must take steps to monitor and enforce the user's conformance to his contractually agreed interactions with the network. This is called *policing* the contract. For example, a telephone network might wish to police its users for an agreed maximum frequency of dialling, so as to prevent an overload of the signalling part of the network.

It is necessary to say what will happen if the user violates his part of the contract. This can be specified in the service contract itself. One possibility is to specify that if there is a violation then a different quality of service will be provided. For example, if conformance of the user's traffic stream is being policed by a leaky bucket, then a simple specification would be that if there is congestion in the network then it can discard any non-conforming cells, either as they enter or traverse the network. The network could mark non-conforming cells, so that they can be the first to be dropped when congestion occurs.

Of course, there is a cost to policing and it is desirable that policing be implemented such that it introduces low overhead and on the basis of measurements that are easy to make. The user has the incentive to produce traffic that conforms to the contract. One way to do this is by policing his own traffic prior to delivering it to the network. When cells are produced that do not conform to the contract, these are placed in a buffer until they become conforming. This *traffic shaping* tends to smooth the traffic.

A simple way to provide a loose form of policing is by charging. The network simply provides an incentive for the user to respect the contract by imposing a very high charge whenever he violates it. This method has the advantage that it acts indirectly, rather than on a cell by cell basis. Of course it can only work if charges are based upon measurements of the user's traffic that capture the contract violations. Although costly to implement, the flexibility allowed by this type of policing may be of great value to those applications that would sometimes prefer to pay a bit more, rather than see their traffic trimmed by the leaky buckets. Multimedia applications may fall in this category. High capacity networks, serving large numbers of contracts, gain from the fact that traffic streams do not always fully utilize their contracts. It is possible that the provision of a more flexible service contract does not change the load of the network significantly, but does greatly increase the value of the service to the customers.

2.2.3   Static and Dynamic Contract Parameters

There are many service quality characteristics for which no explicit guarantees are made in the traffic contract. The network is free to address them in a best-effort way, or according to some other internal policy. For example, an Internet service provider might make no guarantee as to the probability that a user will find a free modem when he dials-in, but operate enough modems such that he receives very few complaints.

When service contracts do make explicit guarantees, they can do so in different ways. These can differ in the parameters in which they are expressed and the commitments that they require from the network. Consider the following three contracts:

Contract A	Contract B	Contract C
There will be no data loss provided the rate of the source stays below 1 Mbps.	There will be no data loss provided the rate of the source stays below h Mbps, where the network can vary h dynamically between 1 and 2 Mbps.	The data loss rate will be less than 0.000001% provided the rate of the source stays below h Mbps, where the user can vary h dynamically between 1 and 2 Mbps.

Contract A expresses a guarantee in terms of *static parameters*, i.e. ones that are set at the time the contract is established and remain constant throughout its life. This guarantee requires the network to make a firm commitment of resources. The network must reserve 1 Mbps for the service at the start of the contract. Contracts B and C express their guarantee in terms of both static and *dynamic parameters*. Dynamic parameters are ones that are updated during a contract's life. Contract B has a static part, guaranteeing 1 Mbps. Contract C has a static part defined in terms of static parameters 1 Mbps and 0.000001%. Both have an extra, purely dynamic part, for extra rate between 0 and 1 Mbps. A significant difference between B and C is that C is lossy even with rate below h Mb/s.

Apart from the loss guarantee in Contract C being statistical, the main difference between Contracts B and C lies in who chooses h:

1. In Contract B it is the network operator who chooses h, perhaps as a function of his fluctuating spare capacity. He is saying to the user that he can always provide a bandwidth of 1 Mbps, and sometimes up to 2 Mbps. This contract would suit a user of an 'elastic application', that is, an application that can adapt its operation to the available network resources.

2. If in Contract B it were the user who were to choose h, then the network would be saying to the user that he can always provide bandwidth between 1 and 2 Mbps. If this is to be done with a guarantee of no data loss, then the network must always be able to provide 2 Mbps to the user. The only advantage to the network is that he can perhaps sell elsewhere any bandwidth that the user does not use. Thus, Contract C is really more suitable when h is to be chosen by the user. Now the network can make use of statistical multiplexing and overbooking to make efficient use of his bandwidth. Note that it would be a good idea for the network to charge the user according to his usage, otherwise there is no incentive for him to do other than take $h = 2$ always.

In summary, we see that a service contract can be decomposed into static and dynamic parts. The network must permanently reserve resources for the static part. It dynamically reserves resources for the dynamic part, in line with the changing values of the dynamic traffic contract parameters. It is important to specify who is responsible for changing the values of the dynamic parameters, or lay down procedures for the user and network to negotiate them. The network can usually influence the value of the dynamic parameters, even if it cannot choose them directly. By pricing the dynamic contract parameters, the network gives users the incentive to purchase smaller values of them. As prices tend to infinity, the network ends up needing to fulfil only the minimal static part of contract. Depending on the pricing mechanisms available, prices may vary over timescales of seconds to months. For example, time-of-day pricing operates over a timescale of hours. The contract parameters must always be available at the interface between the user and the network, that is, 'visible' to them both.

Example 2.6 (A service with a purely dynamic part) An example of a service with a purely dynamic part is the Internet transport service provided by the TCP protocol and introduced in Example 2.2. The TCP software plays the role of a 'trusted third party'. It runs on the user's computer and dynamically controls the maximum rate, h, at which the user is allowed to send. There is an implicit guarantee of small packet loss inside the network if the user sends packets at this rate (although, in any case, the user does not notice packet losses since TCP retransmits lost packets). To control h, the network sends congestion signals to the TCP module. In response to the rate at which these signals are received, TCP increases or decreases the value of h. There is no guarantee on a minimum value for h. The generation of the congestion signals that are sent to the competing connections relies upon the complete control of the network. TCP will produce a fair allocation of h values amongst the users provided the TCP modules on different computers all run the same algorithm. This is a major weakness, since a user could cheat by installing his own version of TCP, which he has designed to obtain for him a greater bandwidth at the expense of other users. Charging connections for their network usage can remedy this weakness since users that insist upon obtaining a greater share of the bandwidth will incur a greater charge. (Note that thus far we have equated packet loss with congestion. The network does not generate explicit signals of congestion, but rather the user receives implicit signals when he detects packet loss. In some networks, such as those using wireless links, packet loss can occur because of transmission errors, rather than congestion. This can cause TCP to reduce its rate even if there is no congestion.)

A service contract for which the network makes no guarantees whatsoever requires no individual resource allocation. This type of contract is used for best-effort services. The network has maximum flexibility, since it can degrade the performance at any time, without even notifying the user. An example is the UDP protocol, and most prominently, the basic IP protocol of Example 2.2. Typically, the network does the best possible for the best-effort services and may even choose to reserve some resources so as to improve the overall level of service they receive. Of course the user may be willing to pay for a better level of service, with less frequent degradation in service level. It is certainly desirable that a user who is willing to pay more should receive a better level of service. This is not possible within the present best-effort model of the Internet where connections are treated equally. We turn to ideas on how this might be achieved in Chapter 10.

In the literature, the phrase 'guaranteed service' has often been used for a service whose performance level is constant during the life of the contract. We would call this a service that has only strictly static parts. Similarly, the term 'elastic service' has sometimes been used for services having purely dynamic parts, i.e. without any *a priori* commitment from the network. We like to reserve the adjective 'elastic' to describe applications rather than services: an *elastic application* is one that can adapt its operation to the available network resources.

Finally, we observe that some parameters of the traffic contract are *measured parameters*: such as the time the connection starts, its duration and the total volume of data sent. These parameters are known only after the measurements have been made. They are not dynamic contract parameters, because they are not negotiated and do not act as constraints. Measured parameters appear in the *accounting records* that summarize the contract's activity. They and the rest of the parameters in the contract are used to compute charges from the tariff specified in the contract.

2.3   Further reading

A simple tutorial on SLAs has been prepared by Visual Networks, Inc and Telechoice (2002) See also the simple article on SLAs at the web site of AT&T (2000). A rather advanced slide presentation on important SLA implementation issues for IP networks is published by Cisco (2002b). Two more interesting white papers are those of Cisco (2001) and (2002e). Some concrete examples of Service Level Agreements for ATM and Frame Relay can be seen at the web sites of Nortel Networks (2002a) and (2002b).

The concepts of network service layering and quality of service are covered in the classic networking textbooks of Walrand (1998), Walrand and Varaiya (2000) and Kurose and Ross (2001). The latter contains an in-depth exposition of the various service layers that are used to provide Internet services, including the application layer. The first two chapters of Walrand (1998) and Walrand and Varaiya (2000) can provide a good introduction to network service types, layering, and various technologies. These also contain in-depth material on leaky buckets and traffic policing.

3

Network Technology

This chapter concerns the generic aspects of network technology that are important in providing transport services and giving them certain qualities of performance. We define a set of generic control actions and concepts that are deployed in today's communication networks. Our aim is to explain the workings of network technology and to model those issues of resource allocation that are important in representing a network as a production plant for service goods.

In Section 3.1 we outline the main issues for network control. These include the timescale over which control operates, call admission control, routing control, flow control and network management. Tariffing and charging mechanisms provide one important type of control and we turn to these in Section 3.2. Sections 3.3 and 3.4 describe in detail many of the actual network technologies in use today, such as Internet and ATM. We relate these examples of network technologies to the generic control actions and concepts described in earlier sections. In Section 3.5 we discuss some of the practical requirements that must be met by any workable scheme for charging for network services. Section 3.6 presents a model of the business relations amongst those who participant in providing Internet services.

3.1 Network control

A *network control* is a mechanism or procedure that the network uses to provide services. The more numerous and sophisticated are the network controls, the greater and richer can be the set of services that the network can provide. Control is usually associated with the procedures needed to set up new connections and tear down old ones. However, while a connection is active, network control also manages many other important aspects of the connection. These include the quality of the service provided, the reporting of important events, and the dynamic variation of service contract parameters.

Synchronous services provided by synchronous networks have the simple semantics of a constant bit rate transfer between two predefined points. They use simple controls and all bits receive the same quality of service. *Asynchronous networks* are more complex. Besides providing transport between arbitrary points in the network, they must handle unpredictable traffic and connections of arbitrarily short durations. Not all bits require the same quality of service.

Some network technologies have too limited a set of controls to support transport services with the quality required by advanced multimedia applications. Even for synchronous services, whose quality is mostly fixed, some technologies have too limited controls to

Pricing Communication Networks   C. Courcoubetis and R. Weber
© 2003 John Wiley & Sons, Ltd   ISBN 0-470-85130-9 (HB)

make it possible quickly to set up new connections on demand. A knowledge of the various network control mechanisms is key to understanding how communication networks work and how service provisioning relates to resource allocation. In the rest of the chapter we mainly focus on the controls that are deployed by asynchronous networks. These controls shape the services that customers experience.

3.1.1   Entities on which Network Control Acts

A network's topology consists of *nodes* and *links*. Its nodes are routers and switches. Its links provide point-to-point connectivity service between two nodes, or between a customer and a node, or amongst a large number of nodes, as in a Metropolitan Gigabit Ethernet. We take the notion of a link to be recursive: a point-to-point link in one network can in fact be a transport service provided by a second network, using many links and nodes. We call this a 'virtual' link. Since links are required to provide connectivity service for bits, cells or packets at some contracted performance level, the network must continually invoke control functions to maintain its operation at the contracted level. These control functions are implemented by hardware and software in the nodes and act on a number of entities, the most basic of which are as follows.

Packets and cells. These are the parcels into which data is packaged for transport in the network. Variable size parcels are called *packets*, whereas those of fixed size are called *cells*. Internet packets may be thousands of bytes, whereas cells are 53 bytes in the ATM technology. Higher level transport services often use packets, while lower-level services use cells. The packets must be broken into cells and then later reconstructed into packets. We will use the term packet in the broad sense of a data parcel, unless specific reasons require the terminology of a cell.

Connections. A connection is the logical concept of binding end-points to exchange data. Connections may be point-to-point, or point-to-multipoint for multicasting, although not all technologies support the latter. A connection may last from a few seconds (as in the access of web pages) to years (as in the connection of a company's network to the Internet backbone). Depending on the technology in use, a connection may or may not be required. The transfer of web page data as packets requires a connection to be made. In contrast, there is no need to make a connection prior to sending the packets of a datagram service. Clearly, the greater is a technology's cost for setting up a connection the less well suited it is to short-lived connections. Once a connection has been set up, the network may have to allocate resources to handle the connection's traffic in accordance with an associated Service Level Agreement.

Flows. The information transported over a connection may be viewed as a continuous flow of bits, bytes, cells or packets. An important attribute of a flow is its rate. This is the amount of information that crosses a point in the network, averaged over some time period. The job of a network is to handle continuous flows of data by allocating its resources appropriately. For some applications, it may have to handle flows whose rates are fluctuating over time. We call such flows 'bursty'. When network resources are shared, instead of dedicated on a per flow basis, the network may seek to avoid congestion by using *flow control* to adjust the rates of the flows that enter the network.

Calls. These are the service requests that are made by applications and which require connections to be set up by the network. They usually require immediate response from the network. When a customer places a call in the telephone network, a voice circuit connection must be set up before any voice information can be sent. In the Internet, requests for web pages are calls that require a connection set-up. Not all transport technologies possess controls that provide immediate response to calls. Instead, connections may be scheduled long in advance.

Sessions. These are higher-level concepts involving more than one connection. For example, a video conference session requires connections for voice, video, and the data to be displayed on a white board. A session defines a context for controlling and charging.

3.1.2 Timescales

One way to categorize various network controls is by the timescales over which they operate. Consider a network node (router) connected to a transatlantic ATM link of speed 155 Mbps or more. The IP packets are broken into 53 byte ATM cells and these arrive every few microseconds. The packets that are reassembled from the cells must be handled every few tens of microseconds. Feedback signals for flow control on the link arrive every few tens of milliseconds (the order of a round trip propagation time, which depends on distance). Requests for new connections (at the TCP layer) occur at the rate of a few per second (or tenths of a second). Network management operations, such as routing table updates, take place over minutes. From milliseconds to a year are required for pricing policies to affect demand and the link's load. (see Figure 3.1).

In the next sections, we briefly review some key network controls.

3.1.3 Handling Packets and Cells

The fastest timescale on which control decisions can be made is of the order of a packet interarrival time. Each time a network node receives a packet it must decide whether the

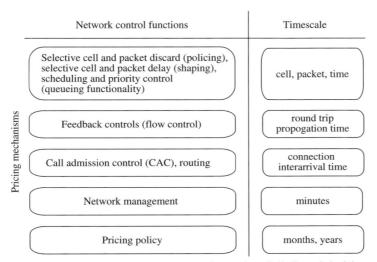

Figure 3.1 Network control takes place on many timescales. Cell discard decisions are made every time a cell is received, whereas pricing policy takes place over months or years. Pricing mechanisms (algorithms based on economic models) can be used for optimizing resource sharing at all levels of network control.

packet conforms to the traffic contract. If it does not, then the node takes an appropriate policing action. It might discard the packet, or give it a lower quality service. In some cases, if a packet is to be discarded, then a larger block of packets may also be discarded, since losing one packet makes all information within its context obsolete. For instance, consider Internet over ATM. An Internet packet consists of many cells. If a packet is transmitted and even just one cell from the packet is lost, then the whole packet will be resent. Thus, the network could discard all the cells in the packet, rather than waste effort in sending those useless cells. This is called 'selective cell discard'.

A crucial decision that a network node must take on a per packet basis is where to forward an incoming packet. In a connectionless network, the decision is based on the destination of the packet through the use of a *routing table*. Packets include network-specific information in their *header*, such as source and destination addresses. In the simplest case of a *router* or *packet switch* the routing table determines the node that should next handle the packet simply from the packet's destination.

In a connection-oriented network, the packets of a given connection flow through a path that is pre-set for the connection. Each packet's header contains a label identifying the connection responsible for it. The routing function of the network defines the path. This is called *virtual circuit switching*, or simply switching. More details are given in Section 3.1.4. Forwarding in a connection-oriented network is simpler than in a connectionless one, since there are usually fewer active connections than possible destinations. The network as a whole has responsibility for deciding how to set routing tables and to construct and tear down paths for connections. These decisions are taken on the basis of a complete picture of the state of the network and so are rather slow to change. Network management is responsible for setting and updating this information.

An important way to increase revenue may be to provide different qualities of service at different prices. So in addition to making routing decisions, network nodes must also decide how to treat packets from different connections and so provide flows with different qualities of packet delay and loss. All these decisions must be taken for each arriving packet. The time available is extremely short; in fact, it is inversely proportional to the speed of the links. Therefore, a large part of the decision-making functionality for both routing and differential treatment must be programmed in the hardware of each network node.

3.1.4 Virtual Circuits and Label Switching

Let us look at one implementation of circuit switching. A network path r between nodes A and B is a sequence of links $l_1, l_2, \ldots, l_n$ that connect A to B. Let $1, \ldots, n+1$ be the nodes in the path, with $A = 1$ and $B = n+1$. A label-switched path r_a over r is a sequence $(l_1, a_1), (l_2, a_2), \ldots, (l_n, a_n)$, with labels $a_i, i = 1, \ldots, n$. Labels are unique identifiers and may be coded by integers. Such a label-switched path is programmed inside the network by

    1. associating r_a at node A with the pair (l_1, a_1), and at node B with (l_n, a_n);

    2. adding to the switching table of each of the intermediate node i the local mapping information $(l_{i-1}, a_{i-1}) \rightarrow (l_i, a_i), i = 2, \ldots, n$.

When a call arrives requesting data transport from A to B, a connection a is established from A to B in terms of a new label-switched path, say r_a. During data transfer, node A breaks the large units of data that are to be carried by the connection a into packets, assigns the label a_1 to each packet, and sends it through link l_1 to node 2. Node $i, i = 2, \ldots, n$, switches arriving packets from input link l_{i-1} with label a_{i-1} to the output link l_i and

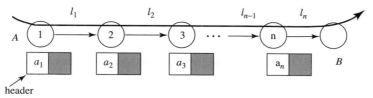

Figure 3.2   A label-switched path implementing a virtual circuit between nodes A and B.

changes the label to the new value a_i, as dictated by the information in its switching table, see Figure 3.2. At the end of the path, the packets of connection a arrive in sequence at node B carrying label a_n. The pair (l_n, a_n) identifies the data as belonging to connection a. When the connection is closed, the label-switched path is cleared by erasing the corresponding entries in the switching tables. Thus, labels can be reused by other connections.

Because a label-switched path has the semantics of a circuit it is sometimes called a *virtual circuit*. One can also construct 'virtual trees' by allowing many paths to share an initial part and then diverge at some point. For example, binary branching can be programmed in a switching table by setting $(l_i, a_i) \rightarrow [(l_j, a_j), (l_k, a_k)]$. An incoming packet is duplicated on the outgoing links, l_j, l_k, with the duplicates possibly carrying different labels. Trees like this can be used to multicast information from a single source to many destinations. Virtual circuits and trees are used in networks of ATM technology, where labels are integer numbers denoting the virtual circuit number on a particular link (see Section 3.3.5). In a reverse way, label-switched paths may be merged inside the network to create reverse trees (called *sink-trees*). This is useful in creating a logical network for reaching a particular destination. Such techniques are used in MPLS technology networks (see Section 3.3.7). Virtual circuits and trees are also used in Frame Relay networks (see Section 3.3.6).

3.1.5   Call Admission Control

We have distinguished best-effort services from services that require performance guarantees. A call that requires a guaranteed service is subject to call admission control to determine if the network has sufficient resources to fulfil its contractual obligations. Once admitted, policing control ensures that the call does not violate its part of the contract. Policing controls are applied on the timescale of packet interarrival times. Call admission control (CAC) is applied on the timescale of call interarrival times. Since call interarrival times can be relatively short, admission decisions must usually be based upon information that is available at the entry node. This information must control the admission policy and reflect the ability of the network to carry calls of given types to particular destinations. (It may also need to reflect the network provider's policy concerning bandwidth reservation and admission priorities for certain call types.) It is not realistic to have complete information about the state of the network at the time of each admission decision. This would require excessive communication within the network and would be impossible for networks whose geographic span means there are large propagation delays. A common approach is for the network management to keep this information as accurately as possible and update it at time intervals of appropriate length.

The call admission control mechanism might be simple and based only on traffic contract parameters of the incoming call. Alternatively, it might be complex and use data from on-line measurements (*dynamic call admission control*). Clearly, more accurate CAC allows for better loading of the links, less blocking of calls, and ultimately more profit for the network operator. To assess the capacity of the network as a transport service 'production facility', we need to know its topology, link capacities and call admission control policy. Together, these constrain the set of possible services that the network can support simultaneously. This is important for the economic modelling of a network that we pursue in Chapter 4. We define for each contract and its resulting connection an *effective bandwidth*. This is a simple scalar descriptor which associates with each contract a resource consumption weight that depends on static parameters of the contract. Calls that are easier to handle by the network, i.e. easier to multiplex, have smaller effective bandwidths. A simple call admission rule is to ensure that the sum of the effective bandwidths of the connections that use a link are no more than the link's bandwidth.

In networks like the Internet, which provide only best-effort services, there is, in principle, no need for call admission control. However, if a service provider wishes to offer better service than his competitors, then he might do this by buying enough capacity to accommodate his customers' traffic, even at times of peak load. But this would usually be too expensive. An alternative method is to control access to the network. For instance, he can reduce the number of available modems in the modem pool. Or he can increase prices. Prices can be increased at times of overload, or vary with the time of day. Customers who are willing to pay a premium gain admission and so prices can act as a flexible sort of call admission control. In any case, prices complement call admission control by determining the way the network is loaded, i.e. the relative numbers of different service types that are carried during different demand periods.

Call admission control is not only used for the short duration contracts. It is also used for contracts that may last days or months. These long duration contracts are needed to connect large customers to the Internet or to interconnect networks. In fact, connection-oriented technology, such as ATM, is today mainly used for this purpose because of its particular suitability for controlling resource allocation.

3.1.6   Routing

Routing has different semantics depending on whether the network technology is connection-oriented or connectionless. In connection-oriented technology, routing is concerned with the logic by which network's routers forward individual packets. In connectionless technology it is concerned with the logic by which the physical paths for connections are chosen. Let us investigate each case separately.

In a connection-oriented network, as depicted in Figure 3.3, routing is concerned with choosing the path that a connection's data is to take through the network. It operates on a slower timescale than policing, since it must be invoked every time a new call arrives. In *source routing*, information at the source node is used to make simultaneous decisions about call acceptance and about the path the call will follow. When the load of the network changes and links that have been favoured for routing are found to have little spare capacity, then the information that is kept at entry nodes can be updated to reflect the change of network state. On the basis of the updated information, the routing control algorithms at the entry nodes may now choose different paths for connections. Again, network management is responsible for updating information about the network state.

Source routing is relevant to networks that support the type of connection-oriented services defined in Section 2.1.4. (It is also defined, but rarely used, in datagram networks, by including in a packet's header a description of the complete path that the packet is to follow in the network.) Connection-oriented networks have the connection semantics of an end-to-end data stream over a fixed path. The basic entity is a connection rather than

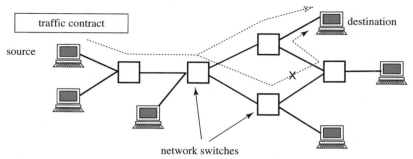

Figure 3.3 In a connection-oriented network each newly arriving call invokes a number of network controls. *Call routing* finds a path from the source to destination that fulfils the user's requirements for bandwidth and QoS. *Call admission control* is applied at each switch to determine whether there are enough resources to accept the call on the output link. *Connection set-up* uses signalling mechanisms to determine the path of the connection, by routing and CAC; it updates switching tables for the new virtual circuit and reserves resources. Above, X marks a possible route that is rejected by routing control. *Flow control* regulates the flow in the virtual circuit once it is established.

individual packets. When a call is admitted, the network uses its *signalling mechanism* to set the appropriate information and reserve the resources that the call needs at each network node along the path. This signalling mechanism, together with the ability to reserve resources for an individual call on a virtual circuit, is a powerful tool for supporting different QoS levels within the same network. It can also be used to convey price information.

During the signalling phase, call admission control functions are invoked at every node along the connection's path. The call is blocked either if the entry node decides that there are insufficient resources inside the network, or if the entry node decides that there may be enough resources and computes a best candidate path, but then some node along that path responds negatively to the signalling request because it detects a lack of resources. A similar operation takes place in the telephone network. There are many possibilities after such a refusal: the call may be blocked, another path may be tried, or some modification may be made to the first path to try to avoid the links at which there were insufficient resources. Blocking a call deprives the network from extra revenue and causes unpredictable delays to the application that places the call. Call blocking probability is a quality of service parameter that may be negotiated at the service interface. Routing decisions have direct impact on such blocking probabilities, since routing calls on longer paths increases the blocking probability compared with routing on shorter paths.

In a connectionless (datagram) network, the reasoning is in terms of the individual packets, and so routing decisions are taken, and optimized, on a per packet basis. Since the notion of a connection does not exist, a user who needs to establish a connection must do so by adding his own logic to that provided by the network, as when the TCP is used to make connections over the Internet. The goal might be to choose routes that minimize transit delay to packet destinations. Routers decide on packet forwarding by reading the packet destination address from the packet header and making a lookup in the *routing table*. This table is different for each router and stores for each possible destination address the next 'hop' (router or the final computer) that the packet should take on the way to its destination. Routing tables are updated by routing protocols on a timescale of minutes, or when an abrupt event occurs. In pure datagram networks the complexity of network controls is reduced because no signalling mechanism is required.

If packets that are destined for the same end node may be roughly described as indistinguishable, as is the case in the present Internet, then there is an inherent difficulty in allocating resources on a per call basis. Admission control on a per call basis does not make sense in this case. A remedy is to add extra functionality; we see this in the architectures of Internet Differentiated Services and Internet Integrated Services, described in Section 3.3.7. The extra functionality comes at the expense of introducing some signalling mechanisms and making the network more complex.

Routing is related to pricing since it defines how the network will be loaded, thus affecting the structure of the network when viewed as a service factory. For example, video connections may use only a subset of the possible routes. One could envisage more complex interactions with pricing. For instance, having priced different path segments differently, a network operator might allow customers to 'build' for themselves the routes that their traffic takes through the network. In this scenario, the network operator releases essential aspects of network control to his customers. He controls the prices of path segments and these directly influence the customers' routing decisions. A challenging problem is to choose prices to optimize the overall performance of the network. Observe that such an approach reduces the complexity of the network, but places more responsibility with the users. It is consistent with the Internet's philosophy of keeping network functions as simple as possible. However, it may create dangerous instabilities if there are traffic fluctuations and users make uncoordinated decisions. This may explain why network operators presently prefer to retain control of routing functions.

3.1.7   Flow Control

Once a guaranteed service with dynamic contract parameters is admitted, it is subject to network control signals. These change the values of the traffic contract parameters at the service interface and dictate that the user should increase or decrease his use of network resources. The service interface may be purely conceptual; in practice, these control signals are received by the user applications. In principle the network can enforce its flow control 'commands' by policing the sources. However, in networks like the Internet, this is not done, because of implementation costs and added network complexity.

In most cases of transport services with dynamic parameters (such as the transport service provided by the TCP protocol in the Internet), the network control signals are congestion indication signals. *Flow control* is the process with which the user increases or decreases his transmission rate in response to these signals. The timescale on which flow control operates is that of the time it takes the congestion indication signals to propagate through the network; this is at most the round trip propagation time. Notice that the controls applied to guaranteed services with purely static parameters are open-loop: once admitted, the resources that are needed are reserved at the beginning of the call. The controls applied to guaranteed services with purely dynamic parameters are closed-loop: control signals influence the input traffic with no need for *a priori* resource reservation.

Flow control mechanisms are traditionally used to reduce *congestion*. Congestion can be recognized as a network state in which resources are poorly utilized and there is unacceptable performance. For instance, when packets arrive faster at routers than the maximum speed that these can handle, packet queues become large and significant proportions of packets overflow. This provides a good motivation to send congestion signals to the sources before the situation becomes out of hand. Users see a severe degradation in the performance of the network since they must retransmit lost information (which further increases congestion), or they find that their applications operate poorly. In any case, congestion results in waste and networks use flow control to avoid it. Of course complete absence of congestion may mean that there is also waste because the network is loaded too conservatively. There are other tools for *congestion control* besides flow control. Pricing policies or appropriate call admission controls can reduce congestion over longer timescales. If prices are dynamically updated to reflect congestion, then they can exert effective control over small timescales. We consider such pricing mechanisms in Chapter 9.

Flow control also has an important function in controlling the allocation of resources. By sending more congestion signals to some sources than others, the network can control the allocation of traffic flow rates to its customers. Thus flow control can be viewed as a mechanism for making a particular choice amongst the set of feasible flows. This is important from an economic perspective as economic efficiency is obtained when bandwidth is allocated to those customers who value it most. Most of today's flow control mechanisms lack the capability to allocate bandwidth with this economic perspective because the part of the flow control process that decides when and to whom to send congestion signals is typically not designed to take it into account. Flow control only focuses on congestion avoidance, and treats all sources that contribute to congestion equally.

Flow control can also be viewed as a procedure for fairly allocating resources to flows. Fairness is a general concept that applies to the sharing of any common good. An allocation is said to be fair according to a given fairness criterion when it satisfies certain fairness conditions. There are many ways to define fairness. For example, proportional fairness emphasizes economic efficiency and allocates greater bandwidth to customers who are willing to pay more. Max-min fairness maximizes the size of the smallest flow. Implicit in a fairness definition for the allocation of bandwidth is a function that takes customer's demands for flows and computes an allocation of bandwidth. The allocation is fair according to the fairness definition and uses as much of the links' bandwidth as possible. Given the way that user applications respond to congestion signals, a network operator can implement his preferred criterion for fair bandwidth allocation by implementing appropriate congestion signalling mechanisms at the network nodes. In Chapter 10 we investigate flow control mechanisms that control congestion and achieve economic fairness.

The use of flow control as a mechanism for implementing fair bandwidth allocation relies on users reacting to flow control signals correctly. If a flow control mechanism relies on the user to adjust his traffic flow in response to congestion signals and does not police him then there is the possibility he may cheat. A user might seek to increase his own performance at the expense of other users. The situation is similar to that in the prisoners' dilemma (see Section 6.4.1). If just one user cheats he will gain. However, if all users cheat, then the network will be highly congested and all users will lose. This could happen in the present Internet. TCP is the default congestion response software. However, there exist 'boosted' versions of TCP that respond less to congestion signals. The only reason that most users still run the standard version of TCP is that they are ignorant of the technological issues and do not know how to perform the installation procedure.

Pricing can give users the incentive to respond to congestion signals correctly. Roughly speaking, users who value bandwidth more have a greater willingness to pay the higher rate of charge, which can be encoded in a higher rate of congestion signals that is sent during congestion periods. Each user seeks what is for him the 'best value for money' in terms of performance and network charge. He might do this using a bandwidth seeking application. It should be possible to keep congestion under control, since a high enough rate of congestion charging will make sources reduce their rates sufficiently.

Sometimes flow control may be the responsibility of the user rather than the network. For instance, if the network provides a purely best-effort service, it may be the responsibility of the user to adjust his rate to reduce packet losses and delays.

3.1.8   Network Management

Network management concerns the operations used by the network to improve its performance and to define explicit policy rules for security, handling special customers, defining services, accounting, and so on. It also provides capabilities for monitoring the traffic and the state of the network's equipment. The philosophy of network management is that it should operate on a slow timescale and provide network elements with the information they need to react on faster timescales as the context dictates.

Network management differs from signalling. Signalling mechanisms react to external events on a very fast timescale and serve as the 'nervous system' of the network. Network management operations take place more slowly. They are triggered when the network administrator or control software detects that some reallocation or expansion of resources is needed to serve the active contracts at the desired quality level. For example, when a link or a node fails, signalling is invoked first to choose a default alternative. At a later stage this decision is improved by the network management making an update to routing tables.

3.2   Tariffs, dynamic prices and charging mechanisms

Network control ensures that the network accepts no more contracts than it can handle and that accepted contracts are fulfilled. However, simple call admission control expresses no preference for the mix of different contracts that are accepted. Such a preference can be expressed through complex call admission control strategies that differentiate contract types in terms of blocking. Or they can also be expressed through tariffing and charging, which may be viewed as a higher-level flow control that operates at the contract level by offering different incentives to users. They not only ensure that demand does not exceed supply, but also that the available capacity is allocated amongst potential customers so as to maximize revenue or be socially efficient (in the sense defined in Section 5.4). Note, however, that for the latter purpose charges must be related to resource usage. We discuss this important concept in Chapter 8. Charges also give users the incentive to release network resources when they do not need them, to ask only for the contracts that are most suited to them, and for those users who value a service more to get more of it. Simplicity and flexibility are arguments for regulating network usage by using tariffing rather than complex network controls. The network operator does not need to reprogram the network nodes, but simply post appropriate tariffs for the services he offers. This pushes some of the decision-making onto the users and leaves the network to carry out basic and simple operations.

Viewed as a long-term control that is concerned with setting tariffs, pricing policy emerges in an iterative manner (i.e. from a tatonnement as described in Section 5.4.1). Suppose that a supplier posts his tariffs and users adjust their demands in response. The supplier reconsiders his tariffs and this leads to further adjustment of user demand. The timescale over which these adjustments take place is typically months or years. Moreover, regulation may prevent a supplier from changing tariffs too frequently, or require that changes make no customer worse off (the so-called 'status-quo fairness' test of Section 10.1). In comparison, dynamic pricing mechanisms may operate on the timescale of a round trip propagation time; the network posts prices that fluctuate with demand and resource availability. The user's software closely monitors the price and optimally adjusts the consumption of network resources to reflect the user's preferences.

Dynamic pricing has an implementation cost for both the network and the customers. A practical approximation to it is *time-of-day pricing*, in which the network posts fixed prices for different periods of the day, corresponding to the average dynamic prices over the given periods. This type of pricing requires less complex network mechanisms. Customers like it because it is predictable.

It is a misconception that it is hard for customers to understand and to react to dynamic prices. One could envision mechanisms that allow customers to pay a flat fee (possibly zero) and the network to adapt the amount of resources allocated at any given time so that each customer receives the performance for which he pays. Or customers might dynamically choose amongst a number of flat rate charging structures (say, gold, silver or bronze) and then receive corresponding qualities of service. In this case prices are fixed but performance fluctuates. Alternatively, a customer might ask for a fixed performance and have a third party pay its fluctuating cost. This is what happens in the electricity market, in which generators quote spot prices, but end-customers pay constant prices per KWh to their suppliers. A customer might buy insurance against severe price fluctuations. All of these new value-added communication service models can be implemented easily since they mainly involve software running as a network application.

Suppose that a network service provider can implement mechanisms that reflect resource scarcity and demand in prices, and that he communicates these to customers, who on the basis of them take decisions. Ideally, we will find that as the provider and users of network services freely interact, a 'market-managed network' emerges, that has desirable stability properties, optimizes global economic performance measures, and allows information to remain local where it is needed for decision-making. The task of creating such a self-managed network is not trivial. The involvement of a large number of entities and complex economic incentives makes security issues of paramount importance. For instance, the network that charges its customers for its services is only the final network in a value chain, which involves many other transport and value-added service providers. Each intermediate network has an incentive to misreport costs and so extract a larger percentage of the customer payment. This means that sophisticated electronic commerce techniques must be used for security and payments. Network may try to provide a worse quality service to customers of other network providers, so as to improve the service offered to its customers or attract the customers of other operators. Networks are no longer trusted parties, as they are in the case of the large state-controlled network monopolies. New security and payment models and mechanisms are required.

3.3 Service technologies

3.3.1 A Technology Summary

The concepts we have mentioned so far are quite general. In this and the following section we discuss some of the data transport services that are standardized and supported by network technologies such as the Internet and ATM. Such services are used to link remote applications and they are differentiated in terms of the quality of the service offered by the network. The reader will recognize most of the generic service interface aspects that we have introduced.

We discussed in Section 2.1.1 the ideas of layering and of synchronous and asynchronous technologies. At a lower layer, synchronous services such as SONET provide for large fixed size containers, called *frames*. We may think of a frame as a large fixed size sequence of bits containing information about the frame itself and the bytes of higher layer service data that are encapsulated in the frame. Synchronous framing services constantly transmit frames one after the other, even if no data are available to fill these frames. Frames may be further subdivided into constant size sub-frames, so allowing multiple synchronous connections of smaller capacities to be set up.

At a higher layer, asynchronous technologies such as IP, ATM and Frame Relay, break information streams into data packets (or cells) that are placed in the frames (or the smaller sub-frames). Their goal is to perform statistical multiplexing, i.e. to efficiently fill these frames with packets belonging to different information streams. At the lowest layer, these *framing services* may operate over fibre by encoding information bits as light pulses of a certain wavelength (the 'λ'). Other possible transmission media are microwave and other wireless technologies. For example, a satellite link provides for synchronous framing services over the microwave path that starts from the sending station and reflects off the satellite to all receivers in the satellite's footprint. In contrast to SONET, Gigabit and 10 Gigabit Ethernet is an example of a framing service that is asynchronous and of variable size. Indeed, an Ethernet frame is constructed for each IP packet and is transmitted immediately at some maximum transmission rate if conditions permit. As we will see, since Ethernet frames may not depart at regular intervals (due to contention resulting from the customers using the same link), Ethernet services may not provide the equivalent of a fixed size bit pipe. Guaranteed bandwidth can be provided by dedicating Ethernet fibre links to single customer traffic. Finally, note that ATM is an asynchronous service that is used by another asynchronous service, namely IP. The IP packets are broken into small ATM cells which are then used to fill the lower-level synchronous frames.

Our discussion so far suggests that customers requiring connections with irregular and bursty traffic patterns should prefer higher layer asynchronous transport services. Asynchronous services then consume lower layer framing services (synchronous or asynchronous), which usually connect the network's internal nodes and the customers to the network. Framing services consume segments of fibre or other transmission media. Observe that a customer whose traffic is both great and regular enough efficiently to use large synchronous containers, might directly buy synchronous services to support his connection. Similarly, large customers with bursty traffic may buy asynchronous container services, e.g. Ethernet services, that allow further multiplexing of the raw fibre capacity.

Figure 3.4 shows a classification of the various transport services that we present in the next sections. For simplicity we assume that the physical transmission medium is fibre. In fact, microwave and wireless are also possible media. This may complicate the picture somewhat, since some of the framing protocols running over fibre may not run over other media.

Services towards the bottom of the diagram offer fixed size bit pipes of coarse granularity, and the underlying controls to set up a call are at the network management layer, i.e. do not work in very fast timescales. By their nature, these are better suited for carrying traffic in the interior of the network where traffic is well aggregated. Ethernet is the only technology offering coarse bit pipes that may be shared. Fibre is 'technology neutral' in the sense that the higher layer protocols dictate the details (speed) of information transmission. Such protocols operate by transmitting light of a certain wavelength. DWDM is a technology that multiplies the fibre throughput by separating light into a large number of wavelengths, each of which can carry the same quantity of information that the fibre was able to carry using a single wavelength.

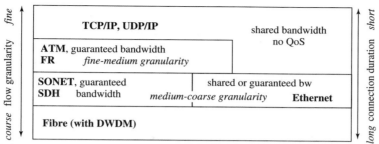

Figure 3.4 Services towards the bottom of this diagram offer fixed size bit pipes of coarse granularity, and the underlying controls for call set-up do not work in very fast timescales. Services towards the top offer flexible pipes of arbitrarily small granularity and small to zero set-up cost that can be established between any arbitrary pair of network edge points. Fibre is 'technology neutral' in the sense that the higher layer protocols dictate the details of information transmission.

Services towards the top of the diagram build flexible pipes of arbitrarily small granularity. These are mainly TCP/IP and UDP/IP pipes, since the dynamic call set-up of the ATM standard is not implemented in practice. (Note, also, that we have denoted ATM and Frame Relay as guaranteed services, in the sense that they can provide bandwidth guarantees by using an appropriate SLA. These service have more general features that allow them to provide best-effort services as well.)

Connections using services at the top of the diagram have little or no set-up cost, and can be established between arbitrary pairs of network edge points. This justifies the use of the IP protocol technology for connecting user applications. In the present client-server Internet model (and even more in future peer-to-peer communications models), connections are extremely unpredictable in terms of duration and location of origin-destination end-points. Hence the only negative side of IP is the absence of guarantees for the diameter of the pipes of the connections. Such a defect can be corrected by extending the IP protocol, or by performing *flow isolation*. This means building fixed size pipes (using any of the fixed size pipe technology) between specific points of the network to carry the IP flows that require differential treatment. This is the main idea in the implementation of Virtual Private Networks described in detail in Section 3.4.1 using the MPLS technology.

We now turn to detailed descriptions of the basic connection technologies.

3.3.2   Optical Networks

Optical networks provide a full stack of connection services, starting from *light path* services at the lowest layer and continuing with framing services, such as SONET and Ethernet, up to ATM and IP services. We concentrate on the lower layer light path services since the higher layers will be discussed in following sections.

Dense Wavelength Division Multiplexing (DWDM) is a technology that allows multiple light beams of different colours (λs) to travel along the same fibre (currently 16 to 32 λs, with 64 and 80 λ in the laboratories). A light path is a connection between two points in the network which is set up by allocating a dedicated (possibly different) λ on each link over the path of the connection. Along such a light path, a light beam enters the network at the entry point, using the λ assigned on the first link, continues through the rest of the links by changing the λ at each intermediate node and finally exits the network at the exit point. This is analogous to circuit-switching, in which the λs play the role of circuit identifiers or of labels on a label-switched path. Lasers modulate the light beam into pulses that encode the bits, presently at speeds of 2.5 Gbps and 10 Gbps, and soon to be 40 Gbps, depending on the framing technology that is used above the light path layer. Optical signals are attenuated and distorted along the light path. Depending on the fibre quality and the lasers, the light pulses need to be amplified and possibly regenerated after travelling for a certain distance. These are services provided internally by the optical network service provider to guarantee the quality of the information travelling along a light path. In an *all-optical network*, the light that travels along a lightpath is regenerated and switched at the optical level, i.e. without being transformed into electrical signals.

In the near future, optical network management technology will allow lightpaths to be created dynamically at the requests of applications (just like dynamic virtual circuits). Even further in the future, optical switching will be performed at a finer level, including switching at the level of packets and not just at the level of the light path's colour. Dynamic light path services will be appropriate for applications that can make use of the vast amounts of bandwidth for a short time. However, the fact that optical technology is rather cheap when no electronic conversion is involved means that such services may be economically sensible even if bandwidth is partly wasted. Presently, lightpath services are used to create virtual private networks by connecting routers of the same enterprise at different locations.

An important property of a lightpath service is *transparency* regarding the actual data being sent over the lightpath. Such a service does not specify a bit rate since the higher layers such as Ethernet or SONET with their electrical signals will drive the lasers which are also part of the Ethernet or SONET specification. A certain maximum bit rate may be specified and the service may carry data of any bit rate and protocol format, even analog data. Essentially the network guarantees a minimum bound on the distortion and the attenuation of the light pulses. In the case of a light path provided over an all-optical network, where there is optical to electrical signal conversion for switching and regeneration, the electro-optical components may pose further restrictions on maximum bit rates that can be supported over the light path.

A *dark fibre* service is one in which a customer is allocated the whole use of an optical fibre, with no optical equipment attached. The customer can make free use of the fibre. For example, he might supply SONET services to his customers by deploying SONET over DWDM technology, hence using more than a single λs.

There is today a lot of dark fibre installed around the world. Network operators claim that their backbones have capacities of hundreds of Gigabits or Terabits per second. Since this capacity is already in place and its cost is sunk, one might think that enormous capacity can be offered at almost zero cost. However, most of the capacity is dark fibre. It is costly to add lasers to light the fibre and provide the other necessary optical and electronic equipment. This means there is a non-trivial variable cost to adding new services. This 'hidden' cost may be one reason that applications such as video on demand are slow to come to market.

3.3.3 Ethernet

Ethernet is a popular technology for connecting computers. In its traditional version, it provides a best-effort framing service for IP packets, one Ethernet frame per IP packet. The framed IP packets are the Ethernet packets which can be transmitted only if no other node of the Ethernet network is transmitting. The transmission speeds are from 10 Mbps to 10 Gbps in multiples of ten (and since the price of a 10 Gbps Ethernet adaptor card is no more than 2.5 times the price of a 1 Gbps card, the price per bit drops by a factor of four). Ethernet technologies that use switching can provide connection-oriented services that are either best-effort or have guaranteed bandwidth. Ethernet can provide service of up to 54 Mbps over wireless and over the twisted-pair copper wires that are readily available in buildings. Twisted-pair wiring constrains the maximum distance between connected equipment to 200 meters. For this reason, Ethernet has been used mainly to connect computers that belong to the same organization and which form a Local Area Network (LAN). It is by far the most popular LAN technology, and more than 50 million Ethernet interface cards are sold each year.

Ethernet service at speeds greater than 100 Mbps is usually provided over fibre; this greatly extends the feasible physical distance between customer equipment. 10 Gigabit Ethernet using special fibre can be used for distances up to 40 km. For this reason and its low cost, Ethernet technology can be effectively used to build Metropolitan Area Networks (MANs) and other access networks based on fibre. In the simplest case, a point-to-point Ethernet service can run over a dedicated fibre or over a light path service provided by an optical network. In this case, distances may extend well beyond 40 Km.

An Ethernet network consists of a central network node which is connected to each computer, router or other Ethernet network node by a dedicated line. Each such edge device has a unique Ethernet address. To send a data packet to device B, device A builds an Ethernet packet which encapsulates the original packet with the destination address of B, and sends it to the central node. This node functions as a *hub* or *switch*. A hub retransmits the packet to all its connected devices, and assumes a device will only keep the packets that were destined for it. A node starts transmitting only if no packet is currently being transmitted. Because two devices may start transmitting simultaneously, the two packets can 'collide', and must be retransmitted. (In fact, propagation delays and varying distances of edge devices mean that collision can occur even if devices start transmitting a little time apart.) Conflict resolution takes time and decreases the effective throughput of the network. The use of switches instead of hubs remedies this deficiency.

A switch knows the Ethernet addresses of the connected edge devices and forwards the packet only to the wire that connects to the device with the destination address. For large Ethernet networks of more than one Ethernet network node, an Ethernet switch will forward the packet to another Ethernet switch only if the destination device can be reached through that switch. In this case the switching tables of the Ethernet switches essentially implement virtual circuits that connect the edge devices. Such a connection may sustain two-way traffic at the maximum rate of the links that connect the edge devices, i.e. 1 Mbps to 1 Gbps (10 Gbps). This maximum rate can be guaranteed at all times if the above physical links are not shared by other virtual circuits. If a number of virtual circuits share some physical links (possibly in the interior of the Ethernet network) then bandwidth is statistically multiplexed among the competing edge devices in a best-effort fashion; see Figure 3.5. This may be a good idea if such a service is provided for data connections that are bursty. Bursty data sources value the possibility of sending at high peak rates, such as 10 Mbps, for short periods of time. Statistical arguments suggest that in high speed links, statistical multiplexing can be extremely effective, managing to isolate each data source from its competitors (i.e. for most of the time each device can essentially use the network at its maximum capability). Proprietary Ethernet switching technologies allow for manageable network resources, i.e. virtual circuits may be differentiated in terms of priority and minimum bandwidth guarantees.

Connectivity providers using the Gigabit and 10 Gigabit Ethernet technology provide services more quickly and in more flexible increments than competitors using the traditional

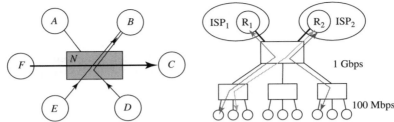

Figure 3.5   The left of the figure shows a simple Ethernet network. *N* is an Ethernet switch, and *A*, *B*, *C*, *D*, *E*, *F* are attached devices, such as computers and routers. Virtual circuit *FC* has dedicated bandwidth. Virtual circuits *EB* and *DB* share the bandwidth of link *NB*. The right of the figure shows the architecture of a simple access network, in which edge customers obtain a 100 Mbps Ethernet service to connect them to the router of their ISP. The 1 Gbps technology is used for links shared among many such customers.

SONET technology that we discuss in Section 3.3.4. Besides a lower cost per Gigabit (almost 10:1 in favour of Ethernet), Ethernet networks are managed by more modern web-based software, allowing these new competitive *bandwidth on demand* features, where bandwidth increments can be as low as a few megabits and can be provided in a short notice. The negative side is that capacity may be shared, as discussed previously.

3.3.4   Synchronous Services

Synchronous services provide end-to-end connections in which the user has a fixed rate of time slots that he can fill with bits. They are the prime example of guaranteed services. Examples of synchronous connection-oriented services are SDH, SONET and ISDN. *SDH* and *SONET* employ similar technologies and are typically used for static connections that are set up by management. The term SONET (Synchronous Optical Network) is used in the US and operates only over fibre, whereas SDH (Synchronous Digital Hierarchy) is used in Europe. They provide synchronous bit pipes in discrete sizes of 51.84 Mbps (only SONET), 155.52 Mbps, 622.08 Mbps, 2.488 Gbps and 9.953 Gbps. It is also possible to subdivide these, to provide smaller rates, such as multiples of 51.84 Mbps. In such services the quality is fixed in a given network and is determined by the bit error rate and the jitter, which are usually extremely small. There is no need for a complex traffic contract and policing since the user has a dedicated bit pipe which operates at a constant bit rate and which he can fill to the maximum. The network has no way to know when such a pipe is not full and when unused capacity could carry other traffic.

We have already explained the operation of SONET and SDH in terms of providing a constant rate of fixed size data frames over the fibre. Such frames may be further subdivided to constant size sub-frames to allow the setting up of multiple synchronous connections of smaller capacities. These smaller frames must be multiples of the basic 155.52 Mbps container. For instance, a 2.488 Gbps SONET link can provide for a single 2.488 Gbps SONET service or four services of 622.08 Mbps, or two 622.08 Mbps and four 155.52 Mbps services. In that sense, SONET and SDH can be seen as multiplexing technologies for synchronous bit streams with rates being multiples of 155.52 Mbps.

An important quality of service provided by SONET and SDH networks is the ability to recover in the event of fibre disruption or node failure. The nodes of SONET and SDH networks are typically connected in a ring topology which provides redundancy by keeping half of the capacity of the ring, the 'protection bandwidth', as spare. If the fibre of the ring is cut in one place, SONET reconfigures the ring and uses the spare capacity to restore full connectivity within 50 ms. The equipment used to build the nodes of such ring topology networks is complex and expensive compared to other technologies such as Gigabit Ethernet.

If one does not want to use the SONET or SDH recovery functionality, these protocols can be seen as simple synchronous framing services over fibre. Internet routers connected with fibre may use SONET to define the fixed size frames to be filled with IP packets and drive the lasers of the fibre. This is the case of *IP over SONET*, a technology used to directly fill the fibre with IP packets by adding little overhead (as the SONET frames add relatively few extra bits). The big gain is that now the complete bandwidth of the fibre is used instead of keeping half of it spare, at the expense of fast failure recovery. In this case one will rely on the higher network layers to do the recovery. In particular, the IP routers will sense the failure and update the routing tables to use other routes for the traffic. This may take much longer than the 50 ms it takes for SONET to recover. A similar concept applies when Ethernet frames are used to fill with IP packets the fibre that connects routers. *Optical Internets* are fibre networks that use SONET and Ethernet in this simple vanilla flavour.

ISDN (Integrated Services Digital Network) provides access services with dynamic end-to-end call set-up capabilities. By dialling an ISDN number, a customer can set up a synchronous constant bit rate pipe to other end-points of the ISDN network. The service interface to the user can provide three types of channels: the B channel (64 kbps) is used for data or digitized voice, the D channel (16 kbps) is used for signalling to the network, and the H channel (384 kbps, 1536 kbps, or 1920 kbps) is used like a B channel but for services requiring greater rates. A user can buy either a *basic access service* or a *primary access service*. The basic service provides a D channel to the network itself and two B channels to any ISDN destination. The primary service provides one 64 kbps D channel to the network, and a larger number of B channels (30 B channels in Europe and 23 B channels in the US).

Today's telephony services are provided by ISDN networks. SDH and SONET are used at the core of the ISDN networks to carry large numbers of voice channels. Indeed, these technologies were initially conceived to carry large numbers of 64 kbps digitized voice circuits. Older telephone networks employed Digital Carrier System (DCS) technology, which allowed network capacity to be divided into logical channels of different bit rates, ranging from 1.5–45 Mbps (2–34 Mbps in Europe). A customer could lease such a logical channel to connect two locations. These are the so-called *leased line services*, the most common of which are T1 (1.544 Mbps) and the T3 (44.736 Mbps). In Europe they are called E1 (2.048 Mbps) and E3 (34.368 Mbps).

3.3.5 ATM Services

ATM (Asynchronous Transfer Mode) technology networks use transport protocols that have been developed by the ATM Forum and ITU-T (International Telecommunications Union Taskforce). The ATM Forum is an industry consortium which performs the standardization activities for ATM. Information is packaged into 53 byte cells, and these are transported through the network over *virtual circuits*, as explained in Section 3.1.4. The small cell size means that it is possible to transport and switch streams of cells with small delay and delay variation.

ATM was originally designed to operate in a similar way to the traditional telephone network. An application places a call to the network that specifies both the address of a remote application and the type of service required. This is specified in the service contract for the call. The network uses signalling to implement controls for accepting the call, choosing the route of the virtual circuit (or tree), reserving resources, and setting parameters in the tables of the intermediate switches. The virtual circuit may be visualized as a virtual pipe (or branching pipe) that is dedicated to the connection's traffic stream. Appropriate resources are allocated for this pipe so that the traffic stream receives the specified quality of service. As we will see, these pipes may not require a fixed bandwidth. Instead, they may 'inflate' and 'deflate' in time, according to the bursts of data sent through the connection. Since such fluctuations cannot be determined *a priori* and occur on fast timescales, their contract traffic parameters bound the maximum duration and frequency of such inflation and deflation and the maximum bandwidth consumed during each period of peak operation. These bounds are expressed by leaky buckets in the traffic contracts of the connections. The network uses statistical models for the behaviour of such pipes to decide how many can be handled simultaneously. These issues are investigated in Chapter 4, where we present a methodology for deriving effective bandwidths for such contracts.

Signalling is the most complex part of ATM. It is common for network operators to disable ATM's full signalling or to use a simpler implementation. It is common to use only permanent virtual circuits or virtual paths (bundles of virtual circuits), which are set by network management rather than by signalling on customer request. These connections remain in place for months or years. They are mainly used to make permanent connections between the networks of an enterprise that has many physical locations, or to connect Internet routers (when Internet is run on top of ATM). This an example of a 'wholesale' service in which bandwidth is sold in large contracts to large customers and other network operators (ISPs). ATM specifies five 'native' service classes for connections; they differ in respect to the traffic descriptors that are used to characterize the carried traffic and the QoS parameters guaranteed by the network. This information is part of the contract for the particular service class. These five classes are as follows. The first three, CBR, VBR-RT and VBR-NRT, are guaranteed services with purely static parameters. ABR has guarantees with both static and dynamic parameters, while UBR is purely best-effort.

CBR (constant bit-rate service) uses the input traffic descriptor of type CBR. This is a simpler version of the VBR descriptor in Example 2.5, in which only the peak rate is policed (by the top leaky bucket). Its QoS parameters are cell loss and delay. This service is appropriate for applications that generate traffic with an almost constant rate and which have specific requirements for cell loss and delay. Examples are leased telephone line emulation and high quality video. In CBR an asynchronous network based on ATM can offer the same set of services as a synchronous network (synchronous bit pipes).

VBR-RT (variable bit-rate, real-time service) uses the input traffic descriptor of type VBR. Its QoS parameters are the same as those of CBR. Real-time services are used for applications such as interactive video and teleconferencing which can tolerate only small delays. Applications with bursty traffic should prefer VBR to CBR if these services have been correctly priced. This is because input traffic with a VBR traffic descriptor can be *statistically multiplexed*, to create a controlled 'overbooking' of resources. As we see in Chapter 8, this makes a difference to the tariffs of VBR and CBR. A CBR contract with peak rate h has an effective bandwidth of h while a VBR contract with the same peak rate generally has a smaller effective bandwidth.

VBR–NRT (variable bit-rate, non-real-time service) uses the input traffic descriptor of type VBR. Its QoS is tight for cell-loss, but relaxed for delay. It can be viewed as a relaxed version of VBR-RT, in which the network is given more flexibility in scheduling the cells of streams. For example, it might assign smaller priority to the cells of some streams or buffer more cells.

ABR (available bit-rate service). This service delivers cells at a minimum rate specified as part of the service contract parameter *MCR* (minimum cell rate). It also provides the user with the value of the maximum allowed rate $h(t)$. This value is updated dynamically by the end-user software in response to congestion signals sent by the network. The user must send at a rate less than $h(t)$ for minimal cell loss to occur inside the network. The network polices the user to prevent him exceeding $h(t)$. Hence the guarantee has a static part of MCR and a dynamic part of $h(t)$-MCR. The network is assumed to fairly share any remaining capacity amongst the competing connections and to deliver cells as fast as possible. Applications which conform to the flow control signals and correctly update $h(t)$ should expect to lose only a small proportion of cells.

UBR (unspecified bit-rate service). This is a purely best-effort service which requires no commitment from either the user or network. There is no feedback information to tell the user to increase or decrease his rate.

Figure 3.6 illustrates how a link is filled with ATM traffic.

3.3.6 Frame Relay

Frame Relay is a packet switched network technology operating at speeds of no more than 45 Mbps and using virtual paths to connect end-points over long time durations (static instead of dynamic connections). The traffic contracts for such virtual paths are similar to ATM-VBR, with the additional feature that they provide minimum throughput guarantees in terms of a Committed Information Rate (CIR) that is specified in the contract. A traffic contract for a virtual path uses parameters (T_c, B_c, B_e), with the following meanings:

- Committed Burst Size (B_c): the network guarantees to transport B_c bytes of data in each interval of duration T_c. This guarantees CIR $= B_c/T_c$.
- Excess Burst Size (B_e): the maximum number of bytes above B_c that the network will attempt to carry during each T_c.

The network operator can statistically multiplex many virtual paths in the core of the network by assuming that customers do not use all their CIR at all times. Hence, in practice, the CIR commitment of the network may be of a statistical nature depending on the overbooking performed by the operator. Operated properly, overbooking should only occur for the B_e part of the contract.

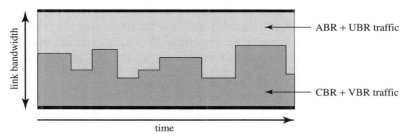

Figure 3.6 An example of how a link is filled by traffic of various services types. CBR and VBR have priority, while ABR and UBR use the remaining bandwidth.

Frame Relay is presently used by many enterprises to connect numbers of local area networks at physically separate locations into a single IP network, and to connect to the Internet. The IP routers of the local area networks are interconnected using Frame Relay virtual paths with the appropriate SLAs. This is a case in which Frame Relay technology is used to provide Virtual Private Network services, as in the case of ATM and MPLS. In many cases, different virtual paths are established for carrying voice. In order to avoid routing voice calls to remote internal locations through the public voice network, such calls are redirected through the private data network (voice is packetized and sent over the Frame Relay network). In this case, an adequate CIR must be reserved in the SLA, and if the same virtual path is used for both voice and data some priority mechanism must be available for the voice traffic, so that it falls into the committed part of the contract, and hence voice packets are rarely discarded due to policing when transmitted together with data packets.

Frame Relay networks are frequently implemented within ATM networks, but used only for the access service to the network, i.e. to connect the customer to the network. In this case, a Frame Relay SLA is translated to an ATM SLA for the virtual path of the connection, and Frame Relay packets sent by the sending end of the connection are broken into ATM cells which are carried further by the ATM network along the virtual path. At the receiving end, the network reassembles the Frame Relay packets from the ATM cells.

3.3.7   Internet Services

The Internet Protocol (IP) is the basic protocol by which packet transport services are provided in the Internet. It operates as a simple packet delivery service. The reader should refer to Example 2.2, where we have already described its basic workings.

TCP and UDP are two transport services that run on top of the IP service. They are denoted as TCP/IP and UDP/IP. These services have representatives (software) that runs only on user machines. Let us now describe these in greater detail than we have in Example 2.2. An application A that wishes to use TCP transport services to send a file to an application B, residing on a different computer (computer B), must take the following steps. First, it must find the IP address of computer B. Next, it must hand the file and the address of B to the local TCP representative. This representative establishes a connection with his peer representative in computer B, which is identified by some new connection identifier, say by choosing an unused tag c. The connection is established by the TCP representatives exchanging special 'signalling' packets using the IP service. Once the connection is established and c is known to both, the local TCP representative breaks the file into smaller packets, tags each packet with the connection identifier c (and a sequence number for detecting losses, see the following discussion), and hands this TCP packet to the IP representative, together with the IP destination address. This representative follows the steps described above, i.e. it builds an IP packet containing the above TCP packet, tagged with the destination IP address, and then forwards it to the IP network. The IP representative at the destination machine eventually receives these IP packets, extracts their content (the TCP packets) and delivers them to the TCP representative. The TCP representative reads the connection identifier, and delivers the data in the packet to the application that is receiving data from the above connection. UDP is simpler than TCP by not requiring the connection set up phase.

A connection using the UDP/IP protocols has no constraints, but also no guarantees. It sends packets (i.e. the UDP representative breaks files into packets and hands them to the IP representative) at a maximum rate, irrespective of congestion conditions, and does not resend lost data. Like TCP, UDP adds some information to the data packets that allows the receiver to detect if some bits where changed, i.e. if the received packet is corrupt. In the Internet, this service is used to send small bursts of data for which, because of their short life, it would not be worthwhile to set up a complete TCP/IP connection. UDP also makes sense when, as for real-time audio and video, there is no value in resending data. UDP is a typical example of a best-effort service with no guarantees. It adds multiplexing services to the basic packet transport service offered by IP.

The TCP Protocol

TCP works as follows. A network connection may send traffic into the network only when the protocol allows. The protocol states that the maximum number of bytes that may be sent without being acknowledged must not exceed the value of the window size W. For simplicity assume that packets each carry the same number of bytes. Each TCP packet carries its own sequence number. When the receiver (which is our shorthand for 'the TCP software at the receiver end of the connection') receives a packet it sends back an acknowledgment packet with the sequence number of the last packet that was received in correct sequence. For instance, suppose packets 0–100 are received in sequence. If packet 101 arrives next, the acknowledgment will be 101, but if packet 102 arrives next, out of sequence, then the acknowledgment will again be 100. This allows the sender to detect packet losses. Indeed, if the sender receives a number of consecutive identical acknowledgments, then it assumes a packet loss and resends the corresponding packet. The size of the window W constraints the number of packets that can be sent beyond those that have been acknowledged. For instance, if the latest acknowledgment received by the sender is 100 and $W = 2$, then the sender is allowed to send packets 101 and 102. The size of W controls the (average) rate h at which packets are sent. It is easy to see that if the round trip delay of the connection (the time for a packet to reach the receiver plus the time of the acknowledgment to travel back to the sender) is T, then the rate of packets is bounded above by W/T. This holds since W is the maximum number of packets that the sender can input to the network during a time of T, which is the time it takes to receives the first acknowledgment.

The actual rate h that is achieved may be less than W/T. This is because at some bottleneck link the network has less bandwidth than h available for the connection. In this case, packets of the connection will queue at the bottleneck link. When this happens, the same rate could be achieved for a smaller W. Thus, if W is chosen too small, it may unnecessarily constrain the rate of the connection. However, if W if chosen too large there will be unnecessary queueing delays inside the network. The ideal value of W achieves the maximum available rate h_{max}, with the minimum possible packet delay. This occurs for $W = h_{max}/T$. However, the problem is to choose W while h_{max} is unknown at the edges of the network. This is where the intelligence of TCP comes in. It searches continuously for the appropriate value of W. It starts with W small and increases it rapidly until it detects that its packets start queueing inside the network. A signal that its packets are queueing is a packet loss. When this occurs, W is decreased to half its previous value. Subsequent to this, W is allowed to increase linearly in time until a new loss occurs. In particular, W increases by approximately $1/W$ packets every time an acknowledgment packet is received. This procedure repeats until the connection runs out of data to send. In many implementations, the routers explicitly send congestion signals, so as to prevent packet losses. A router may detect excessive queue build-up and send packets to signal congestion to the contributing connections, or it may even decide preemptively to discard selected packets before it is crippled by congestion. In any case, the sources running TCP react by halving their window sizes whenever they receive a congestion signal.

The economics of IP

The high economic value of IP is due to its complementarity regarding most other transport services and customer applications. Examples of complementary goods are bread and cheese. The better the quality of the cheese, the more bread is consumed. The reason is that bread complements cheese in most recipes, and hence increasing the value of cheese increases the value of bread. Similarly, if more types of cheese that go well with bread become available, this again increases the economic value of bread. But where are the similarities with IP?

We have already discussed in Figure 3.4 that IP is a protocol (perhaps the only one in practice) that can run on top of all other transport technologies such as ATM, Frame Relay, SONET, Ethernet and pure light paths. In that sense, it is complementary to these technologies. Its added value is the efficient provision of end-to-end connections of arbitrary duration between any end-points on the globe. Once information is converted into IP packets, these can run over any access and link technology connecting the IP routers. This is the definition of a truly *open technology*. Installing IP does not constrain which other technologies should be used in the lower layers. A similar argument holds for applications, i.e., for the layers above IP (implicitly assuming TCP/IP and UDP/IP). Any application that is written to cope with the known IP deficiencies (lack of predictable quality and service guarantees), is a complementary good with IP and enhances its economic value. The more such applications are written, the more valuable IP becomes. The other side of the coin is that a killer application that is incompatible with IP will reduce its economic value by enhancing the value of other protocols that should substitute for IP. However, experience is that IP is well accepted and such incompatible services do not show up at either the application or network layer.

We remind the reader that ATM in its full functionality, which allows the end-to-end connection of customer applications through dynamically switched virtual circuits, was a substitute technology for IP when introduced in the mid-1990s. Unfortunately, it was also a substitute for Ethernet in the local area networks. This was its weakness: the already large installed base of Ethernets, connecting million of computers, and the higher price of ATM network cards made ATM hard to justify. In comparison, IP is a complement to Ethernet. This complementarity has helped IP dominate the market and become the universal standard of end-to-end connectivity. Unfortunately, there are limitation to IP that reduce its economic value, as we see in the next section.

Some limitations of the present Internet

Our discussion so far makes it clear that the present Internet, through TCP and UDP, provides two types of service whose quality in terms of the bandwidth provided to competing connections is unpredictable. The share of bandwidth that a connection obtains at any given time depends on the number of its active competitors. Furthermore, all connections are treated equally by the network in that they receive the same rate of congestion signals when congestion occurs. Such equal treatment is not economically justified and results in a set of services that is rather poorly differentiated. Unless the network happens to be lightly loaded, users cannot use it to run applications that require quality of service guarantees and tight control on either delay or cell loss rate. Such guarantees are needed to transport voice and video, or for a high degree of interactivity. Furthermore, the simple flat pricing structure that is traditionally associated with this sort of resource sharing does not provide any incentives for applications to release expensive resources that can be used by applications that need them more and are willing to pay for them. Basically, the present Internet does not provide the flexibility for a user that needs more bandwidth to get more bandwidth by paying an appropriate amount. Economic theory suggests that service differentiation increases the value of the network to its users by allowing them to choose the services that suit them best, rather than being forced to use a 'one size fits all' service. Increasing the value of the network services to customers is key to increasing revenue and keeping customers loyal.

As an example, consider the problem of transmitting video content at two encoding rates. Suppose that for a low and high quality services one needs bandwidths of 5 kbps and 30 kbps, respectively. How could an ISP provide both services? Assuming that the network treats connections equally, the total load of the network must be kept low enough that any connection can obtain with high probability at least 30 kbps. Suppose most of the video customers request the low quality service, and that the total video traffic is only a part of the overall traffic. If the ISP wants to leave open the possibility of supplying high-quality video, he must allow only a limited number of customers to use the network (by some admission control scheme), even if most of them are not using video. The only way this can be justified is if the revenue of the few high video quality customers is so great that it pays to refuse service to other customers so that the load of the network is kept low enough. In practice, this opportunity cost may be prohibitive, and the ISP will prefer to offer only the low quality service and keep his network highly loaded. But then he loses the revenue from the high-quality customers. The only way to obtain this revenue is if he can offer the high-quality service and also keep the network highly loaded. He can achieve this by using extra network controls that differentiate the resource share that different connections obtain. A crucial and difficult question is whether the cost of such controls can be justified by the extra revenue the network obtains. However, cost is not the only reason that the Internet is slow to adopt changes.

Introducing new mechanisms that may improve the performance of the Internet is complicated for many different reasons. Firstly, they may not provide visible improvements if they are applied in only part of the Internet. No single authority administers the Internet and unanimous decisions may be unrealistic due to the large number of network providers involved. Secondly, there are many doubts about the scalability of various new approaches and about the stability of the network if changes are made. The maxim 'if it's not broken, don't fix it' has many adherents when so many businesses depend on the Internet. Moreover, it is difficult to make small scale experiments in loading new software at the network nodes without switching them off. Finally, some experts believe that capacity will always be so abundant that traditional IP technology will be adequate. However, as we have discussed in Section 1.3.1, there are indications that free bandwidth will not remain unused forever. Bandwidth is consumed by software running on machines rather than by humans, and there is no upper bound on the bandwidth an application may require. Applications are digital goods which cost almost zero to reproduce and distribute.

There exist a number of proposals to enhance present Internet mechanisms to provide services of different QoS. These proposals include architectures for Integrated Services (IS), Differentiated Services (DS) and Multiprotocol Label Switching (MPLS). The procedure for producing such proposals is interesting. At their initial stage the proposals appear in public documents called Internet Drafts. These are discussed and refined by working groups of the Internet Engineering Task Force (IETF). After being discussed openly in the Internet, they become Internet RFCs. These can be required or proposed standards for the Internet community, or simply informational. For example, the IP RFC is a required standard, whereas the ECN RFC is a proposed standard.

Differentiated Services (DS)

Consider the following simple idea. Define a small number of traffic classes, say gold, silver and bronze, expressing the different levels of service (on a per packet basis) available at the network nodes. For instance, routers may have three priority levels for serving incoming IP packets, or may be able to allocate different percentages of the link bandwidth. Each IP packet travelling in the interior of the network is classified when it first enters the network as belonging to one of these classes and receives a tag that identifies its class. Customers that connect to the network specify in their contracts how the data they send to the network should be classified. For instance, the video conferencing traffic might be specified for gold class, web traffic silver class, and all other traffic bronze class. The contract also specifies in terms of leaky buckets the maximum amount that can be sent in each of the above classes. The network knows the average total load in each class and allocates resources inside the network so that the quality of service observed by the traffic in each class is at the desired level. For example, packets in the gold class are delayed by at most 10 ms while travelling on any end-to-end path of the network. Such an architecture presents a clear improvement over the traditional single-class Internet, while avoiding complex network controls such as signalling on a per connection basis.

This is an example of a Differentiated Services (DS) IP network. The network decides on the service differentiation it will support and then posts prices which reflect service quality and demand. Users choose in their contracts how to classify their traffic based on these prices and the average performance provided in each class. Note that this architecture does not provide hard guarantees on performance, but only on an average basis. This is because the network allocates resources to the various classes using some average historical data, rather than on a worst-case basis. If all users decide to send data at the maximum rate allowed by their contracts then the network will be overloaded. The complexity of the approach is kept minimal. Only the routers at the periphery of the network (the *ingress nodes* in the DS terminology) need to classify traffic and establish contracts with customers. DS contracts are established by management and last as long as the customer is connected to the network, rather than for just the time of an individual web connection. In the interior of the network the implementation of DS is simple. A router decides how to route a packet by looking at its destination address and the tag identifying its class. Such a routing policy is easy to implement. This is an important departure from the traditional circuit-switching model, in which a switch applies a different policy on a per connection (virtual circuit) basis. In DS such 'micro' flow information is 'rounded up'. Individual connection flows are aggregated into a small set of much larger flows (the *flow aggregates* in the DS terminology). This coarser information influences control decisions. Complexity is reduced at the expense of control. All micro flows in the same class are treated equally.

The weakness of DS is its inability to offer hard QoS guarantees. A DS service contract with a customer provides a reasonable description of the traffic that will enter the network at the given ingress point, but may not specify its destinations. Hence the network must make informed guesses, based on historic information, as to how each contract will contribute to the traffic of the various network links. This lack of information makes effective resource provisioning extremely difficult. For the same reason, admission control (at the contract level) is difficult. The network may end up being overloaded and, even more interestingly, a low quality class may outperform a higher quality one. This can happen if more customers than anticipated subscribe to the high quality class, for which the network administrator had reserved a fixed amount of resources. Lower quality classes may offer better performance if their load is sufficiently low. Of course, if pricing is done correctly, such situations ought not occur. But the network manager has a complex task. He must construct the right pricing plan, estimate the resulting demand for the various classes, guess the traffic on the various routes of the network, and assign resources. Besides the fact that there are too many control variables (prices, resources, and so on), there are no feedback mechanisms involving the user (aside from TCP). The provider can only measure the network utilization and dynamically increase/decrease capacity to solve temporary overload problems. DS is conceived to be managed in slow timescales relative to the timescale of changes in network load.

Let us investigate in more detail the contract structure and the implementation of DS. In contrast to ATM, in which services are defined for single unidirectional point-to-point connections, the scope of a differentiated service is broader and includes large *traffic aggregates* consisting of:

- multiple connections (i.e. all connections that send web traffic to a particular set of destinations, all Internet telephony calls, and so on);
- traffic generated at an entry point A and going to a set of exit points (possibly singleton, or including all possible destinations).

Hence, a traffic aggregate may be specified by a predicate of the form all packets in connections of types a, b, c that are destined to networks x, y, z. Each DS network, being a *DS domain*, can define its own internal traffic aggregates and the way to handle these in terms of quality of service. This may be part of its business strategy. Traffic aggregates are uniquely identified by IP packets carrying special tags (the 'DS codepoints'). The periphery of the network is responsible for mapping incoming traffic to the traffic aggregates that flow in the interior (the 'core') of the network. This is done by appropriately tagging incoming packets before they enter the core. Such incoming traffic can originate either from end customers or from other DS domains, see Figure 3.7. In either case, there is a service interface and a contract involved.

The service interface specification of DS is called a *Service Level Agreement* (SLA) (see Figure 3.8). It mainly consists of a *Traffic Conditioning Agreement* (TCA) that specifies

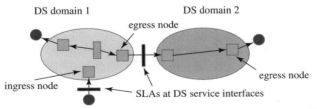

Figure 3.7 The key concepts of the DS architecture. DS domains are responsible for providing service differentiation to the traffic that travels through their core. Incoming flows are assigned by the ingress nodes to the traffic aggregates that travel in the core of the network, according to the contract (the SLA) that specifies how such traffic should be handled. Flows in the same traffic aggregate are treated equally by the network and receive the same QoS. Traffic exits at egress nodes and is either terminated at edge devices or continues its journey through other networks, possibly of the DS type. Different DS domains are free to define their traffic aggregates and the service quality supported.

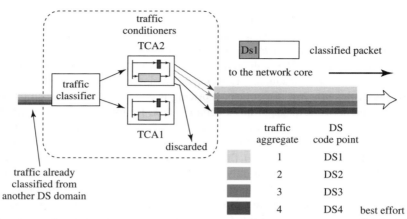

Figure 3.8 Differentiated services architecture. A node of the DS domain performs two basic operations. The first is *classification*: every incoming packet is assigned to the relevant TCA on the basis of DS codepoint. The second is *conditioning*: for every TCA there is logic that uses the leaky bucket descriptors for policing, and assigns the conforming packets to the internal traffic aggregate that meets the QoS requirements of the TCA. This is done physically by tagging packets with the appropriate tag (the DS codepoint). A packet may be marked or discarded. Here there are four such traffic aggregates. Traffic that exceeds its TCA or is not explicitly specified in a TCA, is called default traffic and is mapped to best effort.

the service class to be provided and the part of the input traffic that should receive such service. An example of a TCA is 'video connection traffic at rates less than 2 Mbps should be assigned to the gold traffic aggregate, web traffic at rates less than 25 kbps should be assigned to the silver traffic aggregate, and all other traffic should be assigned to the bronze traffic aggregate'. A TCA for traffic entering from another DS domain could contain the clause 'gold class input traffic not exceeding 4 Mbps should be assigned to the gold traffic aggregate, all other traffic should be assigned to the bronze traffic aggregate'. The SLA also contains other service characteristics such as availability and reliability (rerouting capabilities in case of failures), encryption and authentication, monitoring and auditing, and pricing and billing. The QoS corresponds to the performance parameters offered (delay, loss, throughput), while traffic descriptors in the TCA are again token buckets. Note that QoS requirements may be directly translated to the identity of the internal traffic aggregates that supports such QoS. Part of the TCA specification is the service to be provided to non-conforming packets. The architecture of DS at an ingress node is depicted in Figure 3.8.

SLAs can be static or dynamic, although only static ones are presently implemented. Dynamic SLAs can change frequently because of traffic and congestion level variation or changes in the price offered by the provider. Such dynamically changing SLAs should be configured without human intervention, using the appropriate software and protocols (intelligent agents and bandwidth brokers).

The nodes of the network provide packets with local forwarding services. To reason in an implementation independent fashion, a set of 'high-level' forwarding services has been standardized in the DS context, where such a service is called a Per-Hop Behaviour (PHB). PHBs are characterized in terms of the effects they have on the traffic and not by their implementation details. When a packet arrives at a node, the node looks at the tag of the packet and serves it by using a mapping from tags to PHBs, which is uniquely defined throughout the network. At the network boundary, newly arriving packets of a particular SLA are first policed using the traffic descriptors of the TCA, and then marked with the corresponding tag of the service negotiated in the TCA (the QoS part of the TCA determines the tag and hence the PHB to be received inside the domain). Note that a packet traversing multiple DS domains might need to be re-marked so as to use the services that have been negotiated in a given domain.

Examples of PHBs (a number of which are being standardized) are Expedited Forwarding (EF) (very small delay and loss) and Assured Forwarding (AF). EF guarantees a minimum service rate, say 2 Mbps, at each link in the core. It provides the traffic aggregate that is served by EF with a form of 'isolation' from the other traffic aggregates. The isolation is lost if this traffic aggregate in a given link exceeds 2 Mbps. Then it will have to compete with the other classes for the extra capacity, which may not be available. The network operator can guarantee QoS by keeping the maximum rate in the EF class less than 2 Mbps on every link of the network. AF is more complex. It divides traffic into four service classes, each of which is further subdivided into three subclasses with different drop precedences. Each service class may have a dedicated amount of bandwidth and buffer, and a different priority for service. When congestion occurs in a class, packets are dropped according to their drop precedence value. There are rules for packets changing drop precedence within a class. It is up to the network operator to control the delay and loss rate in each of these classes by varying the amount of dedicated resource and controlling the load by admission control.

In contrast to EF and ATM, the QoS in AF is relative rather than quantitative. A motivation for such qualitative definitions stems from the facts that PHB definitions can be related (in DS this corresponds to a 'PHB group') due to implementation constraints. For example, PHB1 corresponds to providing higher priority link access to the packets, whereas PHB2 provides lower priority access. These PHBs are related since the performance of PHB2 depends on the amount of traffic assigned to PHB1, and only a qualitative differentiation can be made. A TCA can use qualitative definitions of QoS for its conforming and non-conforming traffic respectively, by assigning it to such related PHBs. In order to support a given set of SLAs each node of the network must decide how to allocate its resources to serve the various PHBs. This is a non-trivial problem unless services with quantitative guarantees are only promised for point-to-point traffic aggregates. Only then are the intermediate nodes known and can appropriate resource reservations be made. The management of the resources at the nodes of the network typically occurs on slow timescales (since SLAs should not change frequently) and it is the responsibility of the network manager (or of the 'policy servers' who are meant to have the intelligence to implement a particular management policy within the DS domain).

The strength of DS is *scalability*. Although the number of connections grows with the number of users, the number of traffic aggregates for which services are differentiated need not grow as fast. This is because aggregates correspond to connection types rather than individual connections. The weaknesses of DS are (a) its loose quality guarantees, (b) the difficult task that the network has in reserving resources that can guarantee quality (how can one guarantee a one-to-many contract when 'many' refers to all possible destinations?), and (c) the impossible task for users to check that the network keeps its part of the contract. Basically, DS is the simplest way to differentiate services with the least amount of network control. Network management is involved in setting and activating contracts between the users and the periphery of the network.

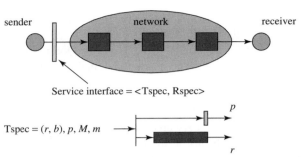

Figure 3.9 Key concepts of integrated services architecture. Two types of service are offered. 'For guaranteed services' there is an upper bound on packet delay. For 'controlled load services' packets receive the same service that they would in an uncongested best-effort network. Tspec consists of a dual leaky bucket VBR traffic descriptor with CDVT = 0 and specifications of the maximum size of datagrams allowed to cross the interface and a minimum datagram size to which smaller datagrams are rounded up for policing purposes. Rspec is usually decided by the receiver and consists of the minimum bandwidth to be reserved by all nodes in the path. This minimum bandwidth is computed so as to provide deterministic guarantees for maximum delay and zero packet loss. Tspec is defined for both Guaranteed Quality and Control Load services. Rspec is defined only for Guaranteed Quality services.

Integrated Services (IS)

The IS architecture is conceptually similar to the end-to-end service architecture of ATM and can similarly provide a controlled level of service to individual network connections (static and dynamic). Presently, two types of services are specified in RFCs, in addition to a default best-effort service. These are Guaranteed Quality service and Controlled Load service. In both, a service contract is agreed at connection set up which follows the general concepts introduced in this chapter. This consists of a traffic descriptor, called Tspec (T for Traffic), and a QoS commitment called Rspec (R for Reserved) (see Figure 3.9).

The QoS provided by Guaranteed Quality services is defined in terms of zero loss and a deterministic upper bound for the end-to-end packet delay (the value of this upper bound being chosen by the individual application). For the Controlled Load services, the QoS is defined as the 'performance visible to applications receiving best-effort service under unloaded conditions'. This is an imprecise definition which leaves room for a network service provider to manage and dimension his network in a way that exploits statistical multiplexing and to load his network sufficiently to compete with other providers who offer similar service. A way to implement Control Load is to combine it with DS. A traffic aggregate in the core of the network is dedicated to control load traffic and is allocated a fixed amount of resource. Using admission control based on the Tspec part of the contract, the network operator makes certain that the load in this 'virtual network' stays below some desired level.

In the case of Guaranteed Quality services, the actual QoS requested (maximum packet delay) is not specified explicitly, but is implicit in the value of B, the minimum bandwidth that should be reserved to all nodes along the path taken by packets of the connection. This is the *Rspec* in the IS terminology, and includes a slack term to allow for some overbooking. The choice of B is made by the receiver using a mathematical formula that relates the maximum delay bound with the values in *Tspec*, B, and some other parameters of the system (which are either known or are guessed). This may be done as follows: the sender sends a message with Tspec towards the destination. This message collects relevant network information from each node in the path, such as propagation delays of the various links. When it reaches the receiver, it contains all the necessary information for the receiver to compute the amount of bandwidth that must be reserved. The receiver explicitly solves the problem 'how much B should be reserved at all nodes in the path so that the worst-case delay is less than d when the source is policed with the leaky buckets in *Tspec* and the links in the path contribute a total propagation delay d_{prop}?' (where clearly we must have $d_{prop} < d$). Note that the receiver is the controlling party for the level of QoS. This is consistent with many applications such as receiving audio or video. Once the value of D is computed, a message with its value is sent back to the sender, suggesting to each node in the path that it reserve the above amount of bandwidth for the connection. Each node can also compute the amount of buffer that must be dedicated to the connection's traffic so that zero loss occurs (which can be done by knowing *Tspec* and B). If the necessary bandwidth and buffer can be allocated, then the node replies positively and the same operation is performed at node next closest to the sender. If some node cannot reserve the necessary resources, the call is blocked (as in ATM). If the resulting delay is unacceptable (due to wrong guesses, for example), then the values of *Tspec* and *Rspec* can be renegotiated. Since IS requires resource reservation and performance guarantees, it must also be subject to policing. At the edge of the network, incoming traffic is policed to conform to *Tspec*, and non-conforming traffic is assigned the default best-effort service.

RSVP (Resource Reservation Protocol) is a signalling protocol that allows for the implementation of the IS service architecture (mainly for the Guaranteed Quality services), by sending messages with *Tspec* towards the receivers and posting the resource reservation requests backwards towards the sender. These messages can carry all the necessary information for IS to work properly. As a signalling protocol it requires less complexity in the network nodes compared to ATM. It does not need to specify routing information for setting up virtual circuits (IS uses the already existing IP routing tables for routing packets). Also, the state of a connection at a router is 'soft', in the sense that it is the responsibility of the receiver to continuously remind routers that the connection is still active, since otherwise the reserved bandwidth is released. There is no explicit connection tear-down signalling phase. Of course, there is the cost that the network must serve all these 'I am alive' messages.

In summary, the strength of IS is its ability to provide strict quality guarantees. The weakness is scalability: when the number of connections grows, the signalling performed by RSVP becomes overly expensive. A possible way to combine DS with IS is to use DS in the backbone of the Internet and IS at the access level. The backbone provides simple service differentiation and is protected from signalling overhead. Signalling at the local level ensures congestion-free access to the backbone and scales better with the size of the network. For this to work, the backbone must be overprovisioned with bandwidth.

Multiprotocol Label Switching

Label switching, introduced in Section 3.1.4, is a network technology for creating label switched paths and trees with dedicated resources. The key idea of *Multiprotocol Label Switching* (MPLS) is to program in the switching fabric of the network, one sink-tree per destination (or set of destinations), and use these trees to carry traffic aggregates that have the same destination, or that travel through some common part of the network. This technique achieves *flow isolation* and reduces the bad effects of uncontrolled statistical multiplexing that is common in IP networks. Using such direct 'tunnels' for sending packets to a destination has the advantage of being able to guarantee performance, since the network can dedicate resources to serving the traffic. Once these tunnels are in place, a router that needs to forward a packet may choose to use such a tunnel instead of forwarding the packet to the next router. Of course such a choice exists only if a tunnel to the particular destination is available, and if the packet is in the traffic aggregate for which the network uses this special service.

Of course, it may not be possible to construct such sink trees for every possible destination in the Internet (although it may be possible in private IP networks). MPLS is mainly used in the core of the Internet where each router i at the periphery of the core is responsible for handling the traffic to and from a specific set of networks N_i. Each such edge router i has an established label-switched path *from* each other edge router, and is also the sink of a tree of paths which connect all members of N_j to it. When an edge router i detects the start of a flow of packets which require a specific quality of transport with a destination of a network in the responsibility of router i, it forwards the packets through the corresponding predefined label switched path (of which there may be more than one for the same destination, with different quality of service parameters). Otherwise, the flow is treated as ordinary IP flow and packets are forwarded to the next IP router inside the core.

In general, these ideas can be used for traffic aggregates of arbitrary definitions (not only those based on the packet's IP destination address), such as video traffic, or other traffic that is considered of higher priority. The network manager must first design these high-quality MPLS tunnels, program them into the network infrastructure, and then specify the traffic aggregate that should use the MPLS service. Hence, at each entry node one must first associate each traffic aggregates with a corresponding label switched path originating at that node. During operation,

- each packet of the traffic aggregate is assigned a forwarding label identifying the entry of the path;
- at each node, the forwarding label of an incoming packet is used to look-up (a) the next-hop node, (b) the service discipline to be used for forwarding the packet (like the PHB in DS), and (c) the replacement label.

This forwarding information that defines the path and the assigned resources is stored in each node by a protocol that is used during path creation. This procedure essentially implements a virtual circuit. Indeed, if the underlying network node technology is ATM, it can directly use the ATM signalling to set and manage the corresponding virtual circuit. (In fact, ATM signalling need not be used; it can be emulated by sending IP packets with the analogous information.) Or since virtual circuits are of a long lasting nature, management procedures may be used instead of signalling. Labels correspond to virtual circuit identifiers. MPLS allows for creating and managing virtual circuits over network node technologies that are not necessarily ATM but are MPLS-capable.

MPLS is a technology for implementing service differentiations within the same network. It is consistent with DS architecture concepts and offers better control of QoS in the network core. This is because if resource allocation is done appropriately then different traffic aggregates that flow through different MPLS tunnels do not interact. However, if resources are shared amongst such tunnels, rather than dedicated, it may not be possible to offer quantitative QoS guarantees. Hence, MPLS technology cannot solve the QoS problem unless properly deployed.

An important application of MPLS is the creation of many *virtual networks* over the same physical network infrastructure. Since routers that are not connected with direct links in the actual IP network may be directly connected through MPLS tunnels of arbitrary capacity, one may design an 'overlay' IP network with links of controllable capacity. This network can be used to provide a single private enterprise having many locations with a *virtual private network*, or to carry high-priority traffic. In principle, a network may support many different quality levels by implementing a number of such parallel virtual networks, one for each quality level.

3.4   Other types of services

3.4.1   Private and Virtual Networks

Enterprises that are spread over geographically remote locations often wish to connect their networks at various locations into one wide area *private network* so computers at all locations can communicate and share applications and information services. Private networks may use internal addressing schemes and exercise complete control over their resources. Presently, private networks are built using IP technology, and can be seen as private Internets. A private network at a local level can be built by installing LANs and interconnecting them with IP routers. Things are rather more interesting at the wide area level.

To create a wide area private network, an enterprise has to interconnect the routers that it owns at different locations. In theory, it might build the necessary communication links itself, for instance, by installing fibre and communication equipment. Although this gives the enterprise complete control of the infrastructure, it is too impractical or expensive. Alternatively, the enterprise can view a link as a communications service and outsource the provision of this link to a network service provider. The outsourcing can take place at different levels. At the lowest level, the network service provider may provide 'raw' infrastructure, such as dark fibre, or even install new fibre in conduit space rented by the enterprise. The enterprise must then provide all the other layers of technology necessary. At a next level, the service provider might provide the link service by offering a lightpath, or a guaranteed bandwidth synchronous services such as SONET, or a leased line. Going even further, he might provide an asynchronous service, such as an ATM or Frame Relay virtual path, or Ethernet over optical. Finally, he could connect the routers of the enterprise to his own IP network and exchange packets using the IP datagram service of his network.

In the list of solutions above the service provider has increasing opportunity to make more efficient use of resources, while the enterprise customer has decreasing control over network resources and the quality of service. For instance, synchronous services require the network service provider to allocate fixed amounts of resources, while Frame Relay and ATM permit statistical multiplexing. At the extreme, best-effort Ethernet and IP connectivity may offer no guarantees on service quality. In practice, the term *Virtual Private Network* (VPN) is used for private networks in which the link outsourcing is substantial and occurs at a level above the use of synchronous services. We refer to such a network as a 'X VPN', where X stands for the link service technology (e.g. an ATM VPN). Of course there are security issues involved in outsourcing link provision, but these can be addressed by the appropriate security protocols.

VPN services have proliferated because it costs a large network operator little to implement VPN services. This is due to the large multiplexing capability of his network. Moreover, instead of requesting constant rate contracts for their virtual paths, customers may buy traffic contracts that take advantage of the bursty nature of their data traffic. There is also a saving in the number of interface cards (see Figure 3.10). Outsourcing the operation of the wide area network can be seen as a step for outsourcing larger parts of the IT of the enterprise to third parties. A high-bandwidth VPN allows for the concentration of critical applications and information (intranet and extranet web servers, data bases) at a

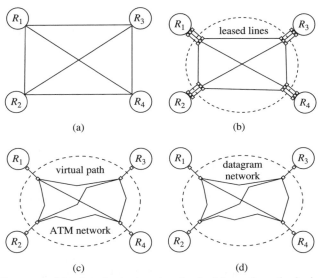

Figure 3.10   Some possible virtual private networks. In (a) we show the logical network that
connects the four routers of the enterprise customer. It consists of six links. In (b) we implement
this network using leased lines; we need six leased lines and 4×6 interface cards (each leased line
needing four cards, i.e. each end of the line, we need one card for the customer's router and one
card for the network equipment to which it is connected). In (c) we implement the VPN using a
connection-oriented service (such as ATM) and replace each leased line with a permanent virtual
path; we now need 4×2 interface cards (one for each router and one for the ATM network
provider's equipment to which this router connects). In part (d) we implement the VPN using a
datagram IP network; we still need 4×2 interface cards. In (c) and (d) the access service to the
network nodes may be obtained from a third party service provider. In theory, the logical network in
(a) can be constructed by using only three circuits in (b) and (c), enough to provide full connectivity.
In practice, graphs with greater connectivity are constructed for reliability and performance.

small number of well-guarded and reliable data centre sites. Observe that bandwidth is a
substitute for storage or processing.

IP VPNs offer great flexibility to the service provider, but may provide no performance
guarantees to the customer using the service. This is simply because the VPN's data traffic
is treated the same as all other IP traffic in the provider's network. There are a number of
solutions that involve flow isolation and service differentiation at the IP level, which need
to be deployed in the network of the provider to offer VPN SLAs with QoS guarantees.
The most popular is MPLS (see Section 3.3.7). VPN SLAs look very similar to the SLAs
used in DS, where one must consider a different SLA for connecting each remote location
to the IP network of the provider. Such SLAs also include upper bounds on packet delays
while travelling in the IP network of the provider, packet loss probabilities, and encryption
services so that no one can read or alter the datagrams. Present network management tools
allow the service provider to offer a visual service interface to its VPN customers, that
allows them to track the performance of their traffic and check the validity of the SLA.

Finally, we remark that the enterprise may itself be a large network operator, but one
whose physical network does not reach certain geographical areas. To be competitive and
offer full coverage, it may be more economical for this enterprise to lease infrastructure
from other providers than to extend his network to cover the areas he does not already reach.
He will outsource his need for links to providers who focus on the wholesale infrastructure
market and who sell existing fibre or install new state-of-the-art fibre on demand. Their

business is one of installing conduits across continents and oceans, each conduit being able to carry a cable of 12–1200 optical fibres. Most of the conduits are empty and can be filled on demand relatively quickly. The infrastructure provider deploys enough optical amplification and regeneration points to allow complete outsourcing of the optical network operation. An infrastructure provider must be 'carrier neutral' since he sells services to competing carriers (i.e. the large telecoms operators who offer transport services to smaller network operators and ISPs). They also run large data centres that are connected to their fibre infrastructure. These data centres host services that can interconnect the different telecoms operator carriers and other bandwidth-critical customer applications.

3.4.2 Access Services

The specific locations at which customers can connect to an ISP or other value-added service provider are called *Points Of Presence* (POPs). The POP contains a router of the ISP's backbone. An *access service* provides a connection from customer x to the POP of service provider X. The customer may not directly pay the access service provider for this service, but may pay the ISP for a bundle consisting of access and valued-added services; the ISP is then responsible for transferring a payment to the access service provider.

In the case of Internet service, the access service connects the customer's computer to the router of the ISP. The access service can be dynamic or 'always on'. It can be of a connection-oriented type (such as an ATM virtual circuit) or of a datagram type (like an Ethernet service). Hence, all the attributes introduced earlier apply; there may be some minimum bandwidth guarantees, or the connection may be purely best-effort. Also the service may be asymmetric in terms of performance. For instance, Internet users tend to receive more information from the network (downstream) than they send to the network (upstream). Thus, they place greater value on services that offer a high downstream bit rate. Other customers may value things differently: for example, a customer who operates a private web site or offers some value-added service. Although access services are conceptually simple, they have many intricacies and play a dominant role in maintaining competition in the communications market.

Consider the case of many access service providers (XSPs) and many ISPs. In a competitive ISP market an end-customer should be able to connect to any of the competing ISPs. In addition, competition in access services should imply that a user can choose both his XSP and the ISP. If ISPs create vertical markets, each with his own XSP, then the quality of the access service may be a decisive factor in a customer's choice of the ISP. In the worst case, a single XSP controls the ISPs to which a customer can connect. Obviously, competition can be assured by having many access service providers, so that no one XSP dominates the market. Unfortunately this is difficult in practice. The infrastructure needed to provide high-quality access services is very expensive and hard to deploy. This is because the total length of the links of the access network is many orders of magnitude greater than the size of the backbones of all network operators added together. Hence, it is highly improbable that more than one operator will ever install an access infrastructure (such as optical fibre) in any one geographic area. Once such infrastructure is in place, even if it is of the older generation of telephone network copper local loop, it deters the introduction of any competitive infrastructure, unless that infrastructure is easy and inexpensive to install. Wireless technologies such as LMDS (local multipoint distribution service) are low cost, fast to deploy, and do compete in performance with the services provided over the local loop.

There are two possible remedies to the lack of competition in the access service market. The first is regulation: the operator of the access infrastructure is required to make it available to his competitors at a reasonable price. This is the well-known 'unbundling of the local loop', which has been applied to the access part of the telephone network, and which could also be applied to access networks of cable, wireless and fibre. The second remedy is the *condominium fibre model*, in which large customers such as communities with schools, hospitals, libraries, and so on, deploy their own common fibre access networks, independently of a carrier. We say more about this in Section 13.4.2. The model becomes complete by having the access network terminate in special carrier-independent locations, so-called *telecom hotels*, which can contain the POPs of many carriers, ISPs and other value-added service providers. The beauty is that the access cost is extremely low, since it is shared by the many parties involved. No single party can control the infrastructure and so artificially raise prices or influence competition.

We have already mentioned that the provision of access service may require purchase of some lower-level services from another party. Let us examine the business model for providing broadband access using the Digital Subscriber Loop (DSL) technology. This technology uses special modems to create a digital two-way pipe of many megabits over the copper wires of the local telephone loop. This pipe operates in parallel with the traditional telephone service, using the same wires. A possible scenario for providing an access service is shown in Figure 3.11.

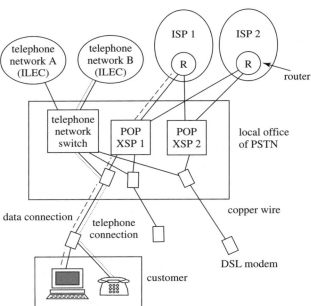

Figure 3.11   An architecture for providing competitive access and value-added services over the local loop. Two competing access service providers (XSPs) connect customers to the POPs of two ISPs. The first part of an access service data connection uses the DSL modems over the local loop (a DSLAM in DSL jargon) to connect to the POP of the customer's XSP. The access service continues to the customer's ISP by sharing the pipe that connects the POPs of the XSP and ISP. The quality of the access service depends on both parts. If the latter part is shared in a best-effort fashion amongst all the connections that terminate to the same ISP, it may be a bottleneck. Similar concepts apply for telephony service. The first dial-tone is provided by the local telephone network switch, which subsequently may continue the connection to the POP belonging to the voice network of the customer's voice service provider. Note that the XSPs' equipment must be located in the same place as the equipment that terminates the local loop.

The XSP must rent from the local telephone company both the local loop and collocation space for his equipment. He must also buy transport services to connect his POP to the ISPs. If the local telephone company is running its own XSP and/or ISP services, it has the incentive to create unfavourable market conditions for the competing XSP. Although the regulator can control the rental price of the local loop, it is hard for him to control other subtle issues. These include the price and true availability of collocation space, the timely delivery of local loop circuits, the maintenance of these circuits and the tracking of malfunctions. These same issues also arise in other access technologies and show the intricacies of the underlying business models. They provide reasons for deploying competing local loop technologies, such as the use of wireless modems to connect users' computers to the POP of their XSPs.

The simplest form of access to the ISPs POP is by dial-up, i.e. a direct telephone network connection. The reader might think that this does not involve any intermediate service provider other than the telephone network. However, to avoid unnecessary waste of telephone network resources, the calls to the ISP's POP are terminated at the periphery of the telephone network, on some access provider's POP (invisible to the user, similar to the architecture in Figure 3.11). These are terminated through a *data network* to the ISP's POP. Such access services are measured by the volume of dial-up call minutes carried, and are provided by third parties to the ISPs. In an even more interesting business model, such third parties deploy the equivalent of a circuit-switched telephone network that is implemented over a pure IP network. This network receives from a local telephone network, telephone calls (or any type of circuit-switched service a telephone network supports, such as T1 and T3), routes them through the data network by transforming voice information into IP packets, and finally terminates them: either directly to the receiving customers' computers if these are connected to the IP network, or converts the IP packets back into telephone calls that are carried through the last part of the telephone network to reach the receiving customer's telephone. The points of conversion between the telephone and the data network are called gateways. This is the business model of voice over IP services. Such a service provider must either run his own IP network or outsource this part to some ISP in the form of an IP VPN with the appropriate SLAs to guarantee low delays for voice packets. Note that this access service architecture allows an ISP to have a small number of POPs, not necessarily located in the vicinity of its customers.

Our business models can be carried further for access network infrastructures other than the local telephone loop. For instance, wireless Ethernet and cellular mobile services can be used instead of the traditional telephone network. A feature shared by most access services is resource scarcity. The XSP's VPN may be restricted in two places. The first is between the end-customer and the XSP's POP. Present access technologies over copper, cable or wireless restrict the available bandwidth to the order of few Mbps. The second is between the POPs of the XSP and ISP. If the market is not competitive, such a provider has the incentive to multiplex a large number of connections and so reduce the bandwidth share of individual users. Suppose a and b are, respectively, the dedicated bandwidths from the XSP's POP to the end-customer and XSP's ISP. If, on average, n customers have active connections (using the Internet service), then as data connections are bursty b may be less than na. Choosing the appropriate b for a given customer base is part of the business strategy of the XSP. However, discouraging users from abusing the service is essential. Any choice of b assumes a statistical pattern of usage. If some users 'overeat' by consuming close to a, then the rest of the users may obtain small bandwidth shares on a regular basis. A policing function can be achieved through usage charges which provide users with the right incentives. Flat access charges may cause unnecessary resource consumption and severe performance degradations.

Users who access the value-added services provided by a server in the ISP's network, observe the end-to-end performance of these services. The performance depends upon many factors, which are divided between the access service, the transport service inside the ISP's network, and the server itself. An ISP who wishes to provide services that are differentiated in terms of transport quality needs to take into account all such factors, not only those he can directly control. Complete control of the quality offered over such a complex value chain is possible if the ISP and the XSP are the same entity, owning also the infrastructure that is used to connect the end customers to the XSP's POP. Such vertically integrated companies, who fill all the blocks of the value chain from content to access, can obtain dominant market position due to the improved service quality they can deliver. Furthermore, they can create strong customer lock-in and refuse other ISPs access to their customers. By controlling interconnection with other networks, they can degrade the performance their customers obtain when accessing servers in other ISP networks. This may create a 'walled garden' environment, controlled by an oligopoly. The high entry cost is a barrier to entry, and further enhances the oligopoly structure. As we have seen, possible remedies are regulation of the access network services and the creation of access networks that are owned by customers.

3.5   Charging requirements

There are several practical requirements that must be met by any workable scheme for charging for network services. We may group these requirements under the three headings of (a) the end-user who pays the charges, (b) the service provider who defines the charges, and (c) the underlying technology that is used to produce the charge. Recipients of charges tend to favour charges that are *predictable*, *transparent* and *auditable*.

A charge is predictable if a user knows in advance what the total cost of using the service will be. For example, many phone customers in the US pay a flat monthly fee for unlimited local telephony usage. Studies show that customers enjoy the fact that they have security against the risk of high bills, and that they use telephone services more than in places where a small usage charges is added to the monthly fee. Although this type of flat pricing can lead to a waste of resources, it does encourage the fast proliferation of other services that generate important social value. Flat rate pricing of Internet access encourages customers to spend more time on the Internet. This has some negative effects in terms of congestion, but it speeds up the acceptance of new electronic commerce services.

It is interesting to compare the way in which consumers prefer to be charged for communications and electricity. Why is it that customers accept usage-based charging for electricity but seem to prefer flat rate charges for communications services such as Internet access and telephony? A possible explanation is that the value of a KWh of electricity is transparent. However, the value of a KByte of Internet data is unknown, since a user consumes it indirectly as a result of higher-level application. Perhaps users would be willing to pay in proportion to the amount of service consumed. Another observation concerns the smaller degree of control a user has over his consumption of communication resources, compared to, say, his consumption of food in a restaurant. In a restaurant, the customer controls the order and can accurately predict the bill. In the case of Internet access with a usage charge, the control of resource consumption is in the hands of the application once it is started. For example, when a conversation starts, the duration cannot be known at the start. This may explain why risk-averse users prefer flat rate charges, even if for the same level of consumption they actually pay more on average. While the above may have convinced the reader that usage charges are to be avoided, technology can provide excellent arguments for them. Consider for example the extreme case of usage charges where prices are dynamic, i.e. change in time to reflect demand. In this context, it may be impractical for end-user to control both his spending rate and his service quality (say measured in terms of the achieved information rate). Does this imply that dynamic pricing should never be considered as a pricing alternative? The answer is not clear. It is very plausible that the computers of users' end-systems could run software that makes optimal choices on their behalf. Such 'intelligent agents' would know their 'master's' preferences, and try to offer them the best price for value. In this context complex charging schemes with better resource control could become practical.

Let us turn to some of the other aspects of charging. Transparent charges are ones that are detailed in an itemized bill, rather than being bundled. The bill explains the total amount spent and helps a user decide if a particular service provides value for money. A charging system is auditable if the provider can, when requested, prove the validity of the charges he has made by tracing them to their origin. Service providers also impose important requirements on charging systems. Since service provisioning often defines a complex value chain, in which many business entities contribute to the end service and so define and share the resulting charges, the charging system must be flexible enough to allow the definition of rich business scenarios. New tariffs and services must be easily programmed while the appropriate service usage parameters must also be easy to access. A good example is multicasting. A sender transmits information to a number of receivers. Depending on the specific business model, the sender must charge the receivers (as in the case of video broadcast), the sender must pay the receivers (as in the case of targeted advertisement), or they must share the charge. The charging system should allow the implementation of all these business models and any degree of service bundling.

3.6 A model of business relations for the Internet

In Sections 2.1.3 and 3.4.2, we described the interesting but complex business relations that can evolve in a large network like the Internet. In this section we provide a simple model which characterizes the various interactions of the business entities that offer services in today's Internet. It can serve as a starting point for understanding the complex service provisioning environment. We begin by describing the hierarchical structure of today's Internet, as shown in Figure 3.12.

This hierarchical access structure allows 'long distance' traffic to flow through the backbone and 'local traffic' to use the regional ISP networks. It also provides for traffic that uses the backbone to be adequately multiplexed so as to fill the large transport containers as described in Section 2.1.3. Network Access Points (NAPs) are the termination points of the access network used by the regional ISPs to access the networks of the Backbone Service Providers (BSPs). BSPs have either installed their own physical fibre optic network backbones, or lease infrastructure from infrastructure providers. The capacity of the already installed fibre infrastructure (assuming fibres lit at 32 wavelengths at 10 Gbps) is estimated to be close to 13 Tbps, and the announced capacity for the near future is 900 Tbps (assuming planned fibres lit at 160 wavelengths at 10 Gbps, as of 2004).[1] The business model of

---

[1] Such calculations may hide important information since they do not include the distance over which such fibre is lit (since the cost of lighting fibre depends on distance). Also, not all planned fibre will be lit due to the high

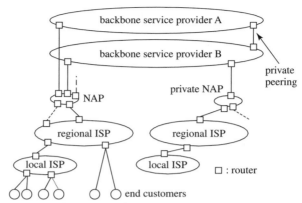

Figure 3.12   The hierarchical structure of the Internet. The concept of a Network Access Point (NAP) was introduced as a prerequisite for the commercialization of the Internet in 1995. A competitive market in the provision of backbone Internet services is achieved through carrier-independent NAPs, at which regional ISPs can freely interconnect to their choice of Backbone Service Provider (BSP).

NAPs is to provide the necessary infrastructure for implementing the SLAs of the transport services sold by BSPs to the ISPs, and amongst BSPs themselves. Peering agreements are interconnection agreements between BSPs that are provided for free on a mutually beneficial basis (see Chapter 12). To provide the above services, a NAP consists of highly secure and reliable local area networks, interconnecting at extremely high speeds the routers (the POPs) of the consumers (the ISPs) and the producers (the BSPs) of the transport services. The NAP manager can connect these POPs with variable size bit-pipes as specified in the SLAs. Such SLAs are usually charged according to the peak rate allowed by the above pipe, and the NAP receives a service fee. In many cases it may act as a bandwidth broker. In this case, the NAP buys such backbone capacity from many BSPs in a wholesale fashion and resells it to its retail ISP customers. One may envision a trend in which NAPs act as virtual BSPs by selling transparent backbone connectivity (or even VPN services) to the ISPs or directly to large customers. Such a business model reduces the market power of the BSPs in favour of the NAPs. SLAs between the NAP and the BSPs, or between the NAP and the ISPs, may be dynamic, reflecting market demand and availability. Such SLAs may also deploy dynamic price mechanisms, such as auctions, to define the market price for bandwidth. To promote competition between NAPs for ISP customers, the access network must make it easy for an ISP to switch NAPs. Metropolitan area networks based on optical network technologies can easily provide this flexibility. Alternatively, the ISP's POP may be located in a carrier-independent facility having fibre connectivity to the various NAPs.

The reader has now all the concepts needed to define a simple model for the value chain in Internet services. We can classify the business entities that contribute to Internet service provisioning in two layers, namely, the *infrastructure layer* and the *Internet service layer*. An entity in the infrastructure layer provides simple services to entities in the Internet service layer, such as the rent or lease of network equipment, point-to-point connectivity (bearer services), and services such as billing, technical support, and call-centre services. By this definition, an ATM network operator is an infrastructure provider who sells bit pipes

---

cost and the questionable demand. A probable figure adjusted by the present spending rate for lighting fibre is 95 Tbps.

(virtual paths) to ISPs. An ISP uses these bit pipes to connect the routers of his network. Note that such an infrastructure provider may buy optical network services from another infrastructure provider who sells point-to-point light paths or dark fibre.

The infrastructure layer also contains an access network service provider who uses DSL technology over copper wires, fibre, satellite, cellular or wireless LAN technology, to provide a bit pipe connection between the end-user equipment and a network node. We have seen examples of such infrastructure services in our previous discussions. An important point is that in the present communications market, such infrastructure services are not provided by vertically integrated monopolies, but by a large number of competing operators. This competitive market is key for the cost-effective provision of continuously upgraded network infrastructures deploying the latest transmission technologies.

The entities in the Internet service layer provide and consume Internet services, where by an Internet service we mean any service that is provided by the Internet software (running on network nodes connected using the infrastructure services), from low level network services (such as IP service, RSVP service and diffserv) to application and value-added information services. The services in this layer might be further subdivided into four types, reflecting whether the nature of the service is distribution or content.

1. *Transport Provider*: provides the infrastructure for forwarding IP packets. Specific cases include

   - *Internet Service Provider*: connects his customers (end-users, end-user networks) to each other and to the Internet backbone. Such a service includes providing customers with network addresses (static or dynamic). More general forms of such services are the Virtual Private Networks. ISPs also offer their customers higher-level information services such as e-mail, electronic commerce, instant messaging and information (portal) or community services. These help to differentiate their service and create customer lock-in. Customers become used to the user-friendly customized way that the ISP provided software allows them to access Internet services. Also an ISP may seek to persuade its customers to use information and e-commerce services supplied by its affiliated content and service providers. Such a preferential treatment can be enforced by designing the network so that these services can be accessed with smaller delays than services offered outside the ISPs network.
   - *Backbone Service Provider*: runs a high-capacity network, has connections to other BSPs through NAPs, and connects ISPs to the backbone. By connecting to a single BSP, an ISP obtains connectivity to the rest of the Internet.

2. *Data Centre Provider*: provides the computing environment that hosts the content and the applications owned by the Information Providers. Examples of such environments are video servers and massive server farms which implement the Internet service layer architecture (client-server, three tier architecture, dynamic content creation) for providing information services upon request. For performance reasons such providers are directly connected to the Internet backbone or to access networks. Since such computing environments can viewed as merely infrastructure, a Data Centre Provider may well be classified as offering infrastructure layer services. Another use of data centres is to host points of presence of ISPs, BSPs, and other telecoms operators. In this case, the data centre may play the role of a NAP and is called a *telecom hotel*. The ability to directly connect (using fibre) the computing environment that hosts the applications to the backbones of the BSPs and ISPs is key to improving access

performance. Also, as already discussed, such architectures allow communication and information service providers to compete for customers. As for infrastructure service for connectivity, data centre services may be layered. At the lowest layer, a customer may rent floor space and simple power reliability. Enhanced services include added security and reliability features, and connectivity to ISPs with backup features. At a higher layer, there are servers and switches that occupy the above floor space, and which the Data Centre Provider can rent to his customers. Different service layers may be provided by different business entities.

3. *Information Provider*: provides the content and the applications broadly described as value-added services. Such a provider rents space and CPU cycles from a Data Centre Provider, and uses one or several Transport Providers to connect with other Service Providers and End-Users. Examples of Information Providers are:

- *Application Service Provider*: leases to customers the use of software applications that he owns or rents. Examples of such applications are www-servers for web hosting, databases, and the complete outsourcing of business IT operations. An ASP rents space from a Data Centre Provider, and often these two types of service are offered by the same business entity.

- *Content Provider*: produces, organizes, manages and manipulates content such as video, news, advertisements and music. When such services are more advanced, including the ability for easily searching and purchasing a broad category of goods and services, they are called *portal services*.

- *Content Distributor*: manages content provided by Content Providers in network caches located near the End-Users. An End-User who accesses the content of a remote web site will receive the same content from the local cache, instead of having to go through the whole Internet. Such services improve the performance of web sites, specially when users access multimedia information that requires high bandwidth, or they access large files. Caches are located as near as possible to the access network, so to avoid bottlenecks and guarantee good performance. A Content Distributor is responsible for regularly updating the information stored in the caches to reflect accurately the content of the primary web site. The quality of a content distribution service improves with the number of cache locations the provider uses. More locations imply a lower average distance from an End-User to such a cache. Note that Content Distributors allow for information to be accessed locally instead of using the Internet backbone. In this respect they are in direct competition to Backbone Service Providers. A local ISP buying services from a large Content Distributor may worry less about transport quality through the backbone. Such competition is greatly influenced by the relative prices of storage and bandwidth.

- *Internet Retailer*: sells products such as books and CDs on the Internet.

- *Communication Service Provider*: runs applications that offer communications services such as Internet Telephony, email, fax and instant messaging.

- *Electronic Marketplace Provider*: runs applications that offer electronic environ-ments for performing market transactions. In such e-commerce environments busi-nesses advertise their products and sell these using market mechanisms simulated electronically.

4. *End-User*: consumes information services produced by the Information Providers, or uses the services of a Transport Provider to connect to other End-Users. He can be an individual user or a private organization.

A Business Perspective

The fundamental reason the Internet has been a catalyst for the generation of such a complex and competitive supply chain for services is that it is an open standard and serves as a common language. It allows new services to be deployed, and no-one has to seek permission from anyone to innovate. There are no owners of the Internet. In that respect it presents a fundamental challenge to the legacy systems such as the telephone network. The basic conceptual difference is that these networks define and restrict the services that can exist. Innovation must come from the network operator instead of the immensely rich community of users and potential entrepreneurs. The Internet is a general purpose language for computers to communicate by exchanging packets, without specifying the service for which these packets are used. This decoupling of networking technology from service creation is fundamental to the Internet revolution and its economic value. The difference between the Internet and the telephone or cable network can be compared to that between highways and railways. The owner of a highway does not constrain beyond very broad limits of size and weight what may travel on it. A vehicle need not file a travel plan and it can enter or leave the highway as it chooses. No central control is exercised. If a traffic jam occurs, vehicles re-route themselves, similarly as do IP packets in the Internet.

A last observation concerns vertical integration. It is natural for a firm that provides services in the above value chain to seek greater control in order to obtain a larger part of the total revenue. The less fragmented is service provisioning, then the more control a firm can obtain. We have already mentioned that another factor that encourages such vertical integration is the provision of end-to-end service quality. The service provider that controls the interaction with the customers may have the most advantageous position due to customer lock-in. This position is mainly held by application and content providers. For other business entities in the value chain a major concern is that their services are not commoditized. So vertical integration between ISPs, access providers and content providers has many advantages. It creates large economies of scope for the content provider by giving him new channels for distributing different versions of his content, for advertisement, and for creating strong customer communities. It also allows him to control the quality of the distribution, and guarantees him some minimum market share (the customers with whom he is vertically integrated).

One way for the access and transport service providers to strengthen their bargaining position with content providers is by increasing their customer base. Access providers using broadband technologies such as cable or wireless can sell their customers a bundle of services consisting of fast Internet access and video. Having a large customer base allows these providers to negotiate low rates for content from content providers such as cable and television channels. In most cases the cost of the content is a substantial part (about 40%) of the operating cost of the access network.

A final issue is the amount of risk involved in deploying new services and generating demand. Certain parts of the value chain, such as the deployment of new fibre-optic networks, involve higher risks. Others are less risky. For example, steady revenues are almost guaranteed to the few telephone companies that control the local loop because of their near monopoly position. However, these companies are often overly risk-averse, due to their past monopoly history, and this reduces their ability to innovate and compete effectively in the new services markets.

3.7 Further reading

References for the Internet and other communication technologies are the classic networking textbooks Walrand (1998), Walrand and Varaiya (2000) and Kurose and Ross (2001). The latter focuses more on the Internet services, whereas the other two cover the complete spectrum of communication technologies and network control mechanisms. Ramaswami and Sivarajan (1998) gives full coverage of optical network technology issues, while Cameron (2001) provides a high-level introduction to issues of modern optical networks, including condominium fibre and access networks.

Substantial information can also be found on-line. For instance, Cisco (2002c) provides a full coverage of major communications technologies (visit Cisco (2002f) for a fuller set of topics), while Cisco (2002d) and Cisco (2002g) serve as a simpler introduction to key networking concepts. We encourage the advanced reader to find in Cisco (2002a) an example of the detailed QoS capabilities of software that runs on network elements and provides Quality of Service. It discusses in depth issues such as congestion control, policing, traffic shaping and signalling.

Excellent starting points for obtaining network technology tutorials are Web Proforum (2002) and the sites of network magazines such as Commweb (2002). Similarly, Webopedia (2002) provides an explanation of most Internet technology concepts, and links for further detailed information. Other useful sites are 'Guide to the Internet' (University of Albany Libraries (2002)), and the web pages of MacKie-Mason and Whittier (2002).

Standards for the Internet are developed by the Internet Engineering Task Force (IETF). The official references are the *Requests for Comments* (RFCs), which are published by the Internet Architecture Board, and start, as their name suggests, as general requests for comments on particular subjects that need standardization. This is precisely the open mentality of the Internet, which can be summarized as: 'rough consensus and running code'. The RFCs can be found in the web pages of RFC Editor (2002) and Internet RFC/STD/FYI/BCP Archives (2002). Two interesting informational RFCs are #1110 (IAB Protocol Standards) and #1118 (The Hitchhikers Guide to the Internet). Between April 1969 and July 2002 there were over 3,300 RFCs. An interesting source for information on the evolution of the Internet telecoms industry is The Cook Report on Internet, Cook (2002).

Information on ATM Forum activities can be found at the web site of the ATM Forum (2002), including approved technical specifications and definitions of services. Information on VPN services is available at the sites of the various equipment vendors and service providers. For example, Cisco (2002h) provides a good introduction to security issues. Information on the Softswitch concepts and the convergence of circuit switched and data network services can be found in the International Softswitch Consortium web page, SoftSwitch (2002).

4

Network Constraints and Effective Bandwidths

This chapter concerns the technological constraints under which networks operate. Just as a manufacturing facility produces goods by consuming input factors, so a communication network provides communications services by consuming factors such as labour and interconnection services, and by leasing equipment and simpler communications services.

We wish to emphasize the importance of timescales in service provisioning. In the short run, a network's size and capabilities for service provisioning are fixed. In the long run, the network can adapt its resources to the amounts of services it wishes to provide. For example, it might purchase and install more optical fibre links. The cost models of Chapter 7 use incremental cost to evaluate the costs of services and are based upon a consideration of network operation over long timescales.

Innovations, such as electronic markets for bandwidth using auctions, are beginning to permit some short run changes in service provisioning through the buying and selling of resources. However, on short timescales of weeks or months, both the size of the network and its costs of operation must usually be taken as fixed. On short timescales, communications services resemble traditional digital goods, in that they have nearly zero marginal cost, but a very large common fixed cost.

Prices can be used as a control to constrain the demand within the production capability of the network: that is, within the so-called technology set. If one does this, then the consumer demand and structure of the technology set determine prices. In this chapter we provide tools that are useful in describing the technology sets of networks that offer the services and service contracts described in Chapter 2. The exact specification of such a technology set is usually not possible. However, by assessing a service's consumption of network resources by its *effective bandwidth*, we can make an accurate and tractable approximation to the technology set.

More specifically, in Section 4.1 we define the idea of a technology set, or acceptance region. Section 4.2 describes the important notion of statistical multiplexing. Section 4.3 concerns call admission control. Section 4.4 introduces the idea of effective bandwidths, using an analogy of filling an elevator with boxes of different weights and volumes. We discuss justifications for effective bandwidths in terms of substitution and resource usage. The general theory of effective bandwidths is developed in Section 4.5. Effective bandwidth theory is applied to the pricing of transport service classes in Section 4.6. Here we

Pricing Communication Networks   C. Courcoubetis and R. Weber
© 2003 John Wiley & Sons, Ltd   ISBN 0-470-85130-9 (HB)

summarize the large N asymptotic, the notion of an operating point, and interpretations of the parameters s and t that characterize the amount of statistical multiplexing that is possible. This section is mathematically technical and may be skipped by reading the summary at the end. In Section 4.7 we work through examples. In Section 4.8 we describe how the acceptance region can be defined by multiple constraints. In Section 4.9 we discuss how various timescales of burstiness affect the effective bandwidth and the effects of traffic shaping. Some of the many subtleties in assigning effective bandwidths to traffic contracts are discussed in Section 4.10. Often, a useful approach is to compute the effective bandwidth of the worst type of traffic that a contract may produce. Some such upper bounds are computed in Section 4.11. The specific case of deterministic multiplexing, in which we require the network to lose no cells, is addressed in Section 4.12. Finally, Section 4.13 presents some extensions to the general network case, and Section 4.14 discusses issues of blocking.

4.1 The technology set

In practice, a network provides only a finite number of different service types. Let x_i denote the amount of service type i that is supplied, where this is one of k types, $i = 1, \ldots, k$. A key assumption in this chapter is that the vector quantity of services supplied, say $x = (x_1, \ldots, x_k)$, is constrained to lie in a *technology set*, X. This set is defined by the provider, who must ensure that he has the resources he needs to provide the services he sells. It is implicit that each service has some associated performance guarantee and so requires some minimum amount of resources. Thus, x lies in X (which we write $x \in X$) if and only if the network can fulfil the service contracts for the vector quantity of services x. Note that here we are concerned only with the constraints that are imposed by the network resources; we ignore constraints that might be imposed by factors such as the billing technology or marketing policy.

Different models of market competition are naturally associated with different optimization problems. This is discussed fully in Chapter 6. In a monopoly market it is natural to consider the problem of maximizing the monopolist's profit. In a market of perfect competition it is natural to consider problems of maximizing social welfare. In both cases, the problems are posed under the constraint $x \in X$. Models of oligopoly concern competition amongst a small number of suppliers and lead to games in which the suppliers choose production and marketing strategies subject to the constraints of their technology sets.

Let x be the vector of quantities of k supplied service types. A general problem we wish to solve is

$$\underset{x \geq 0}{\text{maximize}} \ f(x), \quad \text{subject to } g(x) \leq 0 \tag{4.1}$$

The objective function $f(x)$ might be the supplier's profit, or it might be social welfare. Here $X = \{x : g(x) \leq 0\}$, where the inequality is to be read as a vector inequality, expressing m constraints of the form $g_i(x) \leq 0$, $i = 1, \ldots, m$. It is natural that the technology set be defined in this way, in terms of resource constraints and constraints on guaranteed performance. We suppose that $f(x)$ is a concave function of x. This is mathematically convenient and reasonable in many circumstances. Without loss of generality, we assume that all the service types consume resources and hence that the technology set is bounded.

Note that, for a synchronous network, the technology set is straightforward to define. This is because each service that is provided by the network requires a fixed amount of bandwidth throughout its life on each of the links that it transverses. Therefore, in what follows, we focus on services that are provided over asynchronous networks. In asynchronous networks the links are analogous to conveyor belts with slots, and slots are allocated to services on demand (see the discussion in Section 2.1.4). We suppose that there is finite buffering at the head of each link, where cells can wait for slots in which to be transmitted.

4.2 Statistical multiplexing

Let us consider a service contract with a QoS requirement that the traffic stream should suffer a maximum Cell Loss Probability (CLP). We use the term 'cell loss', instead of 'information loss' or 'packet loss', to make implicit a convenient, but not essential, assumption that information is broken into small cells of equal size. A service provider can guarantee $CLP = 0$ simply by ensuring that on every link the sum of the peak rates of all the connections carried on the link is less than the link's capacity. In other words, for each network link he takes a constraint of the form

$$\sum_{i=1}^{k} x_i h_i \leq C \tag{4.2}$$

where x_i is the number of connections of type i that use the link, h_i is the maximum rate of cells that the service contract allows to service type i, and C is the capacity of the link.

Although such a constraint makes sense for synchronous networks, in which connections are allocated fixed amounts of bandwidth during their lifetimes, equal to their peak rates h_i, it may not make sense for asynchronous networks, where connections are allocated bandwidth only when there is data to carry. If the service provider of such a network uses (4.2) to define the technology set he does not make efficient use of resources. He can do better by making use of *statistical multiplexing*, the idea of which is as follows. Typically, the rate of a traffic stream that uses service type i fluctuates between 0 and h_i, with some mean, of say m_i. At any given moment, the rates of some traffic streams will be near their peaks, others near their mean and others near 0 or small. If there are many traffic streams, then the law of averages states that the aggregate rate is very likely to be much less than $\sum_i x_i h_i$; indeed, it should be close to $\sum_i x_i m_i$. If one is permitted an occasional lost cell, say $CLP = 0.000001$, then it should be possible to carry quantities of services substantially in excess of those defined by (4.2). Instead, we might hope for something like

$$\sum_{i=1}^{k} x_i \alpha_i \leq C \tag{4.3}$$

where $m_i < \alpha_i < h_i$. The coefficient α_i is called an *effective bandwidth*.

Statistical multiplexing is possible when traffic sources are bursty and links carry many traffic streams. A model of a link is shown in Figure 4.1. A link can be unbuffered, or it can have an input buffer, to help it accommodate periods when cells arrive at a rate greater than the link bandwidth, C. Cells are lost when the buffer overflows. If we can tolerate some cell loss then the number of connections that can be carried can be substantially greater

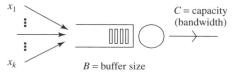

Figure 4.1 The Call Admission Control (CAC) problem. Given the state of the system in terms of the active traffic contracts and a history of load measurements, should a new traffic contract of type i be admitted?

than the number that can be carried if we require no cell loss. If there is just a single type of source then $x_{peak} = C/h_1$ and $x_{stat.} = C/\alpha_1$ would be the number of streams that could be carried without and with statistical multiplexing, respectively. Let us define the *statistical multiplexing gain* for this case as

$$\text{SMG} = \frac{x_{stat.}}{x_{peak}} = \frac{h_1}{\alpha_1}$$

Clearly, it depends upon the CLP. In Example 4.1, the statistical multiplexing gain is a factor of almost 5. The special case of requiring $\text{CLP} = 0$ is usually referred to as *deterministic multiplexing*.

Example 4.1 (Statistical multiplexing) Consider a discrete-time model of an unbuffered link that can carry 950 cells per epoch. There are x identical sources. In each epoch each source produces between 0 and five cells; suppose the number is independently distributed as a binomial random variable $B(5, 0.2)$. Thus, $h = 5$ and $x_{peak} = 950/5 = 190$. The mean number of cells that one source produces is $m = 5 \times 0.2 = 1$, and the number of cells that 900 sources produce is approximately normal with mean 900 and variance $0.2 \times 0.8 \times 900$. From this we calculate that the probability that 900 sources should produce more than 950 cells in a slot is about 0.0000155. Thus for a CLP of 1.55×10^{-5}, we can take $x_{stat.} = 900$ and there is a statistical multiplexing gain of $900/190 = 4.74$. This gain increases as the capacity of the link increases. For example, if C is multiplied tenfold, to 9500, then 9317 sources can be multiplexed with the same CLP of 0.0000155. The SMG is now $9317/1900 = 4.90$. As C tends to infinity the SMG tends to $h/m = 5$.

As we will see in Section 4.12, some multiplexing gain is possible even if we require $\text{CLP} = 0$. For example, if sources are policed by leaky buckets and links are buffered, then it is possible to carry more connections than would be allowed under the peak rate constraint of (4.2).

4.3   Accepting calls

Consider a network comprising only a single link. Suppose that contracts specify exact traffic types and that there are x_i contracts of type i, with $i = 1, \ldots, k$. Suppose that the only contract obligation is the QoS constraint $\text{CLP} \leq p$, for say $p = 10^{-8}$. The technology set A, which we also call the *acceptance region*, is that set of $x = (x_1, \ldots, x_k)$ corresponding to quantities of traffic types that it is possible to carry simultaneously without violating this QoS constraint (see Figure 4.2). Note that the technology set is defined implicitly by the QoS constraint. Later we show how to make explicit approximations of it.

As explained in Sections 2.2.3 and 3.1.5, *Call Admission Control* (CAC) is a mechanism that ensures that x remains in A. It does this by rejecting calls for new service connections through the network that would take the load of active calls outside A. Thus the acceptance region and CAC are intimately related. In practice, however, it is hard to know A precisely and so we must be conservative. In implementing a particular decision rule for CAC, we keep the load x within a region, say A', that lies inside the true acceptance region, A. For instance, a possible rule CAC rule is to accept a call only so long as (4.2) remains satisfied; this would correspond to taking A' as the triangular region near the origin in Figure 4.2. This rule is very conservative. The QoS constraint is easily satisfied, but the network carries fewer calls and obtains less revenue than it would using a more sophisticated CAC. This

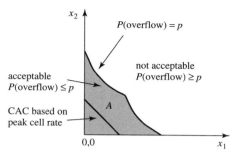

Figure 4.2   The acceptance region problem. Here there are $k = 2$ traffic types and x_i sources of in types i. We are interested in knowing for what (x_1, x_2) is CLP $\leq p$, for say $p = 10^{-8}$. The triangular region close to the origin is the acceptance region defined by $x_1 h_1 + x_2 h_2 \leq C$, which uses the peak cell rates and does not take advantage of the statistical multiplexing.

rule is an example of a static CAC, since it is based only on the traffic contract parameters of calls, in this case $h_1, \ldots, h_k$. In contrast, we say that a CAC is dynamic when it is based both on contract parameters and on-line measurements of the present traffic load. It is desirable that the decision rule for CAC should be simple and that it should keep x within a region that is near as possible to the whole of A, and so there be efficient use of the network. When we define A' in terms of a CAC rule we can call A' the 'acceptance region' of that CAC; otherwise acceptance region means A, the exact technology set where the QoS constraints are met.

Suppose that as new connections are admitted and old ones terminate the mix of traffic remains near a point $\bar{x}$ on the boundary of A. We call $\bar{x}$ the *operating point*. We will shortly see that the acceptance region can be well approximated at $\bar{x}$ by one or more constraints like (4.3), and this constant α_i can be computed off-line as a function of $\bar{x}$, the source traffic statistics, the capacity, buffer size and QoS required.

If a network has many links, connected in an arbitrary topology, then call admission is performed on a per route basis. A route specifies an end-to-end path in the network. A service contract is admitted over that route only if it can be admitted by each link of the route. This may look like a simple extension of the single link case. However, the traffic that is generated by a contract of a certain type is accurately characterized by the traffic contract parameters only at the entrance point of the network. Once this traffic travels inside the network, its shape changes because of interactions with traffic streams that share the same links. In general, traffic streams modelled by stochastic processes are characterized by many parameters. However, for call acceptance purposes, we seek a single parameter characterization, namely the α_i in constraint (4.3). We call α_i an effective bandwidth since it characterizes the resource consumption of a traffic stream of type i in a particular multiplexing context. In the next sections we show how to derive effective bandwidths. We consider their application to networks in Section 4.13. Finally, in Section 4.14, we suppose that a CAC is based on (4.3). What then is the call blocking probability? We discuss blocking in Section 9.3.3.

4.4   An elevator analogy

To introduce some ideas about effective bandwidths we present a small analogy. Suppose an elevator (or lift) can hold a number of boxes, provided their total volume is no greater than V and their total weight is no greater than W. There are k types of boxes. Boxes of type i have volume v_i and weight w_i. Let $v = (v_1, \ldots, v_k)$ and $w = (w_1, \ldots, w_k)$. Suppose $(v_i, w_i) = (2, 5)$ and $(v_j, w_j) = (4, 10)$. Clearly the elevator can equally well carry two boxes of type i as one box of type j, since $(4, 10) = 2 \times (2, 5)$. But what should one say when there is no integer n such that $(v_i, w_i) = n \times (v_j, w_j)$? This is the question posed in Figure 4.3.

It depends upon whether the elevator is full because of volume or because of weight. Suppose that boxes arrive randomly and we place them in the elevator until no more fit. Let x_i denote the number of boxes of type i. If at this point the maximum volume constraint is active, then

$$\sum_{i=1}^{k} x_i v_i = V , \quad \sum_{i=1}^{k} x_i w_i < W$$

and the effective usage is the volume of the box. At such a point we could substitute one small set of boxes for another small set of boxes provided their total volumes are the same. We suppose these sets are small enough that we are in no danger of violating the maximum weight constraint. We then say that a box of type i has effective bandwidth v_i. This is shown in the left of Figure 4.4.

Alternatively, the elevator might fill at a point where the maximum weight constraint is active. Perhaps this is usually what happens in the afternoon, when heavier boxes arrive. Then, again,

$$\sum_{i=1}^{k} x_i v_i < V , \quad \sum_{i=1}^{k} x_i w_i = W$$

and the effective usage is the weight of the box. We then say the effective bandwidth is w_i.

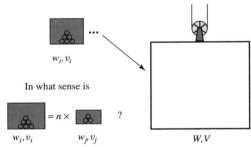

Figure 4.3   The elevator can carry a total weight of at most W and volume at most V. A box of type i has weight w_i and volume v_i. A box of type i has n times the relative effective usage of a box of type j if we are indifferent between packing 1 box of type i or n boxes of type j.

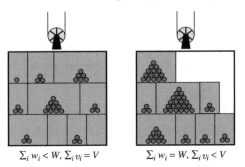

Figure 4.4   At the left the elevator is full because the volume of the boxes is V. The effective resource usage of a box of type i is v_i. At the right the elevator is full because the weight of the boxes equals W. The effective resource usage of a box of type i is w_i.

Thus, the relative effective usage of a box depends on whether the maximum volume or maximum weight constraint is active. We might write these simultaneously as

$$\sum_{i=1}^{k} x_i \alpha_i \leq C^*$$

and define $\alpha(\ldots) = (\alpha_1(\ldots), \ldots, \alpha_k(\ldots))$ and $C^*(\ldots)$ as functions of x, v, w, V and W. If these variables are such that the maximum volume constraint is active then $\alpha = v$ and $C^* = V$. But if they are such that the maximum weight constraint is active, then $\alpha = w$ and $C^* = W$. That is,

$$(\alpha, C^*) = \begin{cases} (v, V) \\ (w, W) \end{cases} \text{ as } \begin{array}{l} \sum_i x_i v_i = V, \ \sum_i x_i w_i < W \\ \sum_i x_i v_i < V, \ \sum_i x_i w_i = W \end{array}$$

At the point of the intersection of the two constraints, both effective bandwidths (volume and weight) are relevant, but not all substitutions are possible. The key point is that the effective bandwidths depend upon known parameters of the box types, (v_i, w_i), and on the capacities, V, W. They also depend on the operating point x, since if we are given the values of x for a full elevator we can determine which constraint is active.

There are various ways this operating point might be reached. It could be, as we have imagined so far, that we simply fill the elevator with boxes as they arrive. Which of the two constraints becomes active depends upon the rates at which the different types of box arrive. This might depend on the time of the day. Alternatively, we might accept and reject offered boxes so as to fill the elevator in a particular way. Alternatively, we might charge boxes for use of the elevator. The more we charge the boxes of type i, the smaller will be their rate of arrival.

Imagine that there are k agents, one associated with each box type. Agent i obtains benefit $u_i(x_i)$ when the elevator carries x_i boxes of type i. Suppose we wish to steer the operating point to maximize the sum of these utilities, i.e. to maximize $f = \sum_i u_i(x_i)$. Let $\bar{x}$ be the point on the boundary of A that does this. Assuming that each u_i is a concave function, one can show that if only the maximum volume constraint is active at $\bar{x}$ then there exists a scalar λ such that $u_i'(\bar{x}_i) = \lambda v_i$ for all i. If only the maximum weight constraint is active then there exists some scalar μ such that $u_i'(\bar{x}_i) = \mu w_i$ for all i. If both constraints are active, then there are λ and μ such that $u_i'(\bar{x}_i) = \lambda v_i + \mu w_i$ for all i.

Let $p_i = \lambda v_i$, $= \mu w_i$ or $= \lambda v_i + \mu w_i$, in line with the three possibilities described above. Then the point $\bar{x}$, at which f is maximized within A, can be characterized as the solution to k problems, the ith of which is to maximize $[u_i(x_i) - p_i x_i]$ over x_i. Note that these k problems decouple and can be solved in a decentralized fashion. The ith problem is to be solved by agent i. He seeks to maximize his net benefit, given that the price per box of type i is p_i. Thus the problem of maximizing the function f can be solved in a decentralized fashion, within a market for services where this optimal price vector will be determined. Observe that in most cases, p_i/p_j equals α_i/α_j, so prices are proportional to effective bandwidths. This is the main motivation for using effective bandwidths in pricing.

Note that in the original model, x_i denoted the number of boxes of type i that are placed in the elevator. We can extend the model and assume that the elevator takes one unit of time for each trip. Now x_i denotes the rate at which boxes of type i are served. We can make the analogy to networks by thinking of services as boxes and the network as an elevator. This is a valid analogy since services consume network resources, of which networks have finite amounts. If it takes time T_i to complete a service of type i and such service requests arrive at a rate of x_i per T_i units of time, then the mean number of services of type i in the network will be x_i. (This follows from *Little's Law*, which says that the mean number of jobs in the system, L, equals the product of arrival rate, λ, and mean time spent in the system, W, i.e., $L = \lambda W$; this translates here to $x_i = (x_i/T_i)T_i$). As above, posting prices that affect arrival rates can solve the problem of maximizing a utility function that captures the value of the services to the customers. A similar observation applies to the interpretation of x in (4.1). If each service is priced with an appropriate price p_i, then the market will find an equilibrium at the solution of this optimization problem. Again, such prices should be proportional to the effective bandwidths of the services.

Note that it is valid to make this simple translation from x_i as a number of boxes to an arrival rate of boxes only if the boxes arrive regularly, i.e., exactly every T_i/x_i time units. If, however, boxes arrive irregularly, say according to a stochastic process, and x_i is only an average arrival rate, then the instantaneous rate of arriving boxes can occasionally exceed the average value. So, if the elevator is full some arriving boxes may be blocked from being served. We discuss models that take account of such blocking effects in Sections 4.14 and 9.3.3.

4.5   Effective bandwidths

We have seen in the elevator example of Section 4.4 that a key notion in assessing resource usage is substitution. Let us explore this in a more general way. Suppose the technology set is defined by $x \geq 0$ and $g_i(x) \leq c_i$, $i = 1, \ldots, m$. Assume $g_i(x)$ is nondecreasing in each component of x. In the elevator example g_i is linear in x. Suppose $g_i(\bar{x}) = c_i$ is the unique binding constraint at point $\bar{x}$. Then (by Taylor's theorem) a change of $\bar{x}$ to $\bar{x} + \epsilon$ changes the value of the left-hand side of this constraint to

$$g_i(\bar{x}) + \epsilon_1 \partial g_i/\partial x_1 + \cdots + \epsilon_k \partial g_i/\partial x_k|_{x=\bar{x}} + o(\epsilon)$$

(where $o(\epsilon)$ denotes a term that is small compared to ϵ: explicitly, $o(\epsilon)/|\epsilon| \to 0$ as $|\epsilon| \to 0$). So, we satisfy the binding constraint to within $o(\epsilon)$ if

$$\epsilon_1 \partial g_i/\partial x_1 + \cdots + \epsilon_k \partial g_i/\partial x_k|_{x=\bar{x}} = 0$$

Thus, it is natural to define the *effective bandwidth* of contract j as $\alpha_j = \partial g_i/\partial x_j|_{x=\bar{x}}$. It can again be viewed as a substitution coefficient, because if we let the number of type 1 contracts, x_1, increase by δ/α_1 and the number of type 2 contracts, x_2, decrease by δ/α_2 and hold all other components of x constant, then the constraints of the technology set are still satisfied to within $o(\delta)$.

The above analysis suggests a method for constructing the effective bandwidths from knowledge of the acceptance region. Unfortunately, it is hard to determine A in practice, since its boundary can be found only by experimentation at a very large number of points. In the next section we present an approach for deriving the $\alpha_1, \ldots, \alpha_k$ from statistical characteristics of the sources.

The previous discussion suggests that we can interpret effective bandwidths as defining a local linear approximation to the boundary of the technology set at the operating point $\bar{x}$. In Figure 4.5 two constraints define A. One is linear and one is nonlinear. Suppose the operating point is on the boundary of the nonlinear constraint, $g_1 \leq c_1$. Then $\sum_j x_j \alpha_j = \sum_j \bar{x}_j \alpha_j$ defines a hyperplane that is tangent to $g_1 = c_1$ at the operating point $\bar{x}$, with $\alpha_j = \partial g_1/\partial x_j|_{x=\bar{x}}$. Here α depends on which constraint is binding and this depends on the operating point $\bar{x}$. Now a local approximation to the boundary of $g_1(x) \leq c_1$

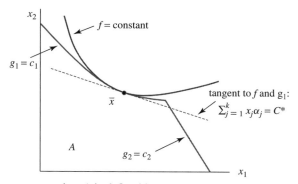

Figure 4.5   The acceptance region A is defined by two constraints. At the operating point $\bar{x}$, which achieves the maximum of f in A, the active constraint is $g_1(x) \leq c_1$ and so the effective bandwidths will be of the form $\alpha_j = \partial g_1 / \partial x_j \big|_{x=\bar{x}}$. Note that the problem of maximizing f subject to $\sum_{j=1}^{k} x_j \alpha_j \leq C^*$, where $C^* = \sum_{j=1}^{k} \bar{x}_j \alpha_j$, is also solved at $\bar{x}$. Thus, we can use simpler effective bandwidth constraints, in place of the actual acceptance region constraints, in posing the optimization problem.

at the operating point $\bar{x}$ is the hyperplane

$$\sum_{j=1}^{k} x_j \alpha_j \leq C^*, \quad \text{defining } C^* := \sum_{j=1}^{k} \bar{x}_j \alpha_j \qquad (4.4)$$

If the operating point $\bar{x}$ was defined by maximizing f over A, then this line is also tangent to a contour of f at this point. The problem of maximizing f subject to (4.4) is solved also at $\bar{x}$. Thus, for the purposes of identifying $\bar{x}$, or motivating users to choose $\bar{x}$ in a decentralized way, the approximation in (4.4) to the boundary of A is as good as the true constraint $g_1(x) \leq c_1$.

4.6   Effective bandwidths for traffic streams

The technology set for transport services depends on the information that is available about the connections. We look first at the case in which we have a full description of each connection's traffic. In subsequent sections, we consider the more realistic case that the only information available about a connection is its service contract. The material of this section is mathematically intricate and some readers may wish to skip to the summary at the end.

We consider the simple problem of determining the number of contracts that can be handled by a single switch. The switch has a buffer of size B and serves C cells per second in a First Come First Serve (FCFS) fashion. In practice, switches may require more sophisticated modelling than FCFS in order to capture the effects of the sophisticated scheduling mechanisms that are used for differentiated services. Suppose the QoS is defined only in terms of the CLP, or equivalently in terms of the probability that the content of the buffer exceeds a certain level. Constraints concerned with exceeding maximum delay bounds can also be modelled this way (see Example 4.7). However, it is reasonable to focus on CLP because in present switch design this is more important than average delay. Present designs use small buffers for real time services. This keeps the maximum delay small. Even if large buffers are used, the CLP is usually already greater than we wish to permit before the size of the delay becomes important.

Suppose that there are k classes of traffic whose statistics are known. We consider QoS constraints that are either deterministic (CLP $= 0$) or probabilistic, say CLP $< 10^{-8}$.

As before x_i denotes the number of sources in class i. The technology set is the set of all $(x_1, \ldots, x_k)$ for which the QoS constraints are not violated. It depends upon the information available in advance (knowledge of the actual source statistics, the leaky bucket constraints of the contracts), dynamic information (on-line measurements), and on the QoS constraints.

Let $X_j[0, t]$ be the number of cells produced by a bursty source of type j in a window of length t seconds. Suppose that at the operating point $\bar{x}$ only a single constraint is binding, and it is of the form $-\log(\text{CLP}) \leq \gamma$. Then the effective bandwidth of a source of type j is defined, (for values of the *space parameter* s and the *time parameter* t, which are defined below), as

$$\alpha_j(s, t) = \frac{1}{st} \log E\left[e^{sX_j[0,t]}\right] \tag{4.5}$$

In Section 4.5, we showed that the binding constraint at the operating point $\bar{x}$ can be approximated by the linear constraint

$$\sum_{j=1}^{k} x_j \alpha_j(s, t) \leq C^* \tag{4.6}$$

where $C^* = \sum \bar{x}_j \alpha_j(s, t)$.

We emphasize that the effective bandwidth of a traffic stream is a function of its multiplexing context. This is fully summarized in the value of the parameters s and t. To determine the effective bandwidth constraint, one must first find the values of these parameters. They depend upon the operating point, $\bar{x}$, the link parameters, (C, B), and the permitted CLP. In fact, defining $\gamma(C, B)$ in (4.9), as an asymptotic value of $\log(\text{CLP})$, it follows that

$$s = \frac{d\gamma}{dB}, \quad st = \frac{d\gamma}{dC} \tag{4.7}$$

Both parameters also have physical interpretations and in principle could be 'observed' in the system. To interpret t, we note that there are many ways in which the buffer of the switch can fill and overflow. An important heuristic, which one can make precise using the mathematical theory of large deviations, is that when a rare event such as buffer overflow occurs, it occurs in its most probable possible way. The time parameter t corresponds to the most probable time over which the buffer fills during a busy period in which overflow occurs (see Figure 4.6). As we have said, this most probable time to overflow depends

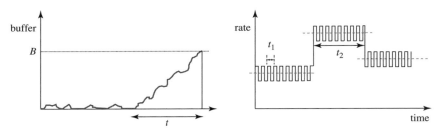

Figure 4.6   The operating point parameter t corresponds to the most probable time over which the buffer fills during a busy period in which overflow occurs. Here the source rate varies on two timescales and t_2 is more relevant to overflow than is t_1. This is because it is when the source produces at a high rate for a relatively long time, of order t_2, that the buffer overflows. During such a long time, fluctuations on the t_1 timescale are evened-out and do not contribute to the overflow.

upon the size of the buffer, the CLP and the precise mix of traffic that is multiplexed at the operating point. If any of these change, then the most probable time to overflow also changes. For example, t tends to zero as the size of buffer, B, tends to zero.

The value of the space parameter s (perhaps measured in kb^{-1}) measures the degree to which advantage can be gained from statistical multiplexing. In particular, for links with capacity much greater than the sum of the mean rates of the multiplexed sources, s tends to zero and $\alpha_j(s, t)$ approaches the mean rate of a source of type j (e.g. as $\lim_{s \to 0}[s^{-1} \log(\frac{1}{2}e^{as} + \frac{1}{2}e^{bs}) = \frac{1}{2}(a + b)$, for a source producing either a or b, with equal probabilities, in a window of length t). For links with capacity not much greater than the sum of the mean rates of the sources, there can be little statistical multiplexing gain. Intuitively, buffering is crucial and increasing the size of the buffer will make a large reduction in the CLP, and thus $s = d\gamma/dB$ will be large. As $s \to \infty$, we find that $\alpha_j(s, t)$ tends to a value, say $\alpha_j(\infty, t) = \bar{X}_j[0, t]/t$, where $\bar{X}_j[0, t] = \sup\{a : P(X_j[0, t] \geq a) > 0\}$, i.e., the least upper bound on the value that $X_j[0, t]$ takes with positive probability (e.g. as $\lim_{s \to \infty}[s^{-1} \log(\frac{1}{2}e^{as} + \frac{1}{2}e^{bs}) = \max\{a, b\}$). For sources that do not have maximum peak rates, such as a Gaussian one, $\bar{X}_j[0, t] = \infty$. Note that this gives the appropriate effective bandwidth for 'deterministic multiplexing' (i.e. for CLP $= 0$), since if $\sum_j x_j \alpha_j(\infty, t) \leq C$, where t is given the value that maximizes the left-hand side of this inequality, then $\sum_j x_j X_j[0, t] \leq Ct$ with probability 1 for all t. We pursue this further in Section 4.12.

There is also a mathematical interpretation for s. Conditional on an overflow event happening, the empirical distributions of the inputs just prior to that event differ from their unconditional distributions. For example, they have greater means than usual and realize a total rate of $C + B/t$ over the time t. The so-called 'exponentially tilted distribution with parameter s', specifies the distribution of the sources' most probable behaviour leading up to an overflow event.

The single constraint (4.6) is a good approximation to the boundary of the acceptance region if the values of s and t remain fairly constant on that boundary of A and so $\alpha_j(s, t)$ does not vary much. In practice, the values of x might be expected to lie within some small part of the acceptance region boundary (perhaps because the network tries to keep x near some point where social welfare or revenue is maximized). In this case it is only important for (4.6) to give a good approximation to A on this part of its boundary.

The motivation for the above approach comes from a large deviations analysis of a model of a single link. Here we simply state the main result. Let C be the capacity of the link and B be the size of its buffer. Suppose the operating point is x (dropping the bar for simplicity). Consider an asymptotic regime in which there are 'many sources', in which link capacity is $C = NC^{(0)}$, buffer size is $B = NB^{(0)}$, the operating point is $x = Nx^{(0)}$, and N tends to infinity. It can be shown that

$$\lim_{N \to \infty} \frac{1}{N} \log(\text{CLP}(N))$$

$$= \sup_{t \geq 0} \inf_{s \geq 0} \left[st \sum_{j=1}^{k} x_j^{(0)} \alpha_j(s, t) - s \left(C^{(0)}t + B^{(0)} \right) \right] \quad (4.8)$$

This holds under quite general assumptions about the distribution of $X_j[0, t]$, even if it has heavy tails. Thus when the number of sources is large we can approximate $(1/N) \log \text{CLP}(N)$ by the right-hand side of (4.8). Making this approximation and then multiplying through by N, we find that a constraint of the form $\text{CLP}(N) \leq e^{-\gamma}$ is approximated by

$$-\gamma(C, B) := \sup_{t \geq 0} \inf_{s \geq 0} \left[st \sum_{j=1}^{k} x_j \alpha_j(s, t) - s(Ct + B) \right] \leq -\gamma \qquad (4.9)$$

Note also that (4.7) is obtained from (4.9) by taking derivatives with respect to B and C. The envelope theorem says that s and t can be treated as constant while taking these derivatives. (It is the theorem that if $F(a) = \max_y f(a, y) \equiv f(a, y(a))$, then $dF(a)/da = \partial f(a, y)/\partial a|_{y=y(a)}$.)

4.6.1 The Acceptance Region

The constraint of (4.9) can be rewritten as the union of an infinite number of constraints, one for each $t \geq 0$, and each taking the form

$$g_t(x) \leq -\gamma \qquad (4.10)$$

where

$$g_t(x) = \inf_{s \geq 0} \left[st \sum_{j=1}^{k} x_j \alpha_j(s, t) - s(Ct + B) \right] \qquad (4.11)$$

We can interpret $g_t(x)$ as the logarithm of the probability that overflow occurs and that it does so over a time t. Hence if x satisfies (4.9) then the logarithm of the probability of overflow during a period of length t is no more than $-\gamma$, for all t.

Let $A_t = \{x : g_t(x) \leq -\gamma\}$. Since it is the minimum of linear functions of x, the right-hand side of (4.11) defines a concave function of x and so each A_t is the complement of a convex set (refer to Appendix A for definitions of concave and convex functions and convex sets). The acceptance region is $A = \cap_t A_t$, as exemplified in Figure 4.7. Note that since (4.9) is an asymptotic approximation of the true CLP, the region A is an asymptotic

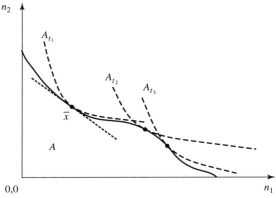

Figure 4.7 The structure of an acceptance region for two types of calls. The acceptance region, A, is the intersection of the complements of the family of convex sets A_t, parts of whose northeast boundaries are shown for three values of t. It may be neither convex nor concave. We illustrate a local approximation at some boundary point $\bar{x}$ using effective bandwidths. Here, the effective bandwidths are defined by the tangent to the boundary of A_{t_1} at $\bar{x}$.

approximation of the true acceptance region for a given finite N. It becomes more exact as N increases. Practical experiments show excellent results for N of the order of 100.

Suppose $\bar{x}$ is the operating point in A and $g_t(x) \leq -\gamma$ is the constraint that is binding at this point. Then t achieves the supremum in (4.9). Let s be the infimizer in the right hand side of (4.11). Then $\partial g_t/\partial x_j\big|_{x=\bar{x}} = st\alpha_j(s,t)$ and so, as above, $g_t(x) \leq -\gamma$ has a linear approximation in the neighbourhood of $\bar{x}$ of

$$st \sum_{j=1}^{k} x_j\alpha_j(s,t) \leq st \sum_{j=1}^{k} \bar{x}_j\alpha_j(s,t) = s(Ct + B) - \gamma$$

Dividing by st, we have

$$\sum_{j=1}^{k} x_j\alpha_j(s,t) \leq C^*, \quad \text{where } C^* = C + \frac{1}{t}\left(B - \frac{\gamma}{s}\right) \tag{4.12}$$

The linear constraint in (4.12) gives a good approximation to the boundary of the acceptance region near $\bar{x}$ if the values of s and t which are optimizing in (4.9) do not change very much as x varies in the neighbourhood of $\bar{x}$. We can extend this idea further to obtain an approximation for the entire acceptance region by approximating it locally at a number of boundary points. Optimizing the selection of such points may be a highly nontrivial task. A simple heuristic when the s and t do not vary widely over the boundary of the acceptance region is to use a single point approximation. One should choose this point to be in the 'interesting' part of the acceptance region, i.e. in the part where we expect the actual operating point to be. Otherwise one may choose some centrally located point such as the intersection of the acceptance region with the ray $(1, 1, \ldots, 1)$.

In practice, points on the acceptance region and their corresponding s and t can be computed using (4.9). We start with some initial point x near 0 and keep increasing all its components proportionally until the target CLP is reached.

Let us summarize this section. We have considered the problem of determining the number of contracts that can be handled by a single switch if a certain QoS constraint is to be satisfied. We take a model of a switch that has a buffer of size B and serves C cells per second in a first come first serve fashion. There are k classes of traffic, and the switch is multiplexing x_j sources of type j, $j = 1, \ldots, k$. We define the 'effective bandwidths' of source type j by (4.5). This is a measure of the bandwidth that the source consumes and depend upon the parameters s and t. As s varies from 0 to ∞, it lies between the mean rate and peak rate of the source, measured over an interval of length t. Arriving cells are lost if the buffer is full. We consider a QoS constraint on the cell loss probability of the form CLP $\leq e^{-\gamma}$, and show that a good approximation to this constraint is given by the inequality in (4.9). The approximation becomes exact as B, C and x_j grow towards infinity in fixed proportions. For this reason, (4.9) is called the 'many sources approximation'. At a given 'operating point', $\bar{x}$, the constraint has an approximation that is linear in x, given by (4.12), where s and t are the optimizing values in (4.9) when we put $x = \bar{x}$ on the right-hand side. The linear constraint (4.12) can be used as an approximation to the boundary of the acceptance region at $\bar{x}$. We can interpret t as the most probable time over which the buffer fills during a busy period in which overflow occurs.

4.7 Some examples

In some cases the acceptance region can be described by the intersection of only a finite number of A_ts. We see this in the first two examples.

Example 4.2 (Gaussian input) Suppose that $X_j[0, t]$ is distributed as a Gaussian random variable with mean $\mu_j t$ and variance $\sigma_j^2 t$. For example, let $X_j[0, t] = \mu_j t + \sigma_j B(t)$, where $B(t)$ is standard Brownian motion. Then

$$\alpha(s, t) = \frac{1}{st} \log E\left[e^{sX[t,0]}\right] = \frac{1}{st} \log e^{\mu_j ts + \sigma_j^2 t s^2/2} = \mu_j + \sigma_j^2 s/2$$

Note that α is independent of t. Also, it tends to infinity as s increases. This is because the Gaussian source does not have a finite peak rate. When sources of k different types are multiplexed, the acceptance region, A, is defined by (4.9), which is

$$\sup_{t \geq 0} \inf_{s \geq 0} \left[st \sum_{j=1}^{k} x_j(\mu_j + \sigma_j^2 s/2) - s(Ct + B)\right] \leq -\gamma$$

The infimum with respect to s occurs at $s = (C + B/t - \sum_j x_j \mu_j)/\sum_j x_j \sigma_j^2$. This gives A to be defined by

$$\sup_{t \geq 0} \left[-\frac{1}{2}t\left(C + B/t - \sum_{j=1}^{k} x_j \mu_j\right)^2 \Big/ \sum_{j=1}^{k} x_j \sigma_j^2\right] \leq -\gamma$$

in which the supremum with respect to t is achieved by

$$t = \frac{B}{C - \sum_j x_j \mu_j}$$

Note that the most likely time over which buffer overflow occurs is the same time that it would take for a full buffer to empty while being fed with new input at the average rate of $\sum_j x_j \mu_j$. So (4.9) is just

$$\sum_{j=1}^{k} x_j\left(\mu_j + \sigma_j^2 s/2\right) \leq C + \frac{1}{t}\left(B - \frac{\gamma}{s}\right) \tag{4.13}$$

The effective bandwidth is $\mu_j + \sigma_j^2 s/2$, where $s = 2(C - \sum_j x_j \mu_j)/\sum_j x_j \sigma_j^2$. Note that the effective bandwidth depends upon C, and on the operating point through the mean and variance of the superimposed sources.

Things are rather special in this example. After substitution for s and t and simplification, (4.13) becomes

$$\sum_{j=1}^{k} x_j\left(\mu_j + \frac{\gamma}{2B}\sigma_j^2\right) \leq C \tag{4.14}$$

Thus, the acceptance region is actually defined by just one linear constraint. Expressions (4.13) and (4.14) are the same because $s = \gamma/B$ is constant on the boundary of the acceptance region. In fact, this acceptance region is exactly the region in which CLP $\leq e^{-\gamma}$, i.e. the asymptotic approximation is exact. This is because the Gaussian input process is infinitely divisible (i.e. $X_j[0, t]$ has the same distribution as the sum of N i.i.d. random variables, each with mean μ_j/N and variance σ_j^2/N — for any N). Therefore the limit in (4.8) is actually achieved.

Example 4.3 (Gaussian input, long range bursts)   Let us calculate the effective bandwidth of a Gaussian source with autocorrelation. This is interesting because positive autocorrelation produces a process with long range bursts. In the previous two examples we have constructed a model in continuous time. Now let us assume that time is discrete, i.e. with epochs $t = 1, 2, \ldots$. Suppose X_i represents the contribution of the source in the ith time interval and $\{X_1, X_2, \ldots\}$ is a sequence of Gaussian random variables with mean μ, variance σ^2 and autocovariance function $\gamma(k)$ (which is not to be confused with the logarithm of the CLP). Then we have $\alpha(s, t) = \mu + \sigma_t^2 s/2$, where

$$t\sigma_t^2 = \mathrm{var}\left(\sum_{i=1}^{t} X_i\right)$$

$$= t\sigma^2 + 2[(t - 1)\gamma(1) + (t - 2)\gamma(2) + \cdots + \gamma(t - 1)]$$

Notice that $\lim_{t\to\infty} \sigma_t^2 = \gamma$, where γ is the so-called 'index of dispersion'; one can show that when the sum converges, $\gamma = \sum_{-\infty}^{\infty} \gamma(k)$. If there is positive autocorrelation then $\sigma_t^2 > \sigma^2$, and so the effective bandwidth is greater than it would be for an uncorrelated Gaussian process with the same variance. Similarly, if $\sigma_t^2 < \sigma^2$ the effective bandwidth is less.

Example 4.4 (Brownian bridge model of periodic sources) In this example the acceptance region is described by just two linear constraints.

Consider a periodic source which produces a burst of size ρ_j at times $U, U + 1, U + 2$, $\ldots$, where U is uniformly distributed on the interval $[0, 1]$. Consider the superposition of x_j such sources, with values of U chosen independently. It is a random process whose value increases from 0 to $x_j \rho_j$ over the interval $[0, 1]$. At each time t between 0 and 1 the probability that any one source has already produced its burst is t. It follows that the traffic produced by the superposition by time $t \leq 1$ has a distribution of ρ_j times a binomial distribution of $B(x_j, t)$; the distribution of this quantity is approximately normal, with mean $x_j \rho_j t$ and variance $x_j \rho_j^2 t(1 - t)$. In fact, the superposition tends to that of ρ_j times a Brownian motion that starts at 0 and is conditioned to reach x_j at time 1. This suggests that we consider a different type of source whose superposition is exactly this. Each of these sources is of the form

$$X_j[0, t] = \rho_j \lfloor t \rfloor + \rho_j Z(t - \lfloor t \rfloor)$$

where $Z(t)$, $0 \leq t \leq 1$, is the standard Brownian bridge having $Z(t) \sim N(t, t(1 - t))$. Superimposing x_j such sources is an approximation for superimposing x_j actual bursty periodic sources. As in Example 4.2, one can compute $\alpha_j(s, t) = \rho_j + \rho_j^2 s f(t)[1 - f(t)]/2t$, where $f(t) = t - \lfloor t \rfloor$ is the fractional part of t. The acceptance region turns out to be $A = \cap_t A_t = A_{0.5} \cap A_\infty$, where A_∞ and $A_{0.5} =$ are the sets of x satisfying the following two constraints:

$$\sum_{j=1}^{k} x_j \rho_j \leq C \tag{4.15}$$

$$\sum_{j=1}^{k} x_j \left(\rho_j + \rho_j^2 \frac{\gamma}{2B}\right) \leq B + C \tag{4.16}$$

Constraints (4.15) and (4.16) correspond to (s, t) values of $(0, \infty)$ and $(2\gamma/B, 1/2)$, respectively. For instance, if (4.15) is the active constraint, there is enough buffer to absorb the temporary bursts (expressed in (4.16)), but these buffers fill infinitely slowly since the average input rate 'barely' exceeds the service capacity. If (4.16) is active, then the most probable time over which the buffer produces overflows is half way through each period.

Brownian bridge inputs are infinitely divisible processes so, as in Example 4.2, the above acceptance region is exact for a simple queue fed by Brownian bridge inputs.

Example 4.5 (On-off sources) Consider a source that alternates between on and off states. When it is on it sends at constant rate h and when it is off it sends at rate zero. The successive lengths of time that it spends in the on and off states are T_{on} and T_{off}, respectively, which can be either deterministic or random. Let p_{on} denote the probability that the source is on. If m is the mean rate of the source, then $p_{on} = m/h$. The effective bandwidth of this on-off source has a simple form when the time parameter t is small compared with T_{on} and T_{off}. This is typical if the buffer is small. In this case, there is only a very small probability that the source is both on and off during a window of length t. Thus, with high probability $X[0, t]$, which is the contribution of the source in a window of size t, takes only two values: zero if the source is off and ht if the source is on, with respective probabilities $1 - m/h$ and m/h. Then (4.5) becomes

$$\alpha_{\text{on-off}}(m, h) = \frac{1}{st} \log\left[\left(1 - \frac{m}{h}\right) + \frac{m}{h} e^{sht}\right] \tag{4.17}$$

This expression illustrates some of the properties of s mentioned in Section 4.6. As s approaches zero, the effective bandwidth approaches the mean rate of source. As s tends to infinity, the effective bandwidth tends to the peak rate of the source. The effective bandwidth is an increasing function of s and thus takes values between the mean and the peak rate. Smaller values of s correspond to more efficient multiplexing.

Example 4.6 (Markov modulated source) Let us take the model of an on-off source in Example 4.5. Let the successive lengths of time that the source spends in the on and off states be i.i.d. exponential random variables with means T_{on} and T_{off}. This model has been used to model voice and video traffic. It can also be used to model the activity during a web browsing session.

We can generalize this model to one with even more than two states; suppose there are m states, $m \geq 2$. Suppose that in state i the source sends at rate μ_i. The state changes according to a continuous-time Markov process with known transition matrix, i.e. the holding time in state i is exponentially distributed, say with mean $1/\lambda_i$, and given that the state is i, the next state will be j with probability P_{ij}.

The nice thing about this class of models is that it is possible to calculate the effective bandwidth (at least numerically). Let $X_i[0, t]$ be the traffic produced over $[0, t]$, conditional on the source starting in state i. In brief, the effective bandwidth is computed from the moment generating function of the $X_i[0, t]$, say $f_i(t) = E \exp(s X_i[0, t])$. Then it is not hard to see that

$$f_i(t) = e^{-\lambda_i t} e^{s\mu_i t} + \int_0^t e^{s\mu_i u} \lambda_i e^{-\lambda_i u} \sum_j P_{ij} f_j(t - u) \, du$$

Let $f_i^*(z)$ be the Laplace transform of $f_i(t)$. The integral above is a convolution integral and so we easily find

$$f_i^*(z) = \frac{1}{z + \lambda_i - s\mu_i} \left(1 + \sum_{j=1}^m P_{ij} f_j^*(z)\right)$$

One can, in principle, solve this set of linear equations and then invert the Laplace transforms.

Consider the special case of an on-off source, with off and on phases that are exponentially distributed with means $1/\lambda_0$ and $1/\lambda_1$, respectively, and taking $\mu_0 = 0$ and peak rate $\mu_1 = h$. We find, after some algebra,

$$f(t) = E\left[\exp\left(s\int_0^t x(s)ds\right)\right] = \frac{\lambda_1}{\lambda_0 + \lambda_1}f_0(t) + \frac{\lambda_0}{\lambda_0 + \lambda_1}f_1(t)$$

$$= \frac{\lambda_1\omega_2 + \lambda_0(\omega_2 - sh)}{(\omega_2 - \omega_1)(\lambda_0 + \lambda_1)}e^{\omega_1 t} + \frac{-\lambda_1\omega_1 + \lambda_0(sh - \omega_1)}{(\omega_2 - \omega_1)(\lambda_0 + \lambda_1)}e^{\omega_2 t}$$

where ω_1, ω_2 are the two roots of $\omega^2 + (\lambda_0 + \lambda_1 - \lambda_0\theta h)\omega - \lambda_0\theta h = 0$.

A discrete time model is even easier. Suppose that the state changes at each epoch according to the transition matrix (p_{ij}). Then the effective bandwidths satisfy the set of linear recurrences

$$f_i(t) = e^{s\mu_i}\sum_j p_{ij}f_j(t-1), \quad t = 1, 2, \ldots$$

These recurrences have been successfully used to make numerical computations of effective bandwidth functions.

4.8 Multiple QoS constraints

As in Section 4.5, the idea of effective bandwidths extends to multiple QoS constraints. The acceptance region is then the intersection of the acceptance regions defined by each constraint. Each constraint might correspond to a different manner of overflow. The effective bandwidth of a stream is defined by the constraint that is active. Example 4.7 demonstrates how multiple constraints may result from the scheduling mechanism of priority queueing.

Example 4.7 (Priority queueing) One way to give different qualities of service to different classes of traffic is by priority queueing. Suppose that traffic classes are partitioned into two sets, J_1 and J_2. Service is FCFS, except that a class in J_1 is always given priority over a class in J_2. For $i \in J_1$ there is a QoS guarantee on delay of the form

$$P(\text{delay} > B_1/C) \le e^{-\gamma_1}$$

For all sources there is a QoS guarantee on CLP of

$$\text{CLP} \le e^{-\gamma_2}$$

This gives the two constraints $g_1(x) \le 0$ and $g_2(x) \le 0$, where x is the vector of the numbers of sources of the different types. The acceptance region is now the intersection of the acceptance regions corresponding to each of the constraints. Assume that on the 'interesting' part of each constraint the values of s and t do not vary widely, being s_i, t_i for constraints $i = 1, 2$. Then, by approximating each constraint globally using (4.12) calculated at a single appropriately chosen point, we obtain the effective bandwidth approximations of the constraints

$$\sum_{j \in J_1} x_j\alpha_j(s_1, t_1) \le K_1 \quad \text{and} \quad \sum_{j \in J_1 \cup J_2} x_j\alpha_j(s_2, t_2) \le K_2 \tag{4.18}$$

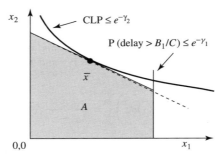

Figure 4.8   An acceptance region defined by two constraints. There are two classes of traffic. The vertical constraint is due to a guarantee on the delay of priority traffic. The second constraint is due to a guarantee on the CLP for both traffic types, and is approximated by a linear constraint at the operating point $\bar{x}$ (shown dotted).

where

$$K_1 := C + \frac{1}{t_1}\left(B_1 - \frac{\gamma_1}{s_1}\right) \quad \text{and} \quad K_2 := C + \frac{1}{t_2}\left(B - \frac{\gamma_2}{s_2}\right)$$

For example, suppose $J_1 = \{1\}$, $J_2 = \{2\}$. Then we have

$$x_1\alpha_1(s_1, t_1) \le K_1 \quad \text{and} \quad x_1\alpha_1(s_2, t_2) + x_2\alpha_2(s_2, t_2) \le K_2$$

If $K_1/\alpha_1(s_1, t_1) < K_2/\alpha_1(s_2, t_2)$ then the acceptance region takes the form illustrated in Figure 4.8. Note that this approximation of the technology set is less accurate if the values of s and t vary significantly along each constraint. Then one might approximate each constraint by tangent hyperplanes at more than one boundary point. The key observation is that our approximations will always be of the form (4.18) but with a larger number of constraints.

4.9   Traffic shaping

It is a characteristic of broadband multimedia and data traffic that its rate can fluctuate widely. These fluctuations can occur at various superimposed frequencies, as illustrated in the right hand part of Figure 4.6. Each frequency of fluctuation defines a timescale of burstiness, i.e. an order of time over which significant changes are observed in the rate of the source, when this is averaged over time periods of the same size. In Figure 4.6, such changes in the rate are observed on timescales of order t_1 and t_2 (and there may be even larger timescales, but these do not show up in the small snapshot taken). Suppose we are at an operating point, where a CLP guarantee is just satisfied and no more traffic can be packed in the link. We can ask, on what timescales are fluctuations most likely to cause buffer overflow? In other words, which aspects of the traffic make it hard to multiplex it and hence contribute to its effective bandwidth? Similar questions were posed in Section 4.4 when we determined effective bandwidths for the boxes that were to be packed in an elevator.

For a constraint on the technology set that is defined in terms of a constraint on CLP, the effective bandwidth of a source depends on the timescales of the source's burstiness that significantly contribute to the event of buffer overflow. Clearly, fluctuations on different timescales do not contribute equally, since those on short timescales may be absorbed by the buffer. This effect is captured in the definition of the effective bandwidth in (4.5). By letting $\alpha(s, t)$ depend upon the total contribution of the source in a window of length of time t, we 'filter out' the fluctuation that occur in timescales smaller than t. For example, in Figure 4.6 the timescale t_1 is absorbed by the buffer and is not reflected in the effective bandwidth of the source. Here t_2 is the dominant timescale of burstiness that constraints the system; within t_2, the timescale t_1 contributes its mean rate. In other words, if we were to replace our source by one obtained by averaging it over a timescale of t_1, this would have no effect on the multiplexing and it would neither increase or decrease the CLP.

Traffic shaping can be used to reduce high frequency fluctuations and produce smoother traffic. A typical traffic shaper consists of a large buffer that is served at a rate smaller than the peak rate of the stream, or of a buffer that is combined with a leaky bucket that holds the part of traffic that is non-conforming with the traffic contract; see Section 2.2.2. One may design the shaper to add delays of the same order of magnitude as the timescales of the fluctuations to be smoothed. Another way to implement a shaper is to collect the traffic that arrives every t time units and then transmit it during the next t time units at a constant rate. This rate will differ during each period, reflecting the variable volume of data to be transmitted. The above discussion explains the effects of shaping mechanisms on the effective bandwidth of the resulting traffic. When buffers in the network are large (and hence t is large), then only substantial traffic shaping can reduce the effective bandwidth of the input traffic. However, for real-time traffic using small buffers, a moderate amount of traffic shaping can drastically reduce the effective bandwidths and thereby increase the multiplexing capability of the network.

Let us look at some data for real traffic. Figure 4.9 shows a 1000 epoch trace from a MPEG-1 encoded video of *Star Wars*, each epoch being 40ms. Note the different timescales of burstiness. Figure 4.10 plots an estimate of the effective bandwidth function for this trace. Observe that as either t becomes small or s becomes large, the effective bandwidth increases; this corresponds to the source becoming more difficult to multiplex. The explanation is simple. The time averaging that takes place in the buffer in which this particular stream is being multiplexed smoothes small traffic bursts. In particular, it smoothes all fluctuation taking place on timescales less than t. The larger is t, the more the fluctuations are absorbed and so the resulting trace can be multiplexed as easily as a smoother trace in which these fluctuations have been averaged-out. Eventually, when t is large enough, the trace is no more difficult to multiplex than a constant bit rate source of the same mean. Small values of t occur when the buffer is small, in which case, the averaging effect is negligible, and so the traffic is more difficult to multiplex. This means a greater effective bandwidth. Similarly, when the link capacity decreases, s increases, and the value of the effective bandwidth

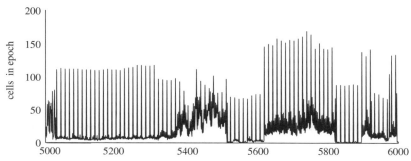

Figure 4.9 Burstiness can be seen in this trace of 1000 epochs of a MPEG-1 encoded video of *Star Wars*. Each epoch is 40 ms.

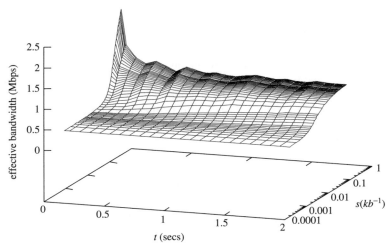

Figure 4.10 Effective bandwidth of MPEG-1 traffic. Note that for different values of s and t, corresponding to different multiplexing contexts, the effective bandwidth of the MPEG traffic stream can differ from about 0.5 to 2.5 Mbps.

increases, again capturing the increased difficulty in multiplexing. Note that for different values of s and t, corresponding to different multiplexing contexts, the effective bandwidth can differ substantially.

4.10   Effective bandwidths for traffic contracts

We have assumed thus far that the source statistics are fully known and so exact effective bandwidths can be computed. In practice, this is not the case. The network knows only the traffic contract of the requested service. This contract only partly characterizes the traffic source, for example, through the leaky bucket constraints. This poses a problem. If only the traffic contract is known, what effective bandwidths should be used for call acceptance? There are several possibilities. Each has its advantages and disadvantages:

1. If the network operator can tell which application generates the traffic, and that application produces traffic with known statistics, then he can use the actual effective bandwidth of that type of traffic. Usually, however, this information is not available.

2. If the traffic contract is used only by applications of a known general type (such as video), then one can use the typical effective bandwidth for the traffic of that type of application. This concept of an 'average' effective bandwidth is used in flat rate charging. The idea is that two applications that use the same contract should be charged the same, i.e. on the basis of an average effective bandwidth for this contract type, irrespective of whether they generate identical traffic.

3. If very little information is known about the source, then it can be reasonable to use the greatest effective bandwidth, say $\bar{\alpha}$, that is possible under the service contract, i.e. the bandwidth of the traffic that is most difficult to multiplex, given the constraints placed upon that traffic by the service contract. This is a conservative approach that results in network resources being underutilized. However, it is the only approach that enables the network to implement hard quality of service guarantees. We pursue this idea in Section 4.11.

4. One can modify the above approach to one of dynamic call acceptance by using information that is obtained by monitoring the actual system. One uses actual measurements of the performance on the links of the network to determine the actual amount of spare capacity available and then decides whether to accept a new contract on the basis of this information and a worst-case model of the new call.

Such a decision can depend on the duration of the new call and the time that it takes for the available capacity to change. It could be that there is available capacity because a majority of the active calls are sending traffic at less than their mean rates. In this case, the existing load will tend to increase as more sources become active at greater rates, although some of them may terminate and depart.

Observe that lack of information about call statistics results in resource underutilization and poor quality of service provisioning. If the network has better information about the resource usage statistics of a new call, then it can better multiplex and load the network more efficiently. This explains why a network operator wishes to have a good idea of the traffic profiles of his customers. But how can he obtain better information than that available through the traffic contract? Since this information is to be used to accept or to reject a call, it must be available at the time the call is set up. One way to obtain more information is through pricing. The network posts a set of possible tariffs for the same traffic contract, each one resulting in a different charge, depending on the traffic that is actually generated during the contract. Assuming the user has some information about the traffic he will generate, he chooses the tariff that minimizes his expected charge. His tariff choice therefore reveals important information to the network operator, who can use this information to obtain a better approximation of the effective bandwidth. This is an example of incentive compatible pricing; when users optimize their tariff choices the network operator gains information that allows him to better load and more efficiently run the network as a whole.

4.11 Bounds for effective bandwidths

Suppose that a connection is policed by multiple leaky buckets, with a set of parameters $\mathbf{h} = \{(\rho_k, \beta_k), k = 1, \ldots, K\}$. Let m be the mean rate of the connection. We are interested in the greatest effective bandwidth, say $\bar{\alpha}(m, \mathbf{h})$, that is possible for connection whose traffic contract has these leaky bucket parameters and whose mean rate is m. In practice, $\bar{\alpha}(m, \mathbf{h})$ can be extremely difficult to calculate exactly, as we are in effect trying to determine a worst-case stochastic process. But we can easily give a simple approximation for $\bar{\alpha}(m, \mathbf{h})$, which nicely shows how various timescales relate to buffer overflow. Since the traffic source is policed by leaky buckets, the maximum amount of traffic $\bar{X}[0, t]$ that could be produced in a time interval of length t is

$$\bar{X}[0, t] \leq H(t) := \min_{k=1,\ldots,K} \{\rho_k t + \beta_k\} \qquad (4.19)$$

One can show that $E \exp(s X[0, t])$ is maximized subject to $E X[0, t] = mt$ and $X[0, t] \leq H(t)$ by the distribution in which $X[0, t]$ equals 0 or $H(t)$ with probabilities $1 - mt/H(t)$ and $mt/H(t)$, respectively. Of course there may not be actual traffic, conforming with the above leaky buckets, for which $X[0, t]$ has this distribution, and this is why we are only obtaining an upper bound on $\bar{\alpha}(m, \mathbf{h})$. This upper bound is

$$\bar{\alpha}(m, \mathbf{h}) \leq \frac{1}{st} \log \left[1 + \frac{tm}{H(t)} \left(e^{sH(t)} - 1 \right) \right] = \tilde{\alpha}_{\text{sb}}(m, \mathbf{h}) \qquad (4.20)$$

We call the right-hand side of (4.20) the *simple bound*.

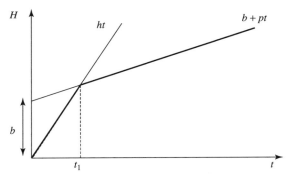

Figure 4.11   A dual leaky bucket policer. Depending on the value of t, a different leaky bucket affects the source's maximum contribution, $H(t)$, and hence the effective bandwidth.

In $\tilde{\alpha}_{sb}(m, \mathbf{h})$ we can see the effects of leaky buckets on the resource usage. Each leaky bucket (ρ_k, β_k) constrains the burstiness of the traffic on a particular timescale. The timescale of burstiness that contributes to buffer overflow is determined by the index k that achieves the minimum in (4.19). We discuss this issue further at the end of this section.

Consider the practical case of a dual leaky bucket $(h, 0)$ and (ρ, β). $H(t)$ is shown in Figure 4.11. If t is small, then $H(t) = ht$ and the bound (4.20) reduces to

$$\tilde{\alpha}_{\text{on-off}}(m, \mathbf{h}) = \frac{1}{st} \log\left[1 + \frac{m}{h}(e^{sht} - 1)\right] \tag{4.21}$$

We refer to this as the *on-off bound*, which we have already met in (4.17). This bound is valid for any value of t, since it is the effective bandwidth of an on-off source with peak rate h and mean rate m which can produce arbitrarily long bursts. Such a source does not comply with the (ρ, β) leaky bucket which restricts the length of such bursts and so is worse than the worst possible source that is compliant with the above traffic contract.

If one wants to obtain more accurate upper bounds on the effective bandwidth, one has to use complex computational procedures to determine the worst-case traffic. In general, the worst-case traffic depends not only on the contract parameters, but also on the parameters s and t. In many cases, the worst-case traffic consists of blocks of an inverted T pattern which repeat periodically or with random gaps, as shown in Figure 4.12. The sizes of the blocks and gaps depend on the values of s and t. This is a general form of extreme traffic which, given the leaky bucket constraints, alternately sends at the maximum rate and at a lesser rate (though not necessarily zero). While sending at the lesser rate it accumulate tokens so that it can again send at the maximum rate.

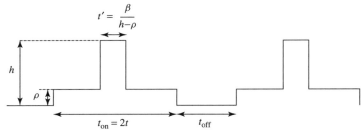

Figure 4.12   Periodic pattern for the inverted T approximation to a worst-case traffic. $t_1 = \beta/(h - \rho)$, $t_{\text{off}} = [(2t - t_1)\rho + t_1 h]/m - 2t$.

As an example, we consider the periodic pattern shown in Figure 4.12. Let $X_L[0, t]$ denote the load produced by the inverted T pattern in t epochs. This gives the *inverted T approximation* for the effective bandwidth bound,

$$\tilde{\alpha}_L(m, \mathbf{h}) = \frac{1}{st} \log E\left[e^{s X_L[0,t]}\right] \tag{4.22}$$

The expected value on the right-hand side of (4.22) can be computed analytically, under the assumption that the start of the inverted T pattern is uniformly distributed within an interval of length $t_{\text{on}} + t_{\text{off}}$, where $t_{\text{on}} = 2t$.

The above inverted T approximation is valid when the time parameter t is large compared to the time for which the leaky bucket permits the source to send at its peak rate h; denote this time by $t_1 = \beta/(h - \rho)$. If t is much smaller than t_1, then worst case traffic is a simple on-off source, for which $t_{\text{on}} = t_1$ and $t_{\text{off}} = t_{\text{on}} h/m - t_{\text{on}}$. In this case, the source operates only at the extremes, sending at full speed or not at all and a reasonable approximation of the worst-case effective bandwidth is given by (4.21).

Which leaky bucket is most constraining of the effective bandwidth produced by a traffic contract? The bound in (4.20) suggests that it depends on the timescale t. This is determined by the network. Given t, the bound for the effective bandwidth depends only upon the leaky bucket that achieves $\min_{k=1,\dots,K}\{\rho_k t + \beta_k\}$ for the particular t that constrains the maximum contribution $H(t)$. This suggests that a worst-case source for the above contract is hard to multiplex because of burstiness that is controlled mainly by this leaky bucket. Changing the value of other leaky buckets will not significantly reduce the difficulty of multiplexing the source. However, smoothening that reduces burstiness can reduce the effective bandwidth. Consider a source that is policed by two leaky buckets $(h, 0), (\rho, \beta)$. In Figure 4.11, we plot $H(t)$. For values of the parameter t less than $t_1 = \beta/(h - \rho)$ the leaky bucket which constrains the peak rate is dominant. For $t > t_1$, the leaky bucket (ρ, β) dominates. The physical explanation is that if burstiness on small timescales is causing overflows, (perhaps because there are small buffers in the network), then reducing the peak rate of these fluctuations will reduce the cell loss. If buffers are large, then rapid fluctuations are absorbed by the buffer and there is no advantage in reducing the peak rate. However, reducing the length of the long bursts will reduce cell loss. These bursts are controlled mainly by the second leaky bucket.

4.12   Deterministic multiplexing

The case of *deterministic multiplexing* is one in which we require no cell loss, so that $\gamma = \infty$. Our effective bandwidth theory suggests that $s = \infty$ (which is consistent with $\gamma = \infty$ in (4.12)). The effective bandwidth of a source of type j is $\alpha_j(\infty, t) = \bar{X}_j[0, t]/t$. Observe that the effective bandwidth of a source does not depend on the complete distribution function, but only on the $\bar{X}_j[0, t]$, the maximum value that $X_j[0, t]$ takes with positive probability.

Let us derive the form of the acceptance region for sources policed by leaky buckets. Suppose that each type of source is policed by a single leaky bucket. A source of type j is guaranteed to satisfy the condition

$$X_j[0, t] \leq \rho_j t + \beta_j, \quad \text{for all } t \tag{4.23}$$

Assume that there is positive probability of arbitrarily near equality in (4.23) for all time windows t. An example of a source that can do this is one that infinitely often repeats the following pattern of three phases: it stays off for time β_j/ρ_j (to empty the token buffer), then produces an instantaneous burst of size β_j, and then stays on at rate ρ_j for a time that is exponentially distributed with mean 1. It is not hard to see that for all $\epsilon > 0$, there is a positive probability that such a source will produce a burst of size β_j within the interval $[0, \epsilon)$ and then remain on at rate ρ_j over $[\epsilon, t]$, so ensuring that in the interval $[0, t]$ the number of cells received are at least $\beta_j + (t - \epsilon)\rho_j$. This implies that $\alpha_j(\infty, t) = \bar{X}[0, t]/t = \rho_j + \beta_j/t$. So, following the notation from Section 4.6.1, as $\gamma \to \infty$ also $s \to \infty$, and $A_t = \{x : g_t(x) \le -\gamma\}$ reduces to

$$\sum_{j=1}^{k} x_j(\rho_j t + \beta_j) \le tC + B$$

This infinite set of hyperplanes is dominated by two extreme ones. That is, $\cap_t A_t = A_0 \cap A_\infty$, where A_∞ and A_0 are regions defined, respectively, by

$$\sum_{j=1}^{k} x_j \rho_j \le C \quad \text{and} \quad \sum_{j=1}^{k} x_j \beta_j \le B \qquad (4.24)$$

There is a nice interpretation of these equations. For each constraint there are some dominant events that cause the quality of service constraint to be critically satisfied. If the first constraint is active, overflow occurs because each source of type j contributes at its maximum allowed average rate ρ_j. If $\sum_j \rho_j$ were to exceed C by a very small amount, say δ, then a buffer of any size would eventually fill, though very slowly. As δ tends to 0 we think of the buffer filling, but over infinite time when $\delta = 0$, hence $t = \infty$. If the second constraint is active, overflow occurs because all sources produce bursts at exactly the same time. In this case, a buffer of size B fills in zero time and hence $t = 0$. Note that the effective bandwidth is defined as ρ_j or β_j, depending upon which constraint is active at the operating point (which can be compared to the w_j and v_j of the elevator analogy).

The case of multiple leaky bucket constraints is more complex, but the results are similar. It turns out that one gets an acceptance region bounded by a finite number of linear constraints, each of which corresponds to a particular way that the buffer can fill. We will briefly investigate the particular case of adding a peak rate constraint, i.e. each source of type j satisfies

$$X_j[0, t] \le \min\{h_j t, \rho_j t + \beta_j\}, \quad \text{for all } t \qquad (4.25)$$

In this case, following the previous reasoning, the effective bandwidth of a source of type j is

$$\alpha_j(\infty, t) = \bar{X}_{jk}[0, t]/t = \begin{cases} h_j & \text{for } 0 \le t \le t_j \\ \rho_j + \beta_j/t & \text{for } t \ge t_j \end{cases} \qquad (4.26)$$

where $t_j = \beta_j/(h_j - \rho_j)$. Note that at time t_j there is a switch in the constraining leaky bucket. At that time, the leaky bucket constraining the maximum contribution $\bar{X}[0, t]$ of the source switches from $(h_j, 0)$ to (ρ_j, β_j).

In the case $j = 1, 2$, and assuming that $t_1 < t_2$, the acceptance region is defined by

$$\begin{array}{llll} x_1 h_1 + x_2 h_2 & \le & C + B/t & \text{for } 0 \le t \le t_1 \\ x_1 \rho_1 + x_2 h_2 & \le & C + (B - x_1 \beta_1)/t & \text{for } t_1 \le t \le t_2 \\ x_1 \rho_1 + x_2 \rho_2 & \le & C + (B - x_1 \beta_1 - x_2 \beta_2)/t & \text{for } t_2 \le t \end{array} \qquad (4.27)$$

These must hold for all t. But as the reader can verify, it is enough that they hold for $t = t_1$, $t = t_2$ and $t = \infty$. The corresponding constraints become

$$x_1 h_1 + x_2 h_2 \leq C + B/t_1$$
$$x_1(\rho_1 + \beta_1/t_2) + x_2 h_2 \leq C + B/t_2$$
$$x_1 \rho_1 + x_2 \rho_2 \leq C$$

Note, again, that depending upon which constraint is active at the operating point, there is a unique description for the way the buffer fills and the time that it takes. If the operating point lies on the first constraint then the buffer fills by all sources producing at their peak rates over $[0, t_1]$, and the buffer becoming full at time t_1 (which is the time at which sources of the first type of must reduce their rates to ρ_1). If the second constraint is active then the buffer first fills at time t_2, by filling at rate $x_1 h_1 + x_2 h_2 - C > 0$ until time t_1, and then at rate $x_1 \rho_1 + x_2 h_2 - C > 0$ from t_1 to t_2. If the last constraint is active, then the long-run average rate of the input equals C and the buffer fills infinitely slowly.

The above examples generalize to more leaky buckets. Again, one has to check a finite set of equations, similar to (4.27), at a finite number of times at which different leaky buckets become active by constraining the maximum contribution of the sources.

4.13 Extension to networks

We have seen how to compute the effective bandwidths of a flow that is multiplexed at one buffered switch. But is this any use in the context of a network? Clearly, the statistics of the flow change as it passes through switches of the network, since the interdeparture times of cells from a switch are not the same as their interarrival times. Can the effective bandwidth still characterize the flow's contribution to rare overflow events in the network? Fortunately, the answer is yes. One can show that in the limiting regime of many sources, in which the number of inputs increases and the service rate and buffer size increase proportionally as in (4.8), the statistical characteristics and effective bandwidth of a traffic stream are essentially unchanged by passage through the switch. To see this intuitively, observe that as the scaling factor N increases, the aggregate of all the traffic streams looks more and more like a constant bit rate source, of a rate less than the switch bandwidth, NC. This means that with a probability approaching 1 the buffer is empty over the fixed interval of time during which any one given traffic source is present. Thus with probability also approaching 1, the input and output processes are identical over the time that a traffic source is present. Note that these are limiting results: they hold in the limit as the capacity of the links become larger. This assumption is realistic in the context of the expanding capacity of today's broadband networks.

Using this result we can describe the technology set of the network as follows. Let L be a set of links and R be a set of routes. Write $j \in r$ if route r uses link j. Assume each route is associated with a unique traffic contract type. (We allow two routes to be identical in their path through the network and differ only in contract type.) Let x_r be the number of contracts using route r. Let B_j and C_j denote as before the resources of link j. The technology set is then defined by a set of constraints like

$$\sum_{r:j\in r} x_r \alpha_r(s_j, t_j) \leq C_j^* = C_j + \frac{1}{t_j}\left(B_j - \frac{\gamma_j}{s_j}\right) \qquad j \in L, \qquad (4.28)$$

which says that the sum of the effective bandwidths of connections that use link j must not exceed C_j^*.

Note that, although the effective bandwidth function $\alpha_r(\cdot, \cdot)$ is the same along the entire route of a traffic stream, this does not imply that the effective bandwidth is the same on all links. This is because parameters of the operating point may vary along the links of the path, so that (s_j, t_j) depend on the link j. But do we expect these parameters to vary widely inside the network? There are economic arguments that suggest not.

Suppose that the marginal cost of adding some extra buffer in a link of the network is b, and the marginal cost of adding extra capacity is a. Then if the most economical switch configuration is used, we must have $t_j = a/b$, independently of the link. To prove this consider the constrained optimization problem

$$\underset{B_j, C_j}{\text{minimize}} \; aC_j + bB_j, \quad \text{subject to } \gamma_j(C_j, B_j) = \gamma_j^*$$

At the optimum we must have $a^{-1}\partial\gamma_i/\partial C_i = b^{-1}\partial\gamma_i/\partial B_i$, or else it would be possible to reduce the cost. The result then follows by (4.7).

So far as s_j is concerned, if total buffer space is allocated through the network to minimize the total CLP, i.e. minimize $\sum_j \gamma_j(C_j, B_j)$ subject to $\sum_j B_j = B$, then the solution is where $s_j = \partial\gamma_j/\partial B_j$ is constant. It is unlikely that things are so optimized, but in any case if all links have substantial capacity, s_j will be uniformly small, as experimental results confirm. The implication of the above is that it is reasonable to assume that each traffic stream in a network can be assigned a unique effective bandwidth, that is independent of it route and of the other flows. A more refined analysis could take account of the precise values of the parameters s and t along the path of the traffic stream, and use values for the effective bandwidth that depend on the particular link.

4.14 Call blocking

At the end of Section 4.3 we wondered what the effects of blocking would be if connections arrive with rates that have stochastic fluctuations and we adopt a call admission control based on (4.3). There is now a cost due to blocked calls. It is reasonable to assume that when a service request is refused there is some cost to the requester, while when it is accepted some positive value in generated. We formulate and analyze such a model in Section 9.3.3, measuring blocking in terms of the call blocking probability. This probability depends on the technology set of the network and the rates of arrival of the various connection types (the service requests). In this section we see how such blocking probabilities may be calculated from the parameters of the system.

Suppose there are J types of connection. A connection of type r is associated with a route r and connections of this type arrive as a Poisson process of rate λ_r and endure for an average time of $1/\mu_r$. Let p_r be the blocking probability for a connection of type r. Let $x_r(t)$ denote the number of connections of type r that are active at time t. These connections place a load on link j of $\alpha_{jr}x_r(t)$. Here α_{jr} is the effective bandwidth of a connection of type r on link j. It equals 0 if the connection does not use link j.

By Little's Law (stated in 4.4), the average number of connections of type r that will be active is the product of the arrival rate for this type and its average holding time, i.e. $(1 - p_r)\lambda_r/\mu_r$. Let $\rho_r = \lambda_r/\mu_r$. Then it is necessary to have, for all links j,

$$E\left[\sum_{r:j\in r} \alpha_{jr}x_r(t)\right] = \sum_{r:j\in r} \alpha_{jr}(1 - p_r)\rho_r \leq C_j$$

This places some restriction on the blocking probabilities. However, it does not take account of the statistical fluctuations.

To be more accurate we could reason as follows. The probability distribution of the number of calls in progress is

$$\pi(x) = G(C)^{-1} \prod_r \frac{\rho_r^{x_r}}{x_r!}, \qquad \text{where } G(C) = \sum_{x:Ax \leq C} \prod_r \frac{\rho_r^{x_r}}{x_r!}$$

Consider the problem of finding the most likely state, over $Ax \leq C$. This is equivalent to maximizing $\sum_r (x_r \log \rho_r - \log r!)$. By Stirling's approximation, this objective function can be approximated by $\sum_r (x_r \log \rho_r - (x_r + 0.5) \log x_r + x_r)$. Taking $x_r + 0/5 \approx x_r$, and using Lagrange multipliers to solve this constrained maximization problem, we can show that there exist B_is such that

$$\sum_{r:j\in r} \rho_r \alpha_{rj} \prod_{i\in r}(1 - B_i)^{\alpha_{ri}} \quad \begin{matrix} = C_j & \text{if } B_j > 0 \\ \leq C_j & \text{if } B_j = 0 \end{matrix} \qquad (4.29)$$

Here, B_i can be interpreted as the blocking probability for a unit of effective bandwidth on link i. This formula has a simple interpretation. It is as if each such unit has a probability of blocking that is independent of other such units on the same link and other links. Clearly, this is not so. However, it motivates a determination of the p_rs as the solution of

$$1 - p_r = \prod_{i\in r}(1 - B_i)^{\alpha_{ri}}, \qquad \text{for all } r \qquad (4.30)$$

$$B_j = E\left(\sum_{r:r\in j} \rho_r \alpha_{rj}(1 - p_r), C_j\right), \qquad \text{for all } j \qquad (4.31)$$

Here, $E(\rho, C)$ is Erlang's formula for the blocking proportion of calls lost at a single link of capacity C when they arrive at rate λ and hold for an average time $1/\mu$, with $\rho = \lambda/\mu$,

$$E(\rho, C) = \frac{\rho^C}{C!} \bigg/ \sum_{n=0}^{C} \frac{\rho^n}{n!} \qquad (4.32)$$

It can be shown that a solution to (4.31)–(4.31) exists and is unique. The approximation becomes more exact the as the routes passing through each link become more diverse and we take a large N limit (i.e. $C = NC^{(0)}$, $x_r = x_r^{(0)}N$).

4.15 Further reading

The first attempt to define effective bandwidths and approximate the cell loss probabilities was in the early 1990s. The most elaborate definitions were based on large buffer asymptotics, that have proved to be less accurate than the definitions using the 'many sources' (large N) asymptotic, which has eventually gained wide acceptance. Large buffer asymptotics did not capture the multiplexing effects due to a large number of independent sources being multiplexed, but only the effects due to the buffers. Relevant references for large buffer asymptotics and the corresponding effective bandwidths are de Veciana, Olivier and Walrand (1993), Elwalid and Mitra (1993), de Veciana and Walrand (1995), Courcoubetis and Weber (1995), Courcoubetis, Kesidis, Ridder, Walrand and Weber (1995).

A classic paper on the calculation of the overflow probabilities in queues handling many sources is that of Anick, Mitra and Sondhi (1982). This paper motivated much subsequent research in the field. Weiss (1986) first derived the large system asymptotic for on-off sources. The large system asymptotic in (4.8) that leads to the effective bandwidth formulas was independently proved by Botvich and Duffield (1995), Simonian and Guilbert (1995) and Courcoubetis and Weber (1996). A refinement of this asymptotic using that Bahadur-Rao approximation is due to Likhanov and Mazumdar (1999). Some early references for the effective bandwidth concept are Hui (1988), Courcoubetis and Walrand (1991), Kelly (1991a) and Gibbens and Hunt (1991). An excellent reference for the theory of the effective bandwidths is Kelly (1996).

For a review of the Brownian bridge model in Section 4.4, see Hajek (1994). The proof that the acceptance region given in (4.16) is exact for a simple queue fed by Brownian bridge inputs can be found in Kelly (1996). The Markov modulated source model of Section 4.6 has played an important role in theoretical and practical developments. Anick, Mitra and Sondhi (1982) show how to calculate the probabilities of buffer overflow. Further details of the derivation of the effective bandwidth for this model can be found in Courcoubetis and Weber (1996). The material in Section 4.9 is taken from Courcoubetis, Siris and Stamoulis (1999) and Courcoubetis, Kelly and Weber (2000). Calculation of the effective bandwidths for real traffic traces first appeared in Gibbens (1996). The use of effective bandwidth concepts for dimensioning network links and for solving other traffic engineering problems is explained in Courcoubetis, Siris and Stamoulis (1999). This includes experimental results that validate our effective bandwidth definition. Siris (2002) maintains a nice web site on large deviation techniques and on-line tools for traffic engineering. The extension of the single link models and the application of the asymptotics to networks in Section 4.13 is due to Wischik (1999). Section 4.14 summarizes ideas from Kelly (1991b) and (1991c). More refined asymptotics for the blocking probabilities are described by Hunt and Kelly (1989). Extensions of the model to include priorities can be found in Berger and Whitt (1998).

Issues of call-admission control are treated in Courcoubetis, Kesidis, Ridder, Walrand and Weber (1995), Gibbens, Key and Kelly (1995), Grossglauser and Tse (1999) and Courcoubetis, Dimakis and Stamoulis (2002).

Part B

Economics

5

Basic Concepts

Economics is concerned with the production, sale and purchase of commodities that are in limited supply, and with how buyers and sellers interact in markets for them. This and the following chapter provide a tutorial in the economic concepts and models that are relevant to pricing communications services. It investigates how pricing depends on the assumptions that we make about the market. For example, we might assume that there is only one sole supplier. In formulating and analysing a number of models, we see that prices depend on the nature of competition and regulation, and whether they are driven by competition, the profit-maximizing aim of a monopoly supplier, or the social welfare maximizing aim of a regulator.

Section 5.1 sets out some basic definitions and describes some factors that affect pricing. It defines types of markets, and describes three different rationales that can provide guidance in setting prices. Section 5.2 considers the problem of a consumer who faces prices for a range of services. The key observation is that the consumer will purchase a service up to an amount where his marginal utility equals the price. Section 5.3 defines the problem of supplier whose aim is to maximize his profit. Section 5.4 concerns the problem that is natural for a social planner: that of maximizing the total welfare of all participants in the market. We relate this to some important notions of market equilibrium and efficiency, noting that problems can arise if there is market failure due to externalities.

Unfortunately, social welfare is achieved by setting prices equal to marginal cost. Since the marginal costs of network services can be nearly zero, producers may not be able to cover their costs unless they receive some additional lump-sum payment. A compromise is to use Ramsey prices; these are prices which maximize total welfare subject to the constraint that producers cover their costs. We consider these in Section 5.5. Section 5.6 considers maximizing social welfare under finite capacity constraints. Section 5.7 discusses how customer demand can be influenced by the type of network externality that we mentioned in Chapter 1.

The reader of this and the following chapter cannot expect to become an expert in all economic theory that is relevant to setting prices. However, he will gain an appreciation of factors that affect pricing decisions and of what pricing can achieve. In later chapters we use this knowledge to show how one might derive some tariffs for communications services.

5.1 Charging for services

5.1.1 Demand, Supply and Market Mechanisms

Communication services are valuable economic commodities. The prices for which they can be sold depend on factors of demand, supply and how the market operates. The key

---

Pricing Communication Networks   C. Courcoubetis and R. Weber
© 2003 John Wiley & Sons, Ltd   ISBN 0-470-85130-9 (HB)

players in the market for communications services are suppliers, consumers, and regulators. The demand for a service is determined by the value users place upon it and the price they are willing to pay to obtain it. The quantity of the service that is supplied in the market depends on how much suppliers can expect to charge for it and on their costs. Their costs depend upon the efficiency of their network operations. The nature of competition amongst suppliers, how they interact with customers, and how the market is regulated all have a bearing on the pricing of network services.

One of the most important factors is competition. Competition is important because it tends to increases economic efficiency: that is, it increases the aggregate value of the services that are produced and consumed in the economy. Sometimes competition does not occur naturally. In that case, regulation by a government agency can increase economic efficiency. By imposing regulations on the types of tariffs, or on the frequency with which they may change, a regulator can arrange for there to be a greater aggregate welfare than if a dominant supplier were allowed to produce services and charge for them however he likes. Moreover, the regulator can take account of welfare dimensions that suppliers and customers might be inclined to ignore. For example, a regulator might require that some essential network services be available to everyone, no matter what their ability to pay. Or he might require that encrypted communications can be deciphered by law enforcement authorities. He could take a 'long term view', or adopt policies designed to move the market in a certain desirable direction.

5.1.2   Contexts for Deriving Prices

In Section 1.4.1 we defined the words 'charge', 'tariff' and 'price'. We said that a customer pays *charges* for network services, and a charge is computed from a *tariff*. This tariff can be a complex function and it can take account of various aspects of the service and perhaps some measurements of the customer's usage. For example, a telephone service tariff might be defined in terms of monthly rental, the numbers of calls that are made, their durations, the times of day at which they are made, and whether they are local or long-distance calls.

A *price* is a charge that is associated with one unit of usage. For example, a mobile phone service provider might operate a *two-part tariff* of the form $a + bx$, where a is a monthly fixed-charge (or access charge), x is the number of minutes of calling per month, and b is the price per minute. For a general tariff of the form $r(x)$, where x is the amount consumed, probably a vector, price may depend on x. Given that x is consumed, the price of one more unit is $p = \partial r(x)/\partial x$. If $r(x) = p^\top x$ for some price vector p, then $r(x)$ is a *linear tariff*. All other tariff forms are *nonlinear tariffs*.[1] For instance, $a + bx$ is a nonlinear tariff (price), while bx is a linear tariff (price).

In thinking about how price are determined, there are two important questions to answer: (a) who sets the price, and (b) with what objective? It is interesting to look at three different answers and the rationales that they give for thinking about prices. The first answer is that sometimes the market that sets the price, and the objective is to match supply and demand. Supply and demand at given prices depend upon the supplier's technological capacities, the costs of supply, and the how consumers value the service. If prices are set too low then there

---

[1] In the economics literature, the terminology *linear price* or *uniform price* is commonly used instead of a linear tariff, and *nonlinear price* instead of a nonlinear tariff. In this case, price refers implicitly to the total amount paid for the given quantity, i.e. the total charge.

will be insufficient incentive to supply and there is likely to be unsatisfied demand. If prices are set too high then suppliers may over-supply the market and find there is insufficient demand at that price. The 'correct' price should be 'market-clearing'. That is, it should be the price at which demand exactly equals supply.

A second rationale for setting prices comes about when it is the producer who sets prices and his objective is to deter potential competitors. Imagine a game in which an incumbent firm wishes to protect itself against competitors who might enter the market. This game takes place under certain assumptions about both the incumbent's and entrants' production capabilities and costs. We find that if the firm is to be secure against new entrants seducing away some of its customers, then the charges that it makes for different services must satisfy certain constraints. For example, if a firm uses the revenue from selling one product to subsidize the cost of producing another, then the firm is in danger if a competitor can produce only the first product and sell it for less. This would lead to a constraint of no cross-subsidization.

A third rationale for setting prices comes about when a principal uses prices as a mechanism to induce an agent to take certain actions. The principal cannot dictate directly the actions he wishes the agent to take, but he can use prices to reward or penalize the agent for actions that are or are not desired. Let us consider two examples. In our first example the owner of a communications network is the principal and the network users are the agents. The principal prices the network services to motivate users to choose services that both match their needs and avoid wasting network resources. Suppose that he manages a dial-in modem bank. If he prices each unit of connection time, then he gives users the incentive to disconnect when they are idle. His pricing is said to be *incentive compatible*. That is, it provides an incentive that induces desirable user response. A charge based only on pricing each byte that is sent would not be incentive compatible in this way.

In our second example the owner of the communications network is now the agent. A regulator takes the role of principal and uses price regulation to induce the network owner to improve his infrastructure, increase his efficiency, and provide the services that are of value to consumers.

These are three possible rationales for setting prices. They do not necessarily lead to the same prices. We must live with the fact that there is no single recipe for setting prices that takes precedence over all others. Pricing can depend on the underlying context, or contexts, and on contradictory factors. This means that the practical task of pricing is as much an art as a science. It requires a good understanding of the particular circumstances and intricacies of the market.

It is not straightforward even to define the cost of a good. For example, there are many different approaches to defining the cost of a telephone handset. It could be the cost of the handset when it was purchased (the historical cost), or its opportunity cost (the value of what we must give up to produce it), or the cost of the replacing it with a handset that has the same features (its modern equivalent asset cost). Although, in this chapter, we assume that the notion of the cost is unambiguously defined, we return to the issue of cost definition in Chapter 7.

In this chapter we review the basic economic concepts that are needed to understand various contexts for defining prices. We focus on defining the various economic agents that interact in a marketplace. In the following chapter we analyze various competition scenarios. We begin by considering the problem that a consumer faces when he must decide how much of each of a number of services to purchase.

5.2 The consumer's problem

5.2.1 Maximization of Consumer Surplus

Consider a market in which n customers can buy k services. Denote the set of customers by $N = \{1, \ldots, n\}$. Customer i can buy a vector quantity of services $x = (x_1, \ldots, x_k)$ for a payment of $p(x)$. Let us suppose that $p(x) = p^\top x = \sum_j p_j x_j$, for a given vector of prices $p = (p_1, \ldots, p_k)$. Assume that the available amounts of the k services are unlimited and that customer i seeks to solve the problem

$$x^i(p) = \arg\max_x \left[u_i(x) - p^\top x \right] \tag{5.1}$$

Here $u_i(x)$ is the *utility* to customer i of having the vector quantities of services x. One can think of $u_i(x)$ as the amount of money he is willing to pay to receive the bundle that consists of these services in quantities $x_1 \ldots, x_k$.

It is usual to assume that $u_i(\cdot)$ is strictly increasing and strictly concave for all i. This ensures that there is a unique maximizer in (5.1) and that demand decreases with price. If, moreover, $u(\cdot)$ is differentiable, then the marginal utility of service j, as given by $\partial u_i(x)/\partial x_j$, is a decreasing function of x_j. We make these assumptions unless we state otherwise. However, we note that there are cases in which concavity does not hold. For example, certain video coding technologies can operate only when the rate of the video stream is above a certain minimum, say x^*, of a few megabits per second. A user who wishes to use such a video service will have a utility that is zero for a rate x that is less than x^* and positive for x at x^*. This is a step function and not concave. The utility may increase as x increases above x^*, since the quality of the displayed video increases with the rate of the encoding. This part of the utility function may be concave, but the utility function as a whole is not. In practice, for coding schemes like MPEG, the utility function is not precisely a step function, but it resembles one. It starts at zero and increases slowly until a certain bit rate is attained. After this point it increases rapidly, until it eventually reaches a maximum value. The first part of the curve captures the fact that the coding scheme cannot work properly unless a certain bit rate is available.

The expression that is maximized on the right-hand side of (5.1) is called the consumer's *net benefit* or *consumer surplus*,

$$\mathrm{CS}_i = \max_x \left[u_i(x) - p^\top x \right]$$

It represents the net value the consumer obtains as the utility of x minus the amount paid for x. The above relations are summarized in Figure 5.1.

The vector $x^i(p)$ is called the *demand function* for customer i. It gives the quantities $x^i = (x_1^i, \ldots, x_k^i)$ of services that customer i will buy if the price vector is p. The *aggregate demand function* is $x(p) = \sum_{i \in N} x^i(p)$; this adds up the total demand of all the users at prices p. Similarly, the *inverse aggregate demand function*, $p(x)$, is the vector of prices at which the total demand is x.

Consider the case of a single customer who is choosing the quantity to purchase of just a single service, say service j. Imagine that the quantities of all other services are held constant and provided to the customer for no charge. If his utility function $u(\cdot)$ is concave and twice differentiable in x_j then his net benefit, of $u(x) - p_j x_j$, is maximized where it is stationary point with respect to x_j, i.e. where $\partial u(x)/\partial x_j = p_j$. At this point, the marginal increase in utility due to increasing x_j is equal to the price of j. We also see that the customer's *inverse demand function* is simply $p_j(x_j) = \partial u(x)/\partial x_j$. It is the price at which

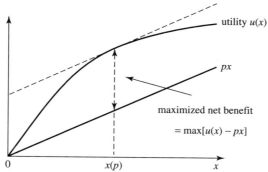

Figure 5.1   The consumer has a utility $u(x)$ for a quantity x of a service. In this figure, $u(x)$ is increasing and concave. Given the price vector p, the consumer chooses to purchase the amount $x = x(p)$ that maximizes his net benefit (or consumer surplus). Note that at $x = x(p)$ we have
$$\partial u(x)/\partial x = p.$$

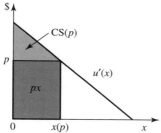

Figure 5.2   The *demand curve* for the case of a single customer and a single good. The derivative of $u(x)$, denoted $u'(x)$, is downward sloping, here for simplicity shown as a straight line. The area under $u'(x)$ between 0 and $x(p)$ is $u(x(p))$, and so subtracting px (the area of the shaded rectangle) gives the consumer surplus as the area of the shaded triangle.

he will purchase a quantity x_j. Thus, for a single customer who purchases a single service j, we can express his consumer surplus at price p_j as

$$CS(p_j) = \int_0^{x_j(p_j)} p_j(x)\, dx - p_j x_j(p_j) \tag{5.2}$$

We illustrate this in Figure 5.2 (dropping the subscript j).

We make a final observation about (5.1). We have implicitly assumed that the (per unit) prices charged in the market are the same for all units purchased by the customer. There are more general pricing mechanisms in which the charge paid by the customer for purchasing a quantity x is a more general function $r(x)$, not of the form $p^\top x$. For instance, prices may depend on the total amount bought by a customer, as part of nonlinear tariffs, of the sort we examine in Section 6.2.2. Unless explicitly stated, we use the term 'price' to refer to the price that defines a linear tariff $p^\top x$.

The reader may also wonder how general is (5.1) in expressing the net benefit of the customer as a difference between utility and payment. Indeed, a more general version is as follows. A customer has a utility function $v(x_0, x)$, where x_0 is his net income (say in dollars), and x is the vector of goods he consumes. Then at price p he solves the problem

$$x^i(p) = \arg\left\{\max_x v(x_0 - p^\top x, x) : p^\top x \le x_0\right\}$$

In the simple case that the customer has a *quasilinear utility function*, of the form $v(x_0, x) = x_0 + u(x)$, and assuming his income is large enough that $x_0 - p^\top x > 0$ at the optimum, he must solve a problem that is equivalent to (5.1). It is valid to assume a quasilinear utility function when the customer's demand for services is not very sensitive to his income, i.e. expenditure is a small proportion of his total income, and this is the case for most known communications services. In our economic modelling, we use these assumptions regarding utility functions since they are reasonable and simplify significantly the mathematical formulas without reducing the qualitative applicability of the results.

5.2.2   Elasticity

Concavity of $u(\cdot)$ ensures that both $x(p)$ and $p(x)$ are decreasing in their arguments, or as economists say, *downward sloping*. As price increases, demand decreases. A measure of this is given by the *price elasticity of demand*. Customer i has elasticity of demand for service j given by

$$\epsilon_j = \frac{\partial x_j(p)/\partial p_j}{x_j/p_j}$$

where, for simplicity, we omit the superscript i in the demand vector x^i, since we refer to a single customer. Thus

$$\frac{\Delta x_j}{x_j} = \epsilon_j \frac{\Delta p_j}{p_j}$$

and elasticity measures the percentage change in the demand for a good per percentage change in its price. Recall that the inverse demand function satisfies $p_j(x) = \partial u(x)/\partial x_j$. So the concavity of the utility function implies $\partial p_j(x)/\partial x_j \leq 0$ and ϵ_j is negative.[2] As $|\epsilon_j|$ is greater or less than 1 we say that demand of customer i for service j is respectively *elastic* or *inelastic*. Note that since we are working in percentages, ϵ_j does not depend upon the units in which x_j or p_j is measured. However, it does depend on the price, so we must speak of the 'elasticity at price p_j. The only demand function for which elasticity is the same at all prices is one of the form $x(p) = ap^\epsilon$. One can define other measures of elasticity, such 'income elasticity of demand', which measures the responsiveness of demand to a change in a consumer's income.

5.2.3   Cross Elasticities, Substitutes and Complements

Sometimes, the demand for one good can depend on the prices of other goods. We define the *cross elasticity of demand*, ϵ_{jk}, as the percentage change in the demand for good j per percentage change in the price of another good, k. Thus

$$\epsilon_{jk} = \frac{\partial x_j(p)/\partial p_k}{x_j/p_k}$$

and

$$\frac{\Delta x_j}{x_j} = \epsilon_{jk} \frac{\Delta p_k}{p_k}$$

---

[2] Authors disagree in the definition of elasticity. Some define it as the negative of what we have, so that it comes out positive. This is no problem provided one is consistent.

But why should the price of good k influence the demand for good j? The answer is that goods can be either *substitutes* or *complements*. Take, for example, two services of different quality such as VBR and ABR in ATM. If the price for VBR increases, then some customers who were using VBR services, and who do not greatly value the higher quality of VBR over ABR, will switch to ABR services. Thus, the demand for ABR will increase. The services are said to be substitutes. The case of complements is exemplified by network video transport services and video conferencing software. If the price of one of these decreases, then demand for both increases, since both are needed to provide the complete video conferencing service.

Formally, services j and k are substitutes if $\partial x_j(p)/\partial p_k > 0$ and complements if $\partial x_j(p)/\partial p_k < 0$. If $\partial x_j(p)/\partial p_k = 0$, the services are said to be independent. Surprisingly, the order of the indices j and k is not significant. To see this, recall that the inverse demand function satisfies $p_j(x) = \partial u(x)/\partial x_j$. Hence $\partial p_j(x)/\partial x_k = \partial p_k(x)/\partial x_j$, and so the demand functions satisfy

$$\frac{\partial x_j(p)}{\partial p_k} = \frac{\partial x_k(p)}{\partial p_j}$$

5.3 The supplier's problem

Suppose that a supplier produces quantities of k different services. Denote by $y = (y_1, \ldots, y_k)$ the vector of quantities of these services. For a given network and operating method the supplier is restricted to choosing y within some set, say Y, usually called the *technology set* or *production possibilities set* in the economics literature. In the case of networks, this set corresponds to the acceptance region that is defined in Chapter 4.

Profit, or *producer surplus*, is the difference between the revenue that is obtained from selling these services, say $r(y)$, and the cost of production, say $c(y)$. An independent firm having the objective of *profit maximization*, seeks to solve the problem of maximizing the profit,

$$\pi = \max_{y \in Y} \left[r(y) - c(y) \right]$$

An important simplification of the problem takes place in the case of *linear prices*, when $r(y) = p^\top y$ for some price vector p. Then the profit is simply a function of p, say $\pi(p)$, as is also the optimizing y, say $y(p)$. Here $y(p)$ is called the *supply function*, since it gives the quantities of the various services that the supplier will produce if the prices at which they can be sold is p.

The way in which prices are determined depends upon the prevailing market mechanism. We can distinguish three important cases. The nature of competition in these three cases is the subject of Chapter 6. If the supplier is a *monopolist*, i.e. the sole supplier in an unregulated *monopoly*, then he is free to set whatever prices he wants. His choice is constrained only by the fact that as he increases the prices of services the customers are likely to buy less of them.

If the supplier is a small player amongst many then he may have no control over p. We say he is a *price taker*. His only freedom is in choice of y. This is a common scenario in practice. In such a scenario, the supplier sells at given linear prices, which are independent of the quantities sold. This is also the case for a *regulated monopoly*, in which the price vector p is fixed by the regulator, and the supplier simply supplies the services that the market demands at the given price p.

A middle case, in which a supplier has partial influence over p, is when he is in competition with just a few others. In such an economy, or so-called *oligopoly*, suppliers compete for customers through their choices of p and y. This assumes that suppliers do not collude or form a cartel. They compete against one another and the market prices of services emerge as the solution to some noncooperative game.

5.4 Welfare maximization

Social welfare (which is also called social surplus) is defined as the sum of all users' net benefits, i.e. the sum of all consumer and producer surpluses. Note that weighted sums of consumer and producer surpluses can be considered, reflecting the reality that a social planner/regulator/politician may attach more weight to one sector of the economy than to another. We speak interchangeably of the goals of social welfare maximization, social surplus maximization, and 'economic efficiency'. The key idea is that, under certain assumptions about the concavity and convexity of utility and cost functions, the social welfare can be maximized by setting an appropriate price and then allowing producers and consumers to choose their optimal levels of production and consumption. This has the great advantage of maximizing social welfare in a decentralized way.

We begin by supposing that the social welfare maximizing prices are set by a supervising authority, such as a regulator of the market. Suppliers and consumers see these prices and then optimally choose their levels of production and demand. They do this on the basis of information they know. A supplier sets his level of production knowing only his own cost function, not the consumers' utility functions. A consumer sets his level of demand knowing only his own utility function, not the producers' cost functions or other customers' utility functions. Individual consumer's utility functions are private information, but aggregate demand is commonly known.

Later we discuss perfectly competitive markets, i.e., a markets in which no individual consumer or producer is powerful enough to control prices, and so all participants must be price takers. It is often the case that once prices settle to values at which demand matches supply, the social welfare is maximized. Thus a perfectly competitive market can sometimes need no regulatory intervention. This is not true, however, if there is some form of market failure, such as that caused by externalities. In Section 5.7 we see, for example, how a market with strong network externality effects may remain small and never actually reach the socially desirable point of large penetration.

In the remainder of this section, we address the problem faced by a social planner who wishes to maximize social welfare. In Sections 5.4.1 and 5.4.2 we show that he can often do this by setting prices. Section 5.4.3 looks at the assumptions under which this is true and what can happen if they do not hold. Section 5.4.4 works through a specific example, that of peak load pricing. Sections 5.4.5 and 5.4.6 are concerned with how the planner's aim can be achieved by market mechanisms and the sense in which a market can naturally find an efficient equilibrium. Social welfare is maximized by marginal cost pricing, which we discuss in Section 5.4.7.

5.4.1 The Case of Producer and Consumers

We begin by modelling the problem of the social planner who by *regulation* can dictate the levels of production and demand so as to maximize social welfare. Suppose there is one producer, and a set of consumers, $N = \{1, \ldots, n\}$. Let x^i denote the vector of quantities of k services consumed by consumer i. Let $x = x^1 + \cdots + x^n$ denote the total demand, and let $c(x)$ denote the producer's cost to produce x. The social welfare (or surplus), S, is the total utility of the services consumed minus their cost of production, and so is written

$$S = \sum_{i \in N} u_i(x^i) - c(x)$$

Since the social planner takes an overall view of network welfare, let us label his problem as

$$\text{SYSTEM} : \underset{x,x^1,\ldots,x^n}{\text{maximize}} \sum_{i \in N} u_i(x^i) - c(x), \qquad \text{subject to } x = x^1 + \cdots + x^n$$

Assume that each $u_i(\cdot)$ is concave and $c(\cdot)$ is convex.[3] Then SYSTEM can be solved by use of a Lagrange multiplier p on the constraint $x = x^1 + \cdots + x^n$. That is, for the right value of p, the solution can be found by maximizing the Lagrangian

$$L = \sum_{i \in N} u_i(x^i) - c(x) + p^\top(x - x^1 - \cdots - x^n)$$

freely over $x^1, \ldots, x^n$ and x. Now we can write

$$L = \text{CS} + \pi \tag{5.3}$$

where

$$\text{CS} = \sum_{i \in N} \left[u_i(x^i) - p^\top x^i \right] \quad \text{and} \quad \pi = p^\top x - c(x)$$

In (5.3) we have written L as the sum of two terms, each of which is maximized over different variables. Hence, for the appropriate value of the Lagrange multiplier p (also called a dual variable), L is maximized by maximizing each of the terms individually. The first term is the *aggregate consumers' surplus*, CS. Following the previous observation, the consumers are individually posed the set of problems

$$\text{CONSUMER}_i : \underset{x^i}{\text{maximize}} \left[u_i(x^i) - p^\top x^i \right], \quad i = 1, \ldots, n \tag{5.4}$$

The second term is the producer's profit, π. The producer is posed the problem

$$\text{PRODUCER} : \underset{x}{\text{maximize}} \left[p^\top x - c(x) \right] \tag{5.5}$$

Thus, we have the remarkable result that the social planner can maximize social surplus by setting an appropriate price vector p. In practice, it can be easier for him to control the dual variable p, rather than to control the primal variables $x, x^1, \ldots, x^n$ directly.

This price controls both production and consumption. Against this price vector, the consumers maximize their surpluses and the producer maximizes his profit. Moreover,

---

[3] This is typically the case when the production facility cannot be expanded in the time frame of reference, and marginal cost of production increases due to congestion effects in the facility. In practice, the cost function may initially be concave, due to economies of scale, and eventually become convex due to congestion. In this case, we imagine that the cost function is convex for the output levels of interest.

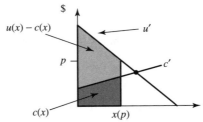

Figure 5.3   A simple illustration of the social welfare maximization problem for a single good. The maximum is achieved at the point where the customer's aggregate demand curve u' intersects the marginal cost curve c'.

from (5.4)–(5.5) we see that provided the optimum occurs for $0 < x_j^i < \infty$, this price vector satisfies

$$\frac{\partial u_i(x^i)}{\partial x_j^i} = \frac{\partial c(x)}{\partial x_j} = p_j.$$

That is, prices equal the supplier's *marginal cost* and each consumer's *marginal utility* at the solution point. We call these prices *marginal cost prices*. A graphical interpretation of the optimality condition is shown in Figure 5.3.

We have called the problem of maximizing social surplus the SYSTEM problem and have seen that price is the catalyst for solving it, through decentralized solution of PRODUCER and CONSUMER$_i$ problems. The social planner, or regulator, sets the price vector p. Once he has posted p the producer and each consumer maximizes his own net benefit (of supplier profit or consumer surplus). The producer automatically supplies x if he believes he can sell this quantity at price p. He maximizes his profit by taking x such that for all j, either $p_j = \partial c(x)/\partial x_j$, or $x_j = 0$ if $p_j = 0$. The social planner need only regulate the price; the price provides a control mechanism that simultaneously optimizes both the demand and level of production. We have assumed in the above that the planner attaches equal weight to consumer and producer surpluses. In this case, the amount paid by the consumers to the producer is a purely internal matter in the economy, which has no effect upon the resulting social surplus.

The same result holds if there is a set M of producers, the output of which is controlled by the social planner to meet an aggregate demand at minimum total cost. Using the same arguments as in the case of a single producer, the maximum of

$$S = \sum_{i \in N} u_i(x^i) - \sum_{j \in M} c_j(y^j)$$

subject to $\sum_{i \in N} x^i = \sum_{j \in M} y^j$, is achieved by

$$p_h = \partial u_i(x^i)/\partial x_h^i = \partial c_j(y^j)/\partial y_h^j, \quad \text{for all } h, i, j \tag{5.6}$$

In other words, consumers behave as previously, and every supplier produces an output quantity at which his marginal cost vector is p.

Iterative price adjustment: network and user interaction

How might the social planner find the prices at which social welfare is maximized? One method is to solve (5.6), if the utilities and the cost functions of the consumers and the producers are known. Another method is to use a scheme of iterative price adjustment. In steps, the social planner adjusts prices in directions that reduce the mismatch between demand and supply. This does not require any knowledge about the utilities and cost functions of the market participants.

Suppose that for price vector p the induced aggregate demand is $x(p)$ and the aggregate supplier output is $y(p)$. Define the *excess demand* as $z(p) = x(p) - y(p)$. Let prices adjust in time according to a rule of the form

$$\dot{p}_i = G_i(z_i(p))$$

where G_i is some smooth sign-preserving function of excess demand. This process is known as *tatonnement*, and under certain conditions p will converge to an equilibrium at which $z(p) = 0$. See Appendix B.1 for a proof that this tatonnement converges.

Tatonnement occurs naturally in markets where producers and consumers are price takers, i.e. in which they solve problems of the form PRODUCER$_j$ and CONSUMER$_i$. Producers lower prices if only part of their production is sold, and raise prices if demand exceeds supply. This occurs in a competitive market as discussed in Section 6.3.

In practice, social planners do not use tatonnement to obtain economic efficiency, due to the high risk of running the economy short of supply, or generating waste due to oversupply. Another issue for the tatonnement mechanism is the assumption that producers and consumers are truthful, i.e., that they accurately report the solutions of their local optimization problems PRODUCER$_j$ and CONSUMER$_i$. More sophisticated approaches that address these issues and may actually be used by a regulator are discussed in Chapter 13.

5.4.2 The Case of Consumers and Finite Capacity Constraints

A similar result can be obtained for a model in which customers share some finite network resources. This is typical for a communication networks in which resources are fixed in the short run. Prices can again be used both to regulate resource sharing and to maximize social efficiency. For the moment, we give a formulation in which the concept of a resource is abstract. Some motivation has already been provided in the elevator analogy of Section 4.4. The ideas are given fuller treatment and made concrete Chapter 8.

Suppose n consumers share k resources under the vector of constraints

$$\sum_{i \in N} x_j^i \le C_j, \quad j = 1, \ldots, k$$

Let us define SYSTEM as the problem of maximizing social surplus subject to this constraint:

$$\text{SYSTEM} : \underset{x^1, \ldots, x^n}{\text{maximize}} \sum_{i \in N} u_i(x^i) \quad \text{subject to} \sum_{i \in N} x_j^i \le C_j, \quad j = 1, \ldots, k$$

Given that $u_i(\cdot)$ is concave, this can be solved by maximizing a Lagrangian

$$L = \sum_{i \in N} u_i(x^i) - \sum_{j=1}^{k} p_j \left(\sum_{i \in N} x_j^i - C_j \right)$$

for some vector Lagrange multiplier $p = (p_1, \ldots, p_n)$. The maximum occurs at the same point as would be obtained if customers were charged the vector of prices p, i.e. if customer i to be posed the problem

$$\text{CONSUMER}_i : \underset{x^i}{\text{maximize}} \left[u_i(x^i) - \sum_j p_j x_j^i \right]$$

Note that $p_j = \partial(\max_x L)/\partial C_j$. That is, p_j equals the marginal increase in aggregate utility with respect to increase of C_j. As above, there is a tatonnement (an iterative method) for computing p (see Section B.2).

5.4.3 Discussion of Assumptions

Relaxing concavity

Thus far in this section we have assumed that utility and cost functions are concave and convex respectively. This ensures that the objective of the SYSTEM problem is a concave function. Hence it can be solved by maximizing a Lagrangian. Fortunately, these assumptions hold in many circumstances. Consumers usually have decreasing marginal utility for a good, and marginal costs of production are usually increasing.

If, however, the utility function is convex, or perhaps sigmoid (a tilted S-shape), then it might be that $u(C/2) + u(C/2) < u(0) + u(C)$. The social welfare maximizing allocation of bandwidth C between two identical users is to give one of them all of the bandwidth and the other nothing. This allocation could be achieved with an auction or by awarding the bandwidth by lottery. However, it cannot be achieved by a classical pricing mechanism.

There are examples in which it is possible to maximize social welfare using nonlinear prices, but not by using linear ones. In Figure 5.4 the user demands the socially optimal quantity C when he faces a nonlinear charge of the form $g(x) = A$ if $x < x_0$ and $g(x) = A + (x - x_0)p$ if $x \geq x_0$, for some x_0 and p equal to the slope of the utility curve at C.

Problems with tatonnement

Certain forms of utility function can give difficulty with the convergence of the tatonnement process. Suppose a user's utility function is concave and increasing. Suppose it is linear between two values, say x_1 and x_2, and p, is the gradient of the line between these two points. Then if the price is p the purchase of any amount between x_1 and x_2 maximizes the user's net benefit. However, a very small change in price from just below p to just above p can flip the demand from above x_2 to below x_1.

Similarly, if a utility function increases linearly from 0 to B and then very quickly becomes nearly flat after B, a small decrease in price can cause the demand to jump from B to 0.

Untruthful declarations

We have assumed thus far that all consumers are too small to affect prices. If this is not the case, then a user who lies about his utility function might have an advantage. Consider a

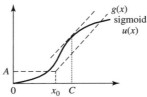

Figure 5.4 If a user has a sigmoid utility function then welfare can be maximized using nonlinear pricing. The user can be made to demand the socially optimal quantity, say C, by being faced with a nonlinear charge of the form $g(x) = A$ if $x < x_0$ and $g(x) = A + (x - x_0)p$ if $x \geq x_0$, for some x_0, where p is equal to the slope of the $u(x)$ at $x = C$. Observe that there is no linear price for which the user will demand C.

market of just two users, whose utilities for x units of bandwidth are both $u(x) = \log(1+x)$. Two units of bandwidth are to be sold. The supplier starts with a high price and decreases it until he finds the price at which all the bandwidth is sold. If both users are truthfully maximizing their net benefits at each price, then the bandwidth is sold at price $p = 1/2$. Each user buys 1 unit of bandwidth and has a net benefit of $\log(2) - 1/2$. However, suppose User 2 cheats. User 1 (who is truthful) buys at price p his net-benefit-maximizing amount, $1/p - 1$. But at each price p that is greater than $1/2 - \epsilon$, User 2 purchases less than the amount required for all two units to be sold, i.e. he deliberately buys less than $3 - 1/p$. The seller keeps decreasing his price. Finally, when the price is $p = 1/2 - \epsilon$, User 1 buys $1/p - 1$ and User 2 now buys $3 - 1/p$. User 2 obtains net benefit of $\log(4 - 1/p) - (3 - 1/p)p$ and this exceeds $\log(2) - 1/2$ for $p = 1/2 - \epsilon$ and ϵ sufficiently small. So User 2 obtains a greater net benefit by lying.

The case of digital goods and network externalities

In our social welfare maximization problem we assumed that marginal utility is decreasing and marginal cost is increasing. This may not be true when selling goods with the sort of network externality effects mentioned in Section 1.1.1. In this case the shape of the demand and cost functions are reversed. First note that the average cost per unit of production decreases with the amount sold, since the marginal cost of producing one more digital copy is zero. Also note that as more items are sold they are of greater value to the customers, and so they may be willing to pay a higher price. For more details on the construction of such demand curves see Section 5.7 and Figure 5.8.

5.4.4 Peak-load pricing

The key result of Section 5.4.1 is that social surplus is maximized by marginal cost pricing. A form of marginal cost pricing is also optimal in circumstances of so-called *peak-load pricing*. Suppose that demand for a service is greater during peak hours, lesser during off-peak hours, and the cost depends on both the amounts consumed and the maximum amount consumed. For example, consider a production facility whose capacity must be great enough to meet demand during the period of maximum demand. The cost of operating the facility during any given period depends both on the level of production during that period and on the maximum of production levels over all periods.

Consider the provision of a single type of service that is consumed during each of T periods. Demand in period t depends upon t and on the prices over all periods

$$x_t = x_t(p_1, \ldots, p_T)$$

Thus, we model the idea that a greater price during one period can shift demand to other periods. Suppose that the total cost of operating the facility takes the form

$$c(x_1, \ldots, x_T) = a \sum_t x_t + b \max_t x_t$$

The problem of finding consumption levels and the corresponding prices that maximize social welfare can be written as

$$\max_{x_1, \ldots, x_T, K} \left[u(x_1, \ldots, x_T) - a \sum_t x_t - bK \right]$$

subject to $x_t \leq K$ for all t.

Suppose that when social welfare is maximized there is a single peak period. A naïve application of the idea of marginal cost pricing suggests that prices should be defined by the relations such as

$$p_t = \begin{array}{ll} a, & x_t < K \\ a+b, & x_t = K \end{array}$$

In other words, the price in each period should reflect the marginal increase in the production cost when the production level in that period is increased. However, the truth is more subtle. There must be a sharing of the rental cost, b, over a number of periods, all of which achieve the peak rate of operation, i.e. the optimal prices take the form

$$p_t = \begin{array}{ll} a, & x_t < K \\ a+y_t, & x_t = K \end{array}$$

where $\sum_t y_t = b$. The fact that social welfare is maximized by prices of this form follows by consideration of the Lagrangian

$$L(x_t, z_t, K, y) = u(x_1, \ldots, x_T) - a \sum_t x_t - bK - \sum_t y_t(x_t + z_t - K)$$

where $z_t \geq 0$, $y_t = 0$ if $x_t \neq K$.

The optimal prices can be constructed in the following way. Start by charging a per unit of capacity in each period. Consider the demand $x_t(a, \ldots, a)$, $t = 1, \ldots, T$, and choose the period with the largest demand, say period i. Start charging this period with some amount y_i in addition to a, i.e. $p = (a, \ldots, a+y_i, \ldots, a)$; let y_i start at zero and gradually increase. The demand x_i will decrease. The demands in other periods will increase or decrease, as they are substitutes or complements. Keep increasing y_i until either $y_i = b$, or the demand in some other period equals x_i. If $y_i = b$ is reached first, then period i should be charged the entire rental cost of b. Otherwise, let M be the set of periods of peak load, i.e. $M = \arg\max\{x_1, \ldots, x_T\}$. Keep increasing the components y_j, $j \in M$, such that demand decreases equally in all periods $j \in M$; add more periods to M as other periods also become peak load periods. Stop when the revenue produced by the peak load periods equals the rental cost, i.e. when $\sum_{i \in M} y_i = b$.

5.4.5 Walrasian Equilibrium

We now turn to two important notions of market equilibrium and efficiency. The key points in this and the next section are the definitions of Walrasian equilibrium and Pareto efficiency, and the fact that they can be achieved simultaneously, as summarized at the end of Section 5.4.6. The reader may wish to skip the proofs and simply read the definitions, summary and remarks about externalities and market failure that introduce the theorems.

We begin with an concept of a market in competitive equilibrium. Suppose that initially each participant in the market is endowed with some amount of each of k goods. Participant i has initial endowments $\omega^i = (\omega_1^i, \ldots, \omega_k^i)$. Suppose the price of good j is p_j, so the monetary value of the participant's endowment is $p^\top \omega^i$. If this participant can sell some of his goods and buy others, he will do this to solve the problem

$$\underset{x^i}{\text{maximize }} u_i(x^i) \quad \text{subject to } p^\top x^i \leq p^\top \omega^i \tag{5.7}$$

where $u_i(x^i)$ is his utility for the bundle x^i. Denote the solution point by $x^i(p, p^\top \omega^i)$, i.e. his preferred bundle of goods, given price vector is p and initial endowment has monetary value $p^\top \omega^i$. Note that we are considering a simplified economy in which there is no production, just exchange. Each participant is effectively both consumer and supplier. Note that actual money may not be used. Prices express simple exchange rules between goods: if $p_i = kp_j$ then one unit of good i can be exchanged for k units of good j. Observe that (5.7) does not depend upon actual price scaling, but only on their relative values.

With $x^i = x^i(p, p^\top \omega^i)$, we say that (x, p) is a *Walrasian equilibrium* from the initial endowment $\omega = \{\omega^i_j\}$ if

$$\sum_i x^i(p, p^\top \omega^i) \leq \sum_i \omega^i \qquad (5.8)$$

that is, if there is no excess demand for any good when each participant buys the bundle that is optimal for him given his budget constraint. It can be proved that for any initial endowments ω there always exists a Walrasian equilibrium for some price vector p. That is, there is some p at which markets clear. In fact, this p can be found by a tatonnement mechanism (a fact we can prove along similar lines as in Appendix B.1). The Walrasian equilibrium is also called a *competitive equilibrium*, since it is reached as participants compete for goods, which become allocated to those participants who value them most. Throughout the following we assume that all utilities are increasing and concave, so that p is certainly nonnegative and the inequalities in (5.7) and (5.8) are sure to be equalities at the equilibrium. Equivalently, under this assumption, (x, p) is a Walrasian equilibrium if

1. $\sum_i x^i = \sum_i \omega^i$.

2. If $\bar{x}^i$ is preferred by participant i to x^i, then $p^\top \bar{x}^i > p^\top x^i$.

5.4.6 Pareto Efficiency

We now relate the idea of Walrasian equilibrium to another solution concept, that of *Pareto efficiency*. We say that a solution point (an allocation of goods to participants) is Pareto efficient if there is no other point for which all participants are at least as well off and at least one participant is strictly better off, for the same total amounts of the goods. In other words, it is not possible to make one participant better off without making at least one other participant worse off. Mathematically, we say as follows.

> The allocation $x^1, \ldots, x^n$ is *not* Pareto efficient if
> there exists $\bar{x}^1, \ldots, \bar{x}^n$, with $\sum_i \bar{x}^i = \sum_i x^i$,
> such that $u_i(\bar{x}^i) \geq u_i(x^i)$ for every i,
> and at least one of these inequalities is strict. $\qquad (5.9)$

Unlike social welfare, Pareto efficiency is not concerned with the sum of the participants' utilities. Instead, it characterizes allocations which cannot be strictly improved 'componentwise'. In the following two theorems we see that Walrasian equilibria can be equated with Pareto efficient points. We assume that the utility functions are strictly increasing and concave and there are no market failures. The following theorem says that a market economy will achieve a Pareto efficient result. It holds under the assumption that (5.7) is truly the problem faced by participant i. In particular, this means that his utility must depend only the amounts of the goods he holds, not the amounts held by others or their utilities. So there must be no unpriced externalities or information asymmetries. These mean there are missing markets (things unpriced), and so-called market failure.

Theorem 1 (first theorem of welfare economics) If (x, p) is a Walrasian equilibrium then it is Pareto efficient.

Proof Suppose x is not Pareto efficient. So there is a $\bar{x}$ for which lines 2–4 of statement (5.9) hold. By assumption that x^i is the preferred bundle of participant i, we have $u_i(\bar{x}^i) - \lambda_i p^\top \bar{x}^i \leq u_i(x^i) - \lambda_i p^\top x^i$ for all i, where $\lambda_i > 0$ is the Lagrange multiplier of the budget constraint in the optimization problem (5.7) that is solved by participant i. Combining these, we have $p^\top \bar{x}^i - p^\top x^i \geq 0$, for all i, with at least one of these a strict inequality. Summing on i, we have $p^\top \sum_i \bar{x}^i > p^\top \sum_i x^i$. Since $p \geq 0$, this can happen only if $\sum_i \bar{x}^i_j > \sum_i x^i_j$ for some j. This contradicts the assumption in the second line of (5.9). ∎

Theorem 2 (second theorem of welfare economics) Suppose ω is a Pareto efficient allocation in which $\omega^i_j > 0$ for all i, j. Then there exists a p such that (ω, p) is a Walrasian equilibrium from any initial endowment $\bar{\omega}$ such that $\sum_i \bar{\omega}^i = \sum_i \omega^i$, and $p^\top \bar{\omega}^i = p^\top \omega^i$ for all i.

Proof If ω is Pareto efficient then any allocation that makes $u_i(x^i) > u_i(\omega^i)$ for some i must make $u_j(x^j) < u_j(\omega^j)$ for some j. So $x = \omega$ solves the problem

$$\text{maximize}_x \left\{ \min_i \left[u_i(x^i) - u_i(\omega^i) \right] \right\}, \qquad \text{subject to} \sum_i x^i = \sum_i \omega^i$$

with a maximized value of 0. Note that $x = \omega$ satisfies $\sum_i x^i \leq \sum_i \bar{\omega}^i$. The above maximization problem is equivalent to

$$\text{maximize}_{t,x} t$$
$$\text{subject to} \sum_i x^i \leq \sum_i \omega_i, \quad \text{and } u_i(x^i) - u_i(\omega^i) \geq t, \quad \text{for all } i.$$

By the concavity of u_i, this problem can be solved by maximizing the Lagrangian

$$L = t + \sum_i \lambda_i [-t + u_i(x^i) - u_i(\omega^i)] + p^\top \left(\sum_i \omega^i - \sum_i x^i \right)$$

over x and t, for some multipliers $\lambda_i \geq 0$ and $p_i \geq 0$. By the assumption that $\omega^i_j > 0$ the maximum with respect to x^i_j must occur at a stationary point, and so at $x = \omega$, we must have

$$\lambda_i \left. \frac{\partial u_i}{\partial x^i_j} \right|_{x^i = \omega^i} - p_j = 0 \tag{5.10}$$

By concavity of u_i, we have that for all i and any x such that $p^\top x^i \leq p^\top \bar{\omega}^i$,

$$u_i(x^i) \leq u_i(\omega^i) + \frac{1}{\lambda_i} p^\top (x^i - \omega^i) = u_i(\omega^i) + \frac{1}{\lambda_i} p^\top (x^i - \bar{\omega}^i) \leq u_i(\omega^i)$$

Thus (ω, p) defines a Walrasian equilibrium. ∎

Note also that (5.10) implies

$$\frac{\partial u_i/\partial x_j^i}{\partial u_i/\partial x_{j'}^i} = \frac{\partial u_{i'}/\partial x_j^{i'}}{\partial u_{i'}/\partial x_{j'}^{i'}}, \quad \text{for all } i, i', j, j'$$

This makes sense; since if it were not so, then participants i and i' could both be better off by exchanging some amounts of goods j and j'.

Alternative proof We have said at the end of Section 5.4.5 that given any initial endowment $\bar{\omega}$ there exists a p, x such (p, x) is a Walrasian equilibrium. Suppose ω is a Pareto point and let $x = x(p, p^\top \bar{\omega}) = x(p, p^\top \omega)$. Now for each i the bundle x^i is preferred to bundle ω^i at prices p. But this preferences cannot be strict for any i, else ω would not be a Pareto point. Thus for all i, ω^i is as just as good for participant i as x^i, and so (p, ω) is a Walrasian equilibrium for initial endowments ω.   ∎

Notice that, given any initial endowment $\bar{\omega}$ such that $\sum_i \bar{\omega}^i = \sum_i \omega^i$, i.e. $\bar{\omega}$ and ω contain the same total quantity of each good, we can support a Pareto efficient ω as the Walrasian equilibrium if we are allowed to first make a lump sum redistribution of the endowments. We can do this by redistributing the initial endowments $\bar{\omega}$ to any $\hat{\omega}$, such that $p^\top \hat{\omega}^i = p^\top \omega^i$ for all i. Of course, this can be done trivially by taking $\hat{\omega} = \omega$, but other $\hat{\omega}$ may be easier to achieve in practice. For example, we might find it difficult to redistribute a good called 'labour'. If there is a good called 'money', then we can do everything by redistributing that good alone, i.e. by subsidy and taxation. We will see this in Section 5.5.1 when we suggest that to maximize social welfare there be a lump-sum transfer of money from the consumers to the supplier to cover his fixed cost.

Let us now return to the problem of social welfare maximization:

$$\underset{x}{\text{maximize}} \sum_i u_i(x^i), \quad \text{subject to } \sum_i x_j^i \leq \omega_j \text{ for all } j \tag{5.11}$$

For $\omega_j = C_j$ this is the problem of Section 5.4.2. We can make the following statement about its solution.

Theorem 3 *Every social welfare optimum is Pareto efficient.*

Proof As we have seen previously, assuming that the u_i are concave functions, this type of problem is solved by posting a price vector p, such that at the optimum $\partial u_i/\partial x_j^i = p_j$ for all j. Suppose the solution point is $\bar{x}$. It satisfies the constraint on the right hand side of (5.11) and also solves for every $i \in N$ the problem of maximizing $u_i(x^i)$ subject to $p^\top x^i \leq p^\top \bar{x}^i$. (Note that in solving such a problem we would take a Lagrangian of $L = u_i(x^i) - \lambda_i(p^\top x^i + z - p^\top \bar{x}^i)$ and then find that taking $x = \bar{x}$, $z = 0$ and $\lambda_i = 1$ maximizes L and satisfies the constraint.) Thus, $(\bar{x}, p)$ is a Walrasian equilibrium for the initial vector of endowments $\bar{x} = (\bar{x}^1, \ldots, \bar{x}^n)$. By Theorem 1, $\bar{x}$ is also a Pareto point. ∎

Finally, note that we can cast the welfare maximization problem of Section 5.4.1 into the above form if we imagine that the producer is participant 1, and the set of consumers is $N = \{2, \ldots, n\}$. In the initial endowment, no consumer has any amount of any good, and producer has a very large endowment of every good, say ω^1. So the constraint in (5.11) is $x^1 + \sum_{i \in N} x^i \leq \omega^1$. We take $u_1(x^1) = -c(\omega^1 - x^1)$, noting that this is a concave increasing function of x^1 if c is convex increasing. At the optimum we will have $x^1 + \sum_{i \in N} x^i = \omega^1$. Thus, (5.11) is simply

$$\underset{x^2,\dots,x^n}{\text{maximize}} \sum_{i \in N} u_i(x^i) - c\left(\sum_{i \in N} x^i\right)$$

Therefore, we can also conclude that for this model with a producer, the same conclusions hold. In summary, these conclusion are that there is a set of relationships between welfare maxima, competitive equilibria and Pareto efficient allocations:

1. Competitive equilibria are Pareto efficient.

2. Pareto efficient allocations are competitive equilibria for some initial endowments.

3. Welfare maxima are Pareto efficient.

The importance of these results is to show that various reasonable notions of what constitutes 'optimal production and consumption in the market' are consistent with one another. It can be found by a social planner who control prices to maximize social welfare, or by the 'invisible hand' of the market, which acts as participants individually seeking to maximize their own utilities.

5.4.7 Discussion of Marginal Cost Pricing

We have seen that marginal cost pricing maximizes economic efficiency. It is easy to understand and is firmly based on costs. However, there can be some problems. First, marginal cost prices can be difficult to compute. Secondly, they can be close to either zero or infinity. This is a problem since communication networks typically have fixed costs which must somehow be recovered.

Think of a telephone network that is built to carry C calls. The main costs of running the network are *fixed costs* (such as maintenance, loans repayments and staff salaries), i.e. invariant to the level of usage. Thus when less than C calls are present the short-run marginal cost that is incurred by carrying another call is near zero. However, it the network is critically loaded (so that all C circuits are busy), then the cost of expanding the network to accommodate another call could be huge. Network expansion must take place in large discrete steps, involving large costs (for increasing the transmission speed of the fibre, or adding extra links and switches). This means that the short-run marginal cost of an extra call can approach infinity.

A proper definition of marginal cost should take account of the time frame over which the network expands. The network can be considered to be continuously expanding (by averaging the expansion that occurs in discrete steps), and the marginal cost of a circuit is then the average cost of adding a circuit within this continuously expanding network. Thus marginal cost could be interpreted as *long-run* marginal cost.

Another difficulty in basing charges on marginal cost is that even if we know the marginal cost and use it as a price, it can be difficult to predict the demand and to dimension the network accordingly. There is a risk that we will build a network that is either too big or too small. A pragmatic approach is to start conservatively and then expand the network only as demand justifies it. Prices are used to signal the need to expand the network. One starts with a small network and adjusts prices so that demand equals the available capacity. If the prices required to do this exceed the marginal cost of expanding the network, then additional capacity should be built. Ideally, this process converges to a point, at which charges equal marginal cost and the network is dimensioned optimally.

Finally, let us comment on the fact that we have been analysing a static model. In practice, the demand for network services tends to increase over time, and so we might like to take a dynamic approach to the problem of building a network and pricing services. If we were to dimension a network to operate reasonably over a period, then we might expect that at the start of the period the network will be under utilized, while towards the end of the period the demand will be greater than the network can accommodate. At this point, the fact that high prices are required to limit the demand is a signal that it is time to expand the network. However, even in a dynamic setting, one can equate competitive equilibrium and with Pareto efficiency.

5.5   Cost recovery

One of the most important issues for a network operator is cost recovery. In many cases, the prices that maximize social welfare may generate income for the supplier that is less than his cost of providing the services. However, if he raises prices in an arbitrary fashion this may significantly reduce the social welfare. In this section we discuss the tradeoff between recovering costs and maximizing social welfare and look at charging methods that take the cost recovery issue into account.

5.5.1   Ramsey Prices

A weakness of marginal cost pricing is that it may not allow the supplier to recover his costs. If he is very large and operates with economies of scale, i.e. costs that increase less than proportionately with output level, then his marginal cost can be very small. The revenue he would obtain under marginal cost pricing could fail to recover his fixed costs of operation (such as property taxes, interest on loans and maintenance). In other words, social welfare is maximized at a point where $\pi < 0$. There are various ways in which to overcome this problem. The simplest is to make a lump-sum transfer of money from the consumers to the supplier that is equal to the supplier's fixed cost and then price services at marginal cost.

A second way is to consider maximization of a weighted objective function, which by taking $0 < \gamma < 1$ places less weight on consumer surplus than supplier profit. Using the notation of Section 5.4.1, we seek to maximize

$$
\begin{aligned}
W &= \pi + (1 - \gamma)\mathrm{CS} \\
&= \gamma \sum_j p_j x_j - c(x) + (1 - \gamma) \sum_i u_i(x^i) \\
&= (1 - \gamma)\left\{ \left[\sum_i u_i(x^i) - c(x) \right] + \frac{\gamma}{1 - \gamma} \left[\sum_j p_j x_j - c(x) \right] \right\}
\end{aligned}
\tag{5.12}
$$

The term in curly brackets is nearly the same as the Lagrangian we would use to solve a problem of maximizing S subject to a constraint $\pi = B$, for some $B > 0$, i.e.

$$
L = \sum_i u_i(x^i) - c(x) + \lambda \left[\sum_j p_j x_j - c(x) - B \right]
$$

for some multiplier λ. The problems are equivalent for $\lambda = \gamma/(1 - \gamma)$.

Now we wish to face the consumers with a price vector p such that W is maximized when each consumer individually maximizes his net benefit. Thus, consumer i will maximize $u_i(x^i) - p^\top x^i$, choosing x^i such that $\partial u_i / \partial x_j^i = p_j$. Differentiating (5.12) with respect to p_h, we have

$$
\begin{aligned}
\frac{\partial W}{\partial p_h} &= \gamma x_h + \gamma \sum_j p_j \frac{\partial x_j}{\partial p_h} - \sum_j \frac{\partial c(x)}{\partial x_j} \frac{\partial x_j}{\partial p_h} + (1 - \gamma) \sum_{i,j} \frac{\partial u_i}{\partial x_j^i} \frac{\partial x_j^i}{\partial p_h} \\
&= x_h \left(\gamma + \sum_j \frac{p_j - \frac{\partial c}{\partial x_j}}{p_j} \epsilon_{hj} \right)
\end{aligned}
\tag{5.13}
$$

where we use $\partial u_i / \partial x_j^i = p_j$ and $\partial x_j / \partial p_h = \partial x_h / \partial p_j$. Here ϵ_{hj} is a cross elasticity of the aggregate demand function, $x(p) = \sum_i x^i(p)$.

For the maximum of W we require $\partial W / \partial p_h = 0$. So in the general case, the p_j are found by solving a complicated set of equations. In the special case that services are independent (i.e. $\epsilon_{ij} = 0$, for $i \neq j$), we have

$$
\frac{p_j - \frac{\partial c}{\partial x_j}}{p_j} = -\frac{\gamma}{\epsilon_i}
\tag{5.14}
$$

Prices of the form of (5.14) are known as *Ramsey prices*. We know that the price elasticity of demand, ϵ_i, is always negative. So if services are independent, as we have imagined above, the Ramsey prices are above marginal cost prices. This means that the demand is reduced below the value at which social welfare is maximized. Recall that an inelastic good is one for which the demand is relatively insensitive to price changes, i.e. $|\epsilon_i|$ is small. Thus, Ramsey pricing has the effect of pricing inelastic goods well above their marginal costs; these goods tend to subsidize goods whose demand is more price sensitive.

If for some value of γ we have $\pi = 0$, then we have found the prices that maximize social welfare, subject to the supplier recovering his costs. Observe that for $\gamma = 1$ we are simply maximizing π, and obtain the prices at which a monopolist maximizes his profit (as we see again in Section 6.2.1).

It is interesting to work through an example for the special case that demand curves are linear and the marginal cost is constant, say $c(x) = x^\top MC$. Suppose that the quantities demanded under marginal cost pricing are $x_1^{MC}, \ldots, x_n^{MC}$. Figure 5.5 shows this for two customers, each demanding a different good, whose demand curves happen to intersect the marginal cost curve at the same point. From (5.14) we have that the Ramsey price p_i, is given by $(p_i - MC_i)/p_i = -\gamma/\epsilon_i$. Since $x_i(p)$ is linear, so that $dx_i/dp = \alpha_i$, say, we have $\epsilon_i = \alpha_i p_i / x_i$. Also, $\alpha_i = -(x_i^{MC} - x_i)/(p_i - MC_i)$. After eliminating ϵ_i and α_i, we find

$$
\frac{x_i - x_i^{MC}}{x_i^{MC}} = \frac{\gamma}{\gamma + 1} \quad \text{for all } i
$$

Thus, the quantities demanded under Ramsey pricing deviate in equal proportion from those demanded under marginal cost pricing. In the special case shown in Figure 5.5, in which two customers have the same demand at marginal cost pricing, i.e., $x_1^{MC} = x_2^{MC}$, the quantities demanded under Ramsey pricing are also equal.

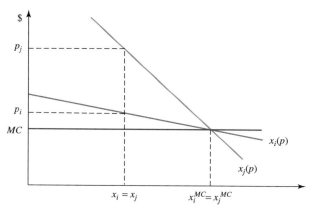

Figure 5.5 Ramsey pricing for two independent goods, with constant marginal cost that is the same for both goods, and linear demand functions. If the quantities demanded under marginal cost pricing are equal, then the quantities demanded under Ramsey pricing are also equal. The Ramsey price for the more inelastic good will be greater.

If services are not independent we have from (5.13) that the Ramsey prices are the solution of the set of equations

$$\sum_j \frac{p_j - \frac{\partial c}{\partial x_j}}{p_j} \epsilon_{hj} = -\gamma , \quad h = 1, \ldots, n \tag{5.15}$$

So Ramsey prices can be below marginal cost if some services are complements (i.e. $\epsilon_{ij} < 0$). For example, if $n = 2$, and the elasticities are constant, with values of $\epsilon_1 = -2$, $\epsilon_2 = -5$ and $\epsilon_{12} = -3$, then we can easily calculate that $p_1 < \partial c/\partial x_1$ and $p_2 > \partial c/\partial x_2$.

As an illustration of this, consider a case of two services, say voice and video. There is demand for voice alone, for video alone, and for voice and video provided as a single teleconferencing service. In this case, voice and video are complements. Let MC_1 and MC_2 be the marginal costs and also the initial prices for voice and video respectively, and suppose that the network needs to recover some fixed cost. Assume that the demand for voice is price inelastic, while the demand for the teleconferencing service is very price elastic. In this case, it is sensible for the network operator to raise the price p_1 for voice above MC_1 to recover a substantial part of the cost. But now the price for the videoconferencing service becomes $p_1 + MC_2 > MC_1 + MC_2$, which will reduce significantly the demand of the price-elastic videoconferencing users. To remedy that, the price p_2 of video should be set below MC_2, so that $p_1 + p_2$ remains close to $MC_1 + MC_2$.

Ramsey prices are linear tariffs and require knowledge of properties of the market demand curves. It turns out that, by using more general nonlinear tariffs, one may be able to obtain Pareto improvements to Ramsey prices, i.e. obtain higher social welfare while still covering costs. For instance, selling additional units at marginal cost can only improve social welfare. Note that in this case prices are non-uniform. As we see in Section 5.5.3, such tariffs are superior, but require more detailed knowledge of the demand, and can only be used under certain market conditions.

5.5.2 Two-part Tariffs

Two other methods by which a supplier can recover his costs while maximizing social welfare are *two-part tariffs* and more general nonlinear prices. A typical two-part tariff is one in which customers are charged both a *fixed charge* and a *usage charge*. Together these cover the supplier's reoccurring *fixed costs* and *marginal costs*. Note the difference between reoccurring fixed costs and nonrecurring *sunk costs*. Sunk costs are those which have occurred once-for-all. They can be included in the firm's book as an asset, but they do not have any bearing on the firm's pricing decisions. For example, once a firm has already spent a certain amount of money building a network, that amount becomes irrelevant to its pricing decisions. Prices should be set to maximize profit, i.e. the difference between revenue and the costs of production, both fixed and variable.

Suppose that the charge for a quantity x of a single service is set at $a + px$. The problem for the consumer is to maximize his net benefit

$$u(x) - a - px$$

He will choose x such that $\partial u / \partial x = p$, unless his net benefit is negative at this point, in which case it is optimal for him to take $x = 0$ and not participate. Thus a customer who buys a small amount of the service if there is no fixed charge may be deterred from purchasing if a fixed charge is made. This reduces social welfare, since although 'large' customers may purchase their optimal quantities of the service, many 'small' customers may drop out and so obtain no benefit. Observe that, when $p = MC$, once a customer decides to participate, then he will purchase the socially optimal amount.

How should one choose a and p? Choosing $p = MC$ is definitely sensible, since this will motivate socially optimal resource consumption. One can address the question of choosing a in various ways. The critical issue is to motivate most of the customers to participate and so add to the social surplus. If one knows the number of customers, then the simplest thing is to divide the fixed cost equally amongst the customers, as in the example of Figure 5.6. If, under this tariff, every customer still has positive surplus, and so continues to purchase, then the tariff is clearly optimal; it achieves maximum social welfare while recovering cost. If, however, some customers do not have positive surplus under this tariff, then their nonparticipating can lead to substantial welfare loss. Participation may be greater if we impose fixed charges that are in proportion to the net benefits that the customers receive, or in line with their incomes.

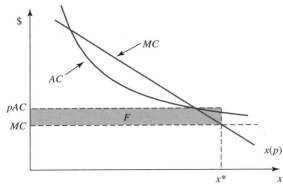

Figure 5.6 In this example the marginal cost is constant and there is a linear demand function, $x(p)$. A two part tariff recovers the additional amount F in the supplier's cost by adding a fixed charge to the usage charge. Assuming N customers, the tariff may be $F/N + x\, MC$. However, a customer will not participate if his net benefit is negative. Observe that if the average cost curve $AC = MC + F/x$ is used to compute prices, then use of the resulting price p_{AC} does not maximize social welfare. Average prices are expected to have worse performance than two part tariffs using marginal cost prices.

Note that such differential charging of customers requires some market power by the operator, and may be illegal or impossible to achieve: a telecoms operator cannot set two customers different tariffs for the same service just because they have different incomes. However, he can do something to differentiate the service and then offer it in two versions, each with a different fixed charge. Customers who are attracted to each of the versions are willing to pay that version's fixed charge. Such price discrimination methods are examined in more detail in Section 6.2.2.

Economists have used various mathematical models to derive optimum values for a and p. They assume knowledge of the distribution of the various customer types and their demand functions. Such models suggest a lower fixed fee and a price above marginal cost. A lower a motivates more small customers to participate, while the extra cost is recovered by the higher p. Remember that small customers do not mind paying more than marginal cost prices, but cannot afford a paying a high fixed fee. Other models assume a fixed cost per customer and a variable cost that depends on usage. This is the case for setting up an access service, such as for telephony or the Internet. Depending on the particular market, a and p may be above or below the respective values of the fixed customer cost and the marginal cost of usage.

5.5.3 Other Nonlinear Tariffs

General nonlinear tariffs can be functions of the form $r(x)$, where x is the amount consumed by the customer. Starting with $r(0) = 0$ they retain the nice property of Ramsey prices, that customers who have low valuations for the service can participate by paying an arbitrarily small amount. The optimal $r(x)$ may have both convex and concave parts. A general property is that the customer purchasing the largest quantity q sees a marginal charge $r'(q)$ equal to marginal cost. In many practical situations, $r(x)$ is a concave function. In this case, the price per unit drops with the quantity purchased, a property known as *quantity discounts*. In practice, smooth tariffs are approximated by *block tariffs*. These are tariffs in which the range of consumption is split into intervals, with constant per unit prices.

A more interesting class of nonlinear tariffs are *optional two-part tariffs*. The customer is offered a choice of tariffs, from which he is free to choose the one from which his charge will be computed. He may required to choose either before or after his use of the service. A set of K optional two-part tariffs is specified by pairs (a_k, p_k), $k = 1, \ldots, K$. Since customers self-select, under plausible assumptions on market demand, these tariffs must satisfy $a_k \leq a_{k+1}$ and $p_k \geq p_{k+1}$, $k = 1, \ldots, K - 1$, thereby defining a concave nonlinear tariff. An optimal choice of these coefficients often has p_K is equal to marginal cost. A nice feature of optional tariffs is the following. Given a K-part optional tariff, we can always construct a $K + 1$-part tariff that is not Pareto inferior. This is because the addition of one more tariff gives customers more choice, and so they can express better their preferences. The coefficients of the new tariff can be easily tuned to cover the supplier's costs. Tariffs of this type are commonly used to charge for fixed-line and mobile telephone services. In the following example, we show how to one can improve on linear prices by adding one optional two-part tariff.

Example 5.1 (Adding an optional two-part tariff) Our initial tariff consists of the Ramsey prices p_R. We assume that customer types have linear parallel demand functions, distributed between a smallest $x_{\min}(p)$ and a largest $x_{\max}(p)$; see Figure 5.7. For simplicity, also assume that we have a constant marginal cost of production MC. Let us first construct a Pareto

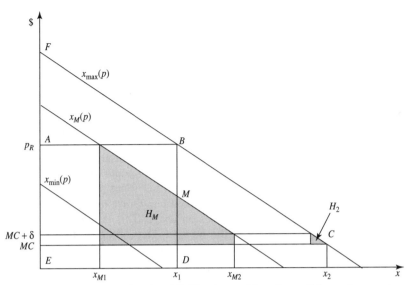

Figure 5.7 The design of an optional tariff. The demand functions of the various customer types are linear, parallel, and distributed between a smallest $x_{min}(p)$ and a largest $x_{max}(p)$. An optional tariff $E_1 + p_1 x$ is added to the Ramsey prices p_R. We start with $E_1 = ABDE$ and $p_1 = MC$. This tariff appeals to customers with demand function of x_M or more, increases social welfare since they consume more and balances costs. By further decreasing E_1, and slightly increasing p_1 to $MC + \delta$, we induce customer type M to use the optional tariff and produce a welfare gain of H_M. This is substantially greater than the welfare loss of H_2 that arises because the larger customer consumer slightly less.

improvement by adding an optional tariff $E_1 + p_1 x$, $p_1 = MC$, targeted at the largest customers. We should compute the E_1 so that such customers are indifferent between the old and new tariffs, while if they switch, their contribution to the common cost (on top of their actual consumption cost) remains the same. Clearly, if we succeed, we obtain a net improvement since the customers using the optional tariff will consume more and hence obtain a larger surplus (after making it a tiny bit more attractive to them). This is in line with a general result, stating that a necessary condition for second degree price discrimination to increase welfare is that output rises as a consequence.

As shown in Figure 5.7, the customers with greatest demand function make a contribution to the common cost, at the consumption level $x_1 = x_{max}(p_R)$, of G, equal to the area of $ABDE$, and they obtain a surplus, CS_1, equal to the area of FAB. By offering the new optional tariff, with $E_1 = G$, these customers find it more profitable to use the new tariff and increase their consumption to x_2, since their surplus becomes $CS_1 + BCD$. They make the same contribution to the common cost, but their surplus increases by the area BCD. Hence, it is a Pareto improvement. But things are even better than that. More large customers will prefer the new tariff since their surplus is greater. The smallest customer type that will switch (being just indifferent between doing so or not) is the one with demand function $x_M(p)$, passing through M, the midpoint of BD. All such customer types have increased their consumption, so there is a clear welfare gains. Moreover, their contributions to the common cost are greater than before, which leaves the network with a net profit. This suggests that E_1 could be further reduced to bring profits to zero while motivating even smaller customers to switch. For simplicity assume that this compensation is already performed and the marginal customer type who is indifferent to switch is M.

Observe that we can make customer type M switch by decreasing E_1 by ϵ, in which case we must increase p_1 by a small amount δ (of the order ϵ) to compensate for the loss of income from the other customers. The net welfare gain by having M switch is the area of the shaded trapezoid H_M (consumption will increase from x_{M1} to x_{M2}. However, there is a welfare loss since customers that have already switched will consume a bit less due to the higher price $MC + \delta$. Each such customer type will produce a welfare loss equal to the shaded triangle H_2. Since H_M is substantially larger than H_2 (which is of order ϵ), it pays to continue decreasing E_1 and increasing p_1 until these effects compensate one other.

5.6 Finite capacity constraints

The problem of recovering costs also arises when there is a finite capacity constraint. Consider the problem of maximizing social surplus subject to two constraints: revenue matches cost and demand does not exceed C. Imagine that there is a set of customers N, a single good, and it is possible to charge different customers different prices for this good. Let x_i be the amount of the good allocated to customer i (so $x_i = x_1^i$, say, where x^i denoted a vector of goods in previous sections). Assuming, for simplicity, that their demand functions are independent, the relevant Lagrangian can be written as

$$\sum_{i \in N} \int_0^{x_i} p_i(y)dy - c\left(\sum_{i \in N} x_i\right)$$

(5.16)

$$+ \lambda_1\left[p^\top x - c\left(\sum_{i \in N} x_i\right)\right] + \lambda_2\left[C - \sum_{i \in N} x_i\right]$$

It is convenient to define write $\gamma = \lambda_1/(1 + \lambda_1)$ and $\mu = \lambda_2/(1 + \lambda_1)$. The first order conditions now become

$$p_i\left(1 + \frac{\gamma}{\epsilon_i}\right) = c' + \mu, \quad i \in N$$

(5.17)

Observe that the price that should be charged to customer i depends on his demand elasticity. The minimum price that he might be charged (when $\gamma = 0$, i.e. no cost recovery is enforced) is the marginal cost augmented by addition of the Lagrange multiplier, $\mu = \lambda_2$ (the shadow price associated with the capacity constraint). Note that $\lambda_2 > 0$ when the capacity constraint is active, i.e. when the optimal total amount allocated would be greater if the capacity C were to increase.

A similar result is obtained when we solve the profit maximization problem under capacity constraints. Although profit maximization is examined in Section 6.2.1, we present the corresponding results to show the similar form of the resulting prices.

In the case of profit maximization, we want to maximize the net profit $p^\top x - c(\sum_{i \in N} x_i)$ subject to the constraint that $\sum_{i \in N} x_i \le C$. For this problem, the Lagrangian is

$$p^\top x - c\left(\sum_{i \in N} x_i\right) + \mu\left(C - \sum_{i \in N} x_i\right)$$

(5.18)

with first order conditions

$$p_i\left(1 + \frac{1}{\epsilon_i}\right) = c' + \mu, \quad i \in N$$

(5.19)

Again, the marginal cost is augmented by addition of the Lagrange multiplier (shadow cost) arising from the capacity constraint. Note that the constraint may not be active if the maximum occurs where $\sum_i x_i < C$. Finally, observe that if the marginal cost is zero, then

$$p_i \left(1 + \frac{1}{\epsilon_i}\right) = \mu \qquad (5.20)$$

Notice that prices are proportional to the shadow cost. The markup in the price charged to customer i is determined by the elasticity of his demand.

5.7 Network externalities

Throughout this chapter, we have supposed that a customer's utility depends only on the goods that he himself consumes. This is not true when goods exhibit network externalities, i.e. when they become more valuable as more customers use them. Examples of such goods are telephones, fax machines, and computers connected to the Internet. Let us analyze a simple model to see what can happen.

Suppose there are N potential customers, indexed by $i = 1, \ldots, N$, and that customer i is willing to pay $u_i(n) = ni$ for a unit of the good, given that n other customers will be using it. Thus, if a customer believes that no one else will purchase the good, he values it at zero. Assume also that a customer who purchases the good can always return it for a refund if he detects that it is worth less to him that the price he paid. We will compute the demand curve in such a market, i.e. given a price p for a unit of the good, the number of customers who will purchase it. Suppose that p is posted and n customers purchase the good. We can think of n as an equilibrium point in the following way: n customers have taken the risk of purchasing the good (say by having a strong prior belief that $n - 1$ other customers will also purchase it), and at that point no new customer wants to purchase the good, and no existing purchaser wants to return it, so that n is stable for the given p. Clearly, the purchasers will be customers $N - n + 1, \ldots, N$. Since there are customers that do not think it is profitable in this situation to purchase the good, there must be such an 'indifferent' customer, for which the value of the good equals the price. This should be customer $i = N - n$, and since $u_i(n) = p$ we obtain that the demand at price p is that n such that $n(N - n) = p$. Note that in general there are two values of n for which this holds. For instance, for $N = 100$ and $p = 1600$, n can be 20 or 80.

In Figure 5.8 we plot such a demand function for $N = 100$. For a p in the range of 0–2500 there are, in general, three possible equilibria, corresponding to the points 0, A and B (here shown for $p = 900$). Point 0 is always a possible equilibrium, corresponding to the prior belief that no customer will purchase the good. Points A and B are consistent with prior beliefs that n_1 and n_2 customers will purchase the good, where $p(n_1) = p(n_2) = p$. Here, $n_1 = 10$, $n_2 = 90$. If $p > 2500$ then only 0 is a possible equilibrium. Simple calculations show that the total value of the customers in the system is $n^2(2N - n + 1)/2$, which is consistent with Metcalfe' Law (that the total value in a system is of the order n^2).

It would lengthen our discussion unreasonably to try to specify and analyse a fully dynamic model. However, it should be clear, informally, what one might expect. Suppose that, starting at A, one more customer (say the indifferent one) purchases the good. Then the value of the good increases above the posted price p. As a result, *positive feedback* takes place: customers with smaller indices keep purchasing the good until point B is reached. This is now a stable equilibrium, since any perturbation around B will tend to make the

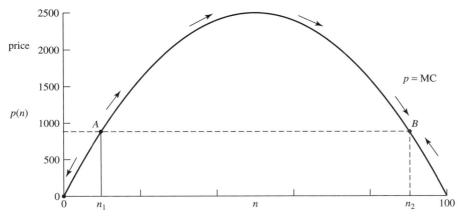

Figure 5.8 An example of a demand curve for $N = 100$ when there are network externalities. Given a price p, there are three possible equilibria corresponding to points 0, A and B, amongst which only 0 and B are stable. Observe that the demand curve is increasing from 0, in contrast to demand curves in markets without network externalities, which are usually downward sloping.

system return to B. Indeed, starting from an initial point n that is below (or above) n_2 will result in customers purchasing (or returning) the good. The few customers left above n_2 have such a small value for the good (including the network externality effects) that the price must drop below p to make it attractive to them. A similar argument shows that starting below n_1 will reduce n to zero.

These simple observations suggest that markets with strong network effects may remain small and never actually reach the socially desirable point of large penetration. This type of *market failure* can occur unless positive feedback moves the market to point B. However, this happens only when the system starts at some sufficiently large initial point above n_1. This may occur either because enough customers have initially high expectations of the eventual market size (perhaps because of successful marketing), or because a social planner subsidizes the cost of the good, resulting in a lower posted price. When p decreases, n_1 moves to the left, making it possible grow the customer base from a smaller initial value. Thus, it may be sensible to subsidize the price initially, until positive feedback takes place. Once the system reaches a stable equilibrium one can raise prices or use some other means to pay back the subsidy.

These conditions are frequently encountered in the communications market. For instance, the wide penetration of broadband information services requires low prices for access services (access the Internet with speeds higher than a few Mbps). But prices will be low for access once enough demand for broadband attracts more competition in the provision of such services and motivates the development and deployment of more cost-effective access technologies. This is a typical case of the traditional 'chicken and egg' problem!

Finally, we make an observation about social welfare maximization. Suppose that in our example with $N = 100$, the marginal cost of the good is p. If we compute the social welfare $S(n)$, it turns out that its derivative is positive at n_2 for any p that intersects the demand curve, and remains positive until N is reached. Hence, it is socially optimal to consume even more than the equilibrium quantity n_2. In this case, marginal cost pricing is not optimal, the optimal price being zero. This suggests that when strong network externalities are present, optimal pricing may be below marginal cost, in which case the social planer should subsidize the price of the good that creates these externalities. Such a subsidy could be recovered from the customers' surplus by taxation.

5.8 Further reading

A good text for the microeconomics presented in this chapter is Varian (1992). A survey of the economics literature on Ramsey pricing and nonlinear tariffs in the telecommunications market is in Mitchell and Vogelsang (1991). Issues related to network externalities and the effects of positive feedback are discussed in Economides and Himmelberg (1995) and Shapiro and Varian (1998). A review of basic results on Lagrangian methods and optimization is in Appendix A

6

Competition Models

Chapter 6 introduces three models of market competition. Their consequences for pricing are discussed in the Sections 6.2–6.4. In Section 6.1 we define three models for a market: monopoly, perfect competition and oligopoly. Section 6.2 looks at the strategies that are available to a monopoly supplier who has prices completely under his control. Section 6.3 describes what happens when prices are out of the supplier's control and effectively determined by 'the invisible hand' of perfect competition. Section 6.4, considers the middle case, called oligopoly, in which there is no dominate supplier, but the competing suppliers are few and their actions can affect prices. Within this section, we present a brief tutorial on some models in game theory that are relevant to pricing problems. Section 6.5 concludes with an analysis of a model in which a combination of social welfare and supplier profit is to be maximized.

6.1 Types of competition

The market in which suppliers and customers interact can be extraordinarily complex. Each participant seeks to maximize his own surplus. Different actions, information and market power are available to the different participants. One imagines that a large number of complex games can take place as they compete for profit and consumer surplus. The following sections are concerned with three basic models of market structure and competition: monopoly, perfect competition and oligopoly.

In a monopoly there is a single supplier who controls the amount of goods produced. In practice, markets with a single supplier tend to arise when the goods have a production function that exhibits the properties of a natural monopoly. A market is said to be a *natural monopoly* if a single supplier can always supply the aggregate output of several smaller suppliers at less than the total of their costs. This is due both to production *economies of scale* (the average cost of production decreases with the quantity of a good produced) and *economies of scope* (the average cost of production decreases with the number of different goods produced). Mathematically, if all suppliers share a common cost function, c, this implies $c(x + y) \leq c(x) + c(y)$, for all vector quantities of services x and y. We say that $c(\cdot)$ is a *subadditive function*. This is frequently the case when producing digital goods, where there is some fixed initial development cost and nearly zero cost to reproduce and distribute through the Internet.

In such circumstances, a larger supplier can set prices below those of smaller competitors and so capture the entire market for himself. Once the market is his alone then his problem is

---

Pricing Communication Networks   C. Courcoubetis and R. Weber
© 2003 John Wiley & Sons, Ltd   ISBN 0-470-85130-9 (HB)

essentially one of profit maximization. In Section 6.2 we show that a monopolist maximizes his profit (surplus) by taking account of the customers' price elasticities. He can benefit by discriminating amongst customers with different price elasticities or preferences for different services. His monopoly position allows him to maximize his surplus while reducing the surplus of the consumers. If he can discriminate perfectly between customers, then he can make a take-it-or-leave it offer to each customer, thereby maximizing social welfare, but keeping all of its value for himself. If he can only imperfectly discriminate, then the social welfare will be less than maximal. Intuitively, the monopolist keeps prices higher than socially optimal, and reduces demand while increasing his own profit.

Monopoly is not necessarily a bad thing. Society as a whole can benefit from the large production economies of scale that a single firm can achieve. Incompatibilities amongst standards, and the differing technologies with which disparate suppliers might provide a service, can reduce that service's value to customers. This problem is eliminated when a monopolist sets a single standard. This is the main reason that governments often support monopolies in sectors of the economy such as telecommunications and electric power generation. The government regulates the monopoly's prices, allowing it to recover costs and make a reasonable profit. Prices are kept close to marginal cost and social welfare is almost maximized. However, there is the danger that such a 'benevolent' monopoly does not have much incentive to innovate.

A price reduction of a few percent may be insignificant compared with the increase of social value that can be obtained by the introduction of completely new and life-changing services. This is especially so in the field of communications services. A innovation is much more likely to occur in the context of a competitive market.

A second competition model is perfect competition. The idea is that there are many suppliers and consumers in the market, every such participant in the market is small and so no individual consumer or supplier can dictate prices. All participants are price takers. Consumers solve a problem of maximizing net surplus, by choice of the amounts they buy. Suppliers solve a problem of maximizing profit, by choice of the amounts they supply. Prices naturally gravitate towards a point where demand equals supply. The key result in Section 6.3 is that at this point the social surplus is maximized, just as it would be if there were a regulator and prices were set equal to marginal cost. Thus, perfect competition is an 'invisible hand' that produces economic efficiency. However, perfect competition is not always easy to achieve. As we have noted there can be circumstances in which a regulated monopoly is preferable.

In practice, many markets consist of only a few suppliers. Oligopoly is the name given to such a market. As we see in Section 6.4 there are a number of games that one can use to model such circumstances. The key results of this section are that the resulting prices are sensitive to the particular game formulation, and hence depend upon modelling assumptions. In a practical sense, prices in an oligopoly lie between two extremes: these imposed by a monopolist and those obtained in a perfectly competitive market. The greater the number of producers and consumers, the greater will be the degree of competition and hence the closer prices will be to those that arise under perfect competition.

We have mentioned that if supply to a market has large production economies of scale, then a single supplier is likely to dominate eventually. This market organization of 'winner-takes-all' is all the more likely if there are network externality effects, i.e. if there are economies of scale in demand. The monopolist will tend to grow, and will take advantages of economies of scope to offer more and more services.

6.2 Monopoly

A monopoly supplier has the problem of *profit maximization*. Since he is the only supplier of the given goods, he is free to choose prices. In general, such (unit) prices may be different depending on the amount sold to a customer, and may also depend on the identity of the customer. Such a flexibility in defining prices may not be available in all market situations. For instance, at a retail petrol station, the price per litre is the same for all customers and independent of the quantity they purchase. In contrast, a service provider can personalize the price of a digital good, or of a communications service, by taking account of any given customer's previous history or special needs to create a version of the service that he alone may use. Sometimes quantity discounts can be offered. As we see below, the more control that a firm has to discriminate and price according to the identity of the customer or the quantity he purchases, the more profit it can make. Before investigating three types of price discrimination, we start with the simplest case, in which the monopolist is allowed to use only linear prices (i.e. the same for all units) uniform across customers.

6.2.1 Profit Maximization

As in Chapter 5, let $x_j(p)$ denote the demand for service j when the price vector for a set of services is p. A monopoly supplier whose goal is *profit maximization* will choose to post prices that solve the problem

$$\underset{p}{\text{maximize}} \left[\sum_j p_j x_j(p) - c(x) \right]$$

The first-order stationarity condition with respect to p_i is

$$x_i + \sum_j p_j \frac{\partial x_j}{\partial p_i} - \sum_j \frac{\partial c}{\partial x_j} \frac{\partial x_j}{\partial p_i} = 0 \tag{6.1}$$

If services are independent, so that $\epsilon_{ij} = 0$ for $i \neq j$, we have, as in (5.14), taking $\gamma = 1$,

$$p_i \left(1 + \frac{1}{\epsilon_i} \right) = \frac{\partial c}{\partial x_i}$$

One can check that this is equivalent to saying that marginal revenue should equal marginal cost. This condition is intuitive, since if marginal revenue were greater (or less) than marginal cost, then the monopolist could increase his profit by adjusting the price so that the demand increased (or decreased). Recall that marginal cost prices maximize social welfare. Since $\epsilon_i < 0$ the monopolist sets a price for service i that is greater than his marginal cost $\partial c(x)/\partial x_i$. At such prices the quantities demanded will be less than are socially optimal and this will result in a loss of social welfare.

Observe that the marginal revenue line lies below the marginal utility line (the demand curve). This is illustrated in Figure 6.1, where also we see that social welfare loss occurs under profit maximization.

If services are not independent then we have, as in (5.15),

$$\sum_j \frac{p_j - \frac{\partial c}{\partial x_j}}{p_j} \epsilon_{ij} = -1, \quad \text{for all } i$$

As already remarked in Section 5.5.1, if some services are complements then it is possible that some of them sold at less than marginal cost.

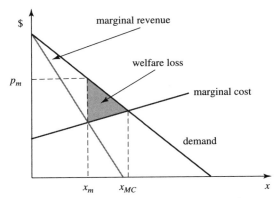

Figure 6.1 A profit maximizing monopolist will set his price so that marginal revenue equals marginal cost. This means setting a price higher than marginal cost. This creates a social welfare loss, shown as the area of the shaded triangle.

6.2.2 Price Discrimination

A supplier is said to engage in *price discrimination* when he sells different units of the same service at different prices, or when prices are not the same for all customers. This enables him to obtain a greater profit than he can by using the same linear price for all customers. Price discrimination may be based on customer class (e.g. discounts for senior citizens), or on some difference in what is provided (e.g. quantity discounts). Clearly, some special conditions should hold in the market to prevent those customers to whom the supplier sells at a low unit price from buying the good and then reselling it to those customers to whom he is selling at a high unit price.

We can identify three types of price discrimination. With *first degree price discrimination* (also called *personalized pricing*), the supplier charges each user a different price for each unit of the service and obtains the maximum profit that it would be possible for him to extract. The consumers of his services are forced to pay right up to the level at which their net benefits are zero. This is what happens in Figure 6.2.

The monopolist effectively makes a 'take it or leave it' offer of the form 'you can have quantity x for a charge of μ'. The customer decides to accept the offer if his net benefit is positive, i.e. if $u(x) - \mu \geq 0$, and rejects the offer otherwise. Hence, given the fact that the monopolist can tailor his offer to each customer separately, he finds vectors x, μ which

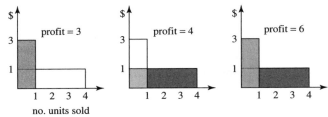

Figure 6.2 A monopolist can increase his profit by price discrimination. Suppose customer A values the service at $3, but customers B, C and D value it only at $1. There is zero production cost. If he sets the price $p = \$3$, then only one unit of the good is (just) sold to customer A for $3. If he sets a uniform price of $p = \$1$, then four units are sold, one to each customer, generating $4. If the seller charges different prices to different customers, then he should charge $3 to customer A, and $1 to customers B, C and D, giving him a total profit of $6. This exceeds $4, which is the maximum profit he could obtain with uniform pricing.

solve the problem

$$\text{maximize}_{x,\mu} \left[\sum_i \mu_i - c(x) \right] \quad \text{subject to } u_i(x_i) \geq \mu_i \text{ for all } i \quad (6.2)$$

At the optimum $\partial c(x)/\partial x_i = u_i'(x_i)$, and hence social surplus is maximized. However, since the consumer surplus at the optimum is zero, the whole of the social surplus goes to the producer. This discussion is summarized in Figure 6.3.

One way a seller can personalize price is by approaching customers with special offers that are tailored to the customers' profiles. Present Internet technology aids such personalization by making it easy to track and record customers' habits and preferences. Of course, it is not always possible to know a customer's exact utility function. Learning it may require the seller to make some special effort (adding cost). Such 'informational' cost is not included in the simple models of price discrimination that we consider here.

In *second degree price discrimination*, the monopolist is not allowed to tailor his offer to each customer separately. Instead, he posts a set of offers and then each customer can choose the offer he likes best. Prices are nonlinear, being defined for different quantities. A supplier who offers 'quantity discounts' is employing this type of price discrimination. Of course his profit is clearly less than he can obtain with first degree price discrimination.

Second degree price discrimination can be realized as follows. The charge for quantity x is set at $\mu(x)$ (where x might range within a finite set of value) and customers *self-select* by maximizing $u_i(x_i) - \mu(x_i)$, $i = 1, \ldots, n$.

Consider the case that is illustrated in Figure 6.4(a). Here customer 1 has high demand and customer 2 has low demand. Assume for simplicity that production cost is zero. If the monopolist could impose first degree price discrimination, he would maximize his revenue by offering customer 1 the deal 'x_1 for $A + B + C$', and offering customer 2 'x_2 for A'. However, under second degree price discrimination, both offers are available to the customers and each customer is free to choose the offer he prefers. The complication is that although the low demand customer will prefer the offer 'x_2 for A', as the other offer is infeasible for him, the high demand customer has an incentive to switch to 'x_2 for A', since he makes a net benefit of B (whereas accepting 'x_1 for $A + B + C$' makes his net benefit zero). To maintain an incentive for the high demand customer to choose a high quantity, the monopolist must make a discount of B and offer him x_1 for $A + C$. It turns out that the

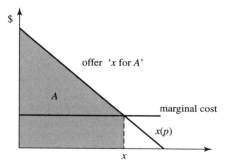

Figure 6.3 In first degree price discrimination the monopolist extracts the maximum profit from each customer, by making each a take-it-or-leave-it offer of the form 'you may have x for A dollars'. He does this by choosing x such that $u'(x) = c'(x)$ and then setting $A = u(x)$. In the example of this figure the demand function is linear and marginal cost is constant. Here A is the area of the shaded region under the demand function $x(p)$.

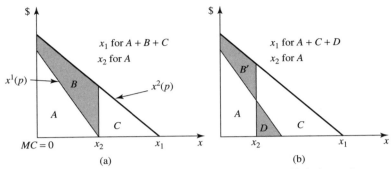

Figure 6.4   Second degree price discrimination for a low and a high demand customer. For simplicity the marginal cost of production is zero. Given the offers in (a), customer 1 (the 'high' demand customer) will choose the offer intended for customer 2 (the 'low' demand customer), unless he is offered 'x_1 for $A + C$ dollars'. The net benefit of customer 1 is the shaded area. This motivates the producer to decrease x_2 and make an offer as in (b), where $B' + D < B$. The optimum value of x_2 achieves the minimum of $B' + D$, which is the amount by which the producer's revenue is less than it would be under first degree price discrimination.

monopolist can do better by reducing the amount that is sold to the low demand customer. This is depicted in Figure 6.4(b), where the offers are x_1 for $A + D + C$, and x_2 for A. There is less profit from the low demand customer, but a lower discount is offered to the high demand customer, i.e. in total the monopolist does better because $B' + D < B$. The optimum value of x_2 achieves the minimum of $B' + D$, which is the amount by which the producer's revenue is less than it would be under first degree price discrimination.

More generally, the monopolist offers two or more of versions of the service, each of which is priced differently, and then lets each customer choose the version he prefers. For this reason second degree price discrimination is also called *versioning*. As illustrated above, one could define the versions as different discrete quantities of the service, each of which is sold at a different price per unit. Some general properties hold when the supplier's creates his versions optimally in this way: (i) the highest demand customer chooses the version of lowest price per unit; (ii) the lowest demand customer has all his surplus extracted by the monopolist; and (iii) higher demand customers receive an *informational rent*. That is, they benefit from having information that the monopolist does not (namely, information about their own demand function).

Quantity is not the only way in which information goods and communications services can be versioned. They can be also versioned by quality. Interestingly, in order to create different qualities, a provider might deliberately degrade a product. He might add extra software to disable some features, or add delays and information loss to a communications service that already works well. Note that the poorer quality version may actually be the more costly to produce. Another trick is to introduce various versions of the products at different times. Versioning allows for an approximation to personalized prices. A version of the good that is adequate for the needs of one customer group, can be priced at what that group will pay. Other customer groups may be discouraged from using this version by offering other versions, whose specific features and relative pricing make them more attractive. Communication services can be price discriminated by the time of day, duration, location, and distance.

In general, if there is a continuum of customer types with growing demand functions the solution to the revenue maximization problem is a nonlinear tariff $r(x)$. Such tariffs can be smooth functions with $r(0) = 0$, where the marginal price $p = r'(x)$ depends on the amount x that the customer purchases. In many cases, $r(x)$ is a concave function and satisfies the property that the greatest quantity sold in the market has a marginal price equal to marginal cost. Observe that this holds in the two customer example above. The largest customer consumes at a level at which his marginal utility is equal to marginal cost, which is zero in this case.

The idea of *third degree price discrimination* is *market segmentation*. By market segment we mean a class of customers. Customers in the same class pay the same price, but customers is different classes are charged differently. This is perhaps the most common form of price discrimination. For example, students, senior citizens and business professionals have different price sensitivities when it comes to purchasing the latest version of a financial software package. The idea is not to scare away the students, who are highly price sensitive, by the high prices that one can charge to the business customers, who are price insensitive. Hence, one could use different prices for different customer groups (the market segments). Of course, the seller of the services must have a way to differentiate customers that belong to different groups (for example, by requiring sight of a student id card). This explains why third degree price discrimination is also called *group pricing*.

Suppose that customers in class i have a demand function of $x_i(p)$ for some service. The monopolist seeks to maximize

$$\max_{\{x_i(\cdot)\}} \sum_{i=1}^{n} p_i x_i(p_i) - c \left(\sum_{i=1}^{n} x_i(p_i) \right)$$

Assuming, for simplicity, that the market segments corresponding to the different classes are completely separated, the first order conditions are

$$p_i(x_i) + p_i'(x_i)x_i = c' \left(\sum_{i=1}^{n} x_i \right)$$

If ϵ_i is the demand elasticity in market i, then these conditions can be written as

$$p_i(x_i) \left(1 + \frac{1}{\epsilon_i} \right) = c' \left(\sum_{i=1}^{n} x_i \right) \tag{6.3}$$

These results are intuitive. The monopolist will charge the lowest price to the market segment that has the greatest demand elasticity. In Figure 6.5 there are two customers classes, with demand functions $x_1(p) = 6 - 3p$ and $x_2(p) = 2 - 2p$. The solution to (6.3) with the right hand side equal to $1/2$ is $p_1 = 5/4$ and $p_2 = 3/4$, with $x_1 = 9/2$ and $x_2 = 1/4$. At these points, $\epsilon_1 = -5/3$, $\epsilon_2 = -3$.

The market segment that is most price inelastic will be charged the highest price. Similar results hold when the markets are not independent and prices influence demand across markets.

A simple but clever way to implement group pricing is through discount coupons. The service is offered at a discount price to customers with coupons. It is time consuming to collect coupons. One class of customers is prepared to put in the time and another is not. The customers are effectively divided into two groups by their price elasticity. Those with a greater price elasticity will collect coupons and end up paying a lower price.

It is interesting to ask whether or not the overall economy benefits from third degree price discrimination. The answer is that it can go either way. Price discrimination can only take place if different consumers have unequal marginal utilities at their levels of consumption,

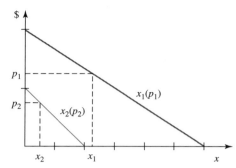

Figure 6.5   In third degree price discrimination customers in different classes are offered different prices. By (6.3) the monopolist maximizes his profits by charging more to customer classes with smaller demand elasticity, which in this example is customer class 1.

which is (generally) bad for welfare. But it can increase consumption, which is good for welfare. A necessary condition for there to be an increase in welfare is that there should be an increase in consumption. This happens in the example of Figure 6.5. There are two markets and one is much smaller than the other. If third degree price discrimination is not allowed, then the monopolist will charge a high price and this will discourage participation from the small market. However, if third degree price discrimination is allowed, he can set the same price for the high demand market, and set a low price for the low demand market, so that this market now participates. If his production cost is zero, the monopolist increases his surplus and users in the second market obtain a nonzero surplus; hence the overall surplus is increased.

6.2.3   Bundling

We say that there is *bundling* when a number of different products are offered as a single package and at a price that differs from the sum of the prices of the individual products. Bundling is a form of versioning.

Consider two products, A and B, for which two customers C_1 and C_2 have different willingness to pay. Suppose that C_1 is prepared to pay $100 and $150 for A and B, respectively, and C_2 is prepared to pay $150 and $100 for A and B, respectively. If no personalized pricing can be exercised, then the seller maximizes his revenue by setting prices of $100 for each of the products, resulting in a total revenue of $400. Suppose now that he offers a new product that consists of the bundle of products A and B for a price of $250. Now both customers will buy the bundle, making the revenue $500. Essentially, the bundle offers the second product at a smaller incremental price than its individual price. Note that $500 is also the maximum amount the seller could obtain by setting different prices for each customer, i.e. by perfect price discrimination.

It is interesting that bundling reduces the dispersion in customers' willingness to pay for the bundle of the goods. For each of the goods in our example, there is a dispersion of $50 in the customers' willingness to pay. This means that overall lower prices are needed to sell the goods to both customers. Now there is no dispersion in the customers' willingness to pay for the bundle. Both are willing to pay the same high price. This is the advantage of creating the new product. In general, optimal bundles are compositions of goods that reduce the dispersion in customers' willingness to pay.

Bundling is common in the service offerings of communication providers. For instance, it is usual for an ISP to charge its subscribers a monthly flat fee that includes an email account, the hosting of a web page, some amount of on-line time, permission to download some quantity of data, messaging services, and so on. If each service were priced individually, there would be substantial dispersion in the users' willingness to pay. By creating a bundle, the service provider decreases the dispersion in pay and can obtain a greater revenue.

6.2.4 Service Differentiation and Market Segmentation

We have discussed the notion of market segmentation, in which the monopolist is able to set different prices for his output in different markets. But can the monopolist always segment the market in this way? Sometimes there is nothing to stop customers of one market from buying in another market. At other times the monopolist can construct a barrier to prevent this. As we have said, he might sell discounted tickets to students, but require proof of student status. Let us investigate these issues further.

We have said that one way to create market segmentation is by *service differentiation* and versioning. This is accomplished by producing versions of a service that cannot fully substitute for one another. Each service is specialized for a targeted market segment. For example, think of a company that produces alcohol. The markets consist of customers that use alcohol as a pharmaceutical ingredient and customers that use it as fuel to light lamps. The manufacturer can segment the market by adding a chemical adulterant to the alcohol that prevents its use as a pharmaceutical. If this market is the least price-elastic, then he will be able to charge a greater price for the pharmaceutical alcohol than for the lamp fuel. Note that the marginal cost of producing the products is nearly the same. The lamp alcohol might actually be a bit more expensive, since it involves addition of the adulterant.

This type of price discrimination is popular in the communications market. The network operator posts a list of services and tariffs and customers are free to choose the service-tariff pair they like better. Versioning of communication services requires care and must take account of substitution effects such as arbitrage and traffic splitting. Arbitrage occurs when a customer can make a profit by buying a service of a certain type and then repackaging and reselling it as a different service at market prices. For instance, if the price of a 2 Mbps connection is less than twice the price of a 1 Mbps connection, then there may be a business opportunity for a customer to buy a number of 2 Mbps connections and become a supplier of 1 Mbps connections at lower prices. Unless there is a substantial cost in reselling bandwidth, such a pricing scheme has serious flaws since no one will ever wish directly to buy a 1 Mbps connection. A similar danger can arise from traffic splitting. This takes place when a user splits a service into smaller services, and pays less this way than if he had bought the smaller services at market prices. In our simple example, the price of a popular 2 Mbps service could be much higher than twice the price of 1 Mbps services. In general, there is cost to first splitting and then later reconstructing the initial traffic. One must take these issues into account when constructing prices for service contracts. Finally, we remark upon the role of content in price discrimination. Usually, it is practically impossible to make prices depend on the particular content that a network connection carries, for instance, to differently price the transport of financial data and leisure content. The network operator is usually not allowed to read the information that his customers send. In any case, data can be encrypted at the application layer. This means that it is usually not possible to price discriminate on the basis of content.

In general, service contracts are characterized by more parameters than just the peak rate, such as the mean rate and burstiness. This weakens the substitution effects since it is not always clear how to combine or split contracts with arbitrary parameters. But the most effective way to prevent substitution is by quality of service differentiation. Consider a simple example. A supplier might offer two services, one with small delay and losses, and one with greater delay. This will divide the market into two segments. One segment consists of users who need high quality video and multimedia services. The other consists of users who need only e-mail and web browsing. Depending on the difference in the two market's demand elasticities, the prices that the supplier can charge per unit of bandwidth can differ by orders of magnitude, even though the marginal costs of production might be nearly the same (proportional to the effective bandwidth of the services, see Section 4.6). Even if the supplier can provide the lower quality services at a quality that is not too different from the high quality ones, it can be to his benefit artificially to degrade the lower quality service, in order to maintain a segmentation of the market and retain the revenue of customers in the first market, who might otherwise be content with the cheaper service.

How about the consumer? Can he benefit from service differentiation, or is it only a means for a profit-seeking producer to increase his profit? The answer is that it depends. A good rule of thumb is to look at the change in the quantity of services consumed. If the introduction of new versions of a service stimulates demand and creates new markets, then both consumer surplus and producer surplus are probably increased. The existence of more versions of service helps consumers express their true needs and preferences, and increases their net benefits. However, the cost of differentiating services must be offset against this.

In networks, service differentiation is often achieved by giving some customers priority, or reserving resources for them. What makes the problem hard is that the service provider cannot completely define the versions of the services *a priori*, since quality factors may depend on the numbers of customers who end up subscribing for the services. In the next example we illustrate some of these issues.

Example 6.1 (Loss model with service differentiation) Consider the following model, which we will meet again in Section 9.4.1. Suppose the users of a transmission channel are divided in two classes, each of size n. Each user produces a stream of packets, as a Poisson processes of rate λ. Time is divided into unit length slots, so that in any given slot the number of packets that a user produces is distributed as a Poisson random variable with mean λ. At most $3C$ packets can be served per slot by the channel and excess packets are lost. Let $C = nc$, for some given c. Users of the two classes have different costs for lost packets, of a_1 and a_2 per packet respectively, where $a_1 > a_2$. Let X_1 and X_2 be Poisson random variables of mean $n\lambda$. If users share the channel and are treated on an equal basis then the expected cost per lost packet is $(a_1 + a_2)/2$ and so the expected cost per slot is

$$(1/2)(a_1 + a_2)E[X_1 + X_2 - 3C]^+$$

where $[x]^+$ denotes $\max\{x, 0\}$.

Suppose a greater part of the channel is reserved for the high-cost users. If channels of sizes $2C$ and C are reserved for the users of class 1 and 2 respectively, the cost is

$$a_1 E[X_1 - 2C]^+ + a_2 E[X_2 - C]^+$$

If n is large and a_1 is very large compared to a_2 then this scenario has smaller cost. This follows from the fact that (for large n): $2E[X_1 - 2C]^+ < E[X_1 + X_2 - 3C]^+$. (Proof of this fact is tedious and we omit it here.) Hence, creating two versions of the service and having each customer class use the appropriate service version increases social welfare. But how can we discourage low-cost customers from using the higher quality service? We assume there is no higher authority to dictate customers' choices; each customer simply chooses whichever service he wants.

A price can be used to provide the right incentives compatibility constraints. The network sets up the two channels and puts a price of p on the channel with capacity $2C$ (the other is free). This p is chosen such that

$$a_1 E[X_1 - 2C]^+ + n\lambda p < a_1 E[X_2 - C]^+$$

and

$$a_2 E[X_1 - 2C]^+ + n\lambda p > a_2 E[X_2 - C]^+$$

Assuming that n is large, these conditions ensure that it is a Nash equilibrium (as defined in Section 6.4.1) for all class 1 users to select the first channel and all class 2 users to select the second channel. (Note that at this operating point, the rate of cost for a single customer of class 1 who uses the first channel is $a_1 E[X_1 - 2C]^+/n + \lambda p$, and if he switches to the second channel his cost becomes $a_1 E[X_2' - C]^+/(n + 1)$, where X_2' is Poisson with mean $(n + 1)\lambda$. For large n this cost is very close to $a_1 E[X_2 - C]^+/n$.)

That is, there is no incentive for any class 1 user to use the second channel, or for a class 2 user to use the first channel. This can be compared with the similar Paris Metro pricing of Section 10.8.1.

Now consider a priority model. There is a single channel that can serve up to C packets per slot. By paying p per packet a user can ensure that his packets are not lost unless all packets that are not lost belong to users who are also paying p per packet. One can check that there is a p which makes it a Nash equilibrium for all class 1 users to pay for this priority treatment, while class 2 users do not. Social welfare is improved if

$$(1/2)(a_1 + a_2)E[X_1 + X_2 - C]^+$$
$$> a_1 E[X_1 - C]^+ + a_2 \left(E[X_1 + X_2 - C]^+ - E[X_1 - C]^+\right)$$

i.e. if

$$(1/2)(a_1 - a_2)\left(2E[X_1 - C]^+ - E[X_1 + X_2 - C]^+\right) < 0$$

This is again true as n gets large.

6.3 Perfect competition

We have discussed the case of a market that is in the control of a profit-seeking monopolist. The job of a regulator is to obtain for the market the benefits of the monopolist's low costs of production, but while maximizing social surplus. A regulator could impose prices in a market with the aim of maximizing social surplus, subject to suppliers being allowed to make certain profits. However, regulation can be costly and imperfect. It may also be difficult for a regulator to encourage a monopolist to innovate and to offer new services and products. Interestingly, the goals of the regulator can be achieved by increasing the competition in the market.

If there is no supplier or customer in the market who is so dominant that he can dictate prices, then social surplus can still be maximized, but by the effect of *perfect competition*. Every participant in the market is small. As a consequence he assumes that prices are determined by the market and cannot be influenced by any of his decisions, i.e. he is a price taker. Consumer i solves a problem of maximizing his net surplus, by demanding x_i, where $\partial u_i(x^i)/\partial x^i = p$. Supplier j solves a problem of maximizing his profit, by supplying y^j, where $\partial c_j(y^j)/\partial y^j = p$.

When equilibrium prices are reached, the aggregate demand, say $\sum_{i=1}^{n} x^i$, equals the aggregate output, say $\sum_{j=1}^{m} y^j$. If this were not so, then some supplier would not be able to sell all he produces and would have the incentive to find a customer for his surplus by reducing his price to just below the market price; or he could produce less, reducing his cost. The circumstance in which demand equals supply is called *market clearance*. At the prices at which the market clears, $\partial u_i / \partial x^i = \partial c_j / \partial y^j = p$; we recognize this as precisely the condition for maximization of the social surplus

$$S = \sum_{i=1}^{N} u_i(x^i) - \sum_{j=1}^{M} c_j(y^j)$$

subject to the constraint $\sum_{i=1}^{N} x^i \leq \sum_{j=1}^{M} y^j$. As we have seen in Section 5.4.1, the prices at which markets clear can be obtained by a tatonnement, i.e. an iterative price adjustment.

6.3.1   Competitive Markets

In a market with perfect competition, suppliers act as *competitive firms*. A competitive firm takes prices as given and decides whether to participate in the market at the given prices. Given a market price p, the firm computes the optimal level of output $y^* = \arg\max_y[py - c(y)]$ and participates by producing y^* if it makes a positive profit, i.e. if $\max_y[py - c(y)] > 0$.

Suppose the cost function has the form $c(y) = F + c_v(y)$. The participation condition becomes $py^* \geq F + c_v(y^*)$, and since $p = c_v'(y^*)$, the firm will participate if the optimal production y^* is such that

$$c_v'(y^*) \geq \frac{F + c_v(y^*)}{y^*}$$

i.e. if at y^* the marginal cost is at least as great as the average cost. This is shown in Figure 6.6. The minimum value of p for which such a condition is met is called the *minimum participation price* of the firm. Note that the participation price depends on whether $c(\cdot)$ denotes the firm's short-run or long-run cost function. In the long-run, a firm can reorganize its production processes optimally for a given production level and find it profitable to participate at a lower price than is profitable in the short-run.

How many firms will participate in a competitive market? Clearly, as more firms enter, more output is produced, and for this extra output to be consumed prices must decrease. This suggests that the number of firms will reach an equilibrium in which if one more firm were to participate in the market the price would drop below the minimum participation price of the firms. The effect of entry on prices is shown in Figure 6.7.

6.3.2   Lock-in

In practice, the perfect competition conditions may not be achieved because of *lock-in* effects. Lock-in occurs both because customers pay a *switching cost* to change providers, and because it is costly for providers to set up to serve new customers. Hence, even though an alternative provider may offer prices and quality more attractive than those of a customer's existing provider he may choose not to switch since he will not gain overall. The effect of lock-in is that prices will be higher than marginal cost, and so allow

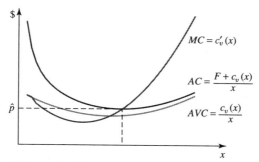

Figure 6.6 In a competitive market firms must take prices as given. Here MC, AC and AVC are respectively the marginal cost, average cost and average variable cost curves. Suppose that if firm participates in the market it has a fixed cost F, plus a variable cost of $c_v(x)$ for producing x. Given a price p, the firm computes its optimal production level, x, by maximizing its profit $px - c_v(x) - F$. This gives $MC = c'(x) = p$. The firm starts producing only if it can make a profit. This gives a participation condition of $px \geq F + c_v(x)$, or equivalently, $MC \geq AC$, or $p \geq \hat{p}$. If the fixed cost F is sunk, i.e., has already occurred, then the participation condition is $AVC \geq AC$.

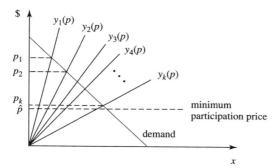

Figure 6.7 For a given demand, the equilibrium price decreases as more firms enter the market. Here, $y_i(p)$ is the total amount that will be supplied to the market when i identical firms compete and the offered price is p. When i firms are in the market the price p_i that prevails occurs at the intersection of $y_i(p)$ and the demand curve. This limits the number of firms that enter, since they do so only if the price is sufficiently high. Here, at most k firms will enter the market.

service providers to obtain positive profits from customers. Examples of switching costs in communications include the cost of changing a telephone number, an email account, or web site address; the costs of installing new software for managing network operation; the costs of setting up to provide access service.

In mass markets, such as telephony and Internet services, even small switching costs can be extremely significant. A provider can make significant profits from lock-in and network externalities, and so may seek to grow his network rapidly in order to obtain a large customer base. Since lock-in reduces the effects of competition and discourages new firms from entering the market, a regulator may seek to reduce its effects. Two examples of regulatory measures that do this are the requirements that telephone numbers be portable (i.e. that customers can keep their telephone number when they switch providers), and that customers may choose a long-distance service provider independently of their local access provider.

The effects of lock-in can be quantified by observing that, in addition to the cost of providing service, a producer can obtain from a customer extra revenue that is equal to his switching cost. We can see this with the following simple argument. Suppose that in equilibrium there are many service providers, each with his own customer base. Provider

i charges customers p_i per month of subscription and has variable monthly cost of c per customer. It costs a customer s to switch providers. Suppose that to entice customers to switch, provider i offers a one-time discount of d_i to a newly acquired customer. Let $100r\%$ be the monthly interest rate. Given a provider i, suppose j is the provider to whom it is best for customers of i to switch if they do switch. In equilibrium, no customer can benefit by switching from i to j, and i cannot increase his charge above p_i without losing customers to j. So we must have,

$$p_i + \frac{p_i}{r} = p_j - d_j + s + \frac{p_j}{r}$$

Also, j must be profitable if customers switch to him, but he cannot lower p_j (which would entice customers to switch to him from i) without becoming unprofitable. So

$$(p_j - c) - d_j + \frac{p_j - c}{r} = 0$$

These imply

$$(p_i - c) + \frac{p_i - c}{r} = s$$

for all i. This says that the that the present value of a customer equals his switching cost, and $p_i = c + rs/(1+r)$ for all i.

One can easily generalize this simple result to the case where cost and quality also differ. Suppose that q_i is the value obtained by a customer using the service in network i (which differs with i because of service quality), and c_i is the cost of service provisioning in this particular network. Now, the discount d_{ij} offered by network j may depend on the network i to which the customer initially belongs. Simple calculations along the previous lines show that

$$p_i - c_i + \frac{p_i - c_i}{r} = s + \left(q_i - q_j + \frac{q_i - q_j}{r} \right) - \left(c_i - c_j + \frac{c_i - c_j}{r} \right)$$

The second term on the right-hand side is the present value of the quality difference of the services provided by networks i and j, and the third term is the present value of the difference in their operating costs. Observe that if the quality difference equals the cost difference, then networks i and j makes the same net profit per customer.

6.4 Oligopoly

In practice, markets are often only partly regulated and partly competitive. A competitive market of a small number of suppliers is called an *oligopoly*. The theory of games is widely used as a tool to study and quantify interactions between a small number of competing firms. In this section we describe a few simple models. The theory of oligopoly involves ideas of equilibria, cartels, punishment strategies and limit pricing. An important methodology for oligopolies is auctions, a subject we take up in Chapter 14.

6.4.1 Games

The reader should not be surprised when we say that many of the ideas and models in this book can be viewed as games. The players of these games are network service suppliers, customers and regulators. Suppliers compete with suppliers for customers. Customers compete with suppliers to obtain services at the best prices. Regulators compete with suppliers over the division of social surplus.

The simplest sort of game has just two players. Each player chooses a strategy, i.e. a rule for taking the action(s) that are available to him in the game. As a function of the players' strategy choices, each player obtains a *pay-off*, i.e. a reward (which may be positive or negative). In a *zero-sum game* one player's reward is the other player's loss. The well-known scissors-stone-paper game is an example of a zero-sum game. In this game each player chooses one of three *pure strategies*: scissors, stone or paper. Scissors beats paper, which beats stone, which beats scissors. If they bet $1 the winner gains $1 and the loser loses $1.

It is well known that a player's expected reward in the scissors-stone-paper game is maximized by using a *randomized strategy*, in which with probabilities of 1/3 he randomly chooses each of the three possible *pure strategies*: scissors, stone or paper. Denote this strategy as σ. Because the situations of the players are symmetric and the game is zero-sum, the expected reward of each player is zero when each uses the optimal strategy σ.

Many real life games have more than two players and are not zero-sum. In markets, both suppliers and customers obtain a positive reward, otherwise the market could not exist. Thus, a more general type of game is one in which many players choose strategies and then rewards are allocated as a function of these strategy choices. The sum of these rewards may be positive or negative, and different for different strategy choices. For two players, the theory of such games is simple. Assuming that players may randomize over their pure strategies, with arbitrary probabilities, then there always exists a unique pair of strategy choices (possibly of randomized strategies), such that neither player can do better if he deviates from his strategy. This is the idea of a *Nash equilibrium*, which extends to games with more than two players, (although with more than two players an equilibrium may not exist). Formally, $(\sigma_1, \ldots, \sigma_n)$ is a Nash equilibrium of a n-player game, if player i cannot do better by deviating from strategy σ_i so long as player j uses strategy σ_j, for all $j \neq i$. For example, in the scissors-stone-paper game, (σ, σ) is the Nash equilibrium. If one player adopts the strategy σ, then the other player has an expected reward of 0 under all possible pure strategies, and so there is no incentive for him to do other than also use the strategy σ.

An example of a game that is not zero-sum is the *prisoners' dilemma* game. In this game two burglars, who have together committed a robbery, have been arrested and imprisoned by the police, and each can choose whether or not to betray the other when interviewed. Each of the two prisoners i, j has available two pure strategies: 'cooperate' and 'defect'. A prisoner whose strategy to cooperate with the other prisoner refuses to give evidence during the interrogation. However, if his strategy is to defect, he helps the police incriminate the other prisoner and is rewarded by being granted some better treatment. Each prisoner must choose his strategy prior to the interrogation.

The *game matrix* describing the possible outcomes shown in Table 6.1. An element (a, b) indicates that prisoner 1 obtains a benefit of a whereas prisoner 2 obtains b. Prisoner 1 chooses the actions indexing the rows while prisoner 2 chooses the actions indexing the columns of the matrix. For instance if prisoner 1 chooses to cooperate while prisoner 2 defects, they obtain 0 and 3 units of benefit respectively. Observe that the strategy 'defect-defect' is the only Nash equilibrium. Although the joint strategy 'cooperate-cooperate' generates a higher benefit to both, it is not a Nash equilibrium, since if Prisoner 1 knows that Prisoner 2 will cooperate then he will do better by defecting. In fact, this game is

Table 6.1 The game matrix of the
prisoners' dilemma game. The only
Nash equilibrium is for both prisoners to
defect

$i \setminus j$	cooperate	defect
cooperate	2,2	0,3
defect	3,0	1,1

dominance solvable, i.e. each player has a pure strategy, namely 'defect', that is better for him than all his other pure strategies, regardless of what pure strategy is chosen by his opponent.

An interesting case of the prisoners' dilemma occurs in the case of a *public goods*. Such goods have the property that one customer's consumption does not reduce the amount available to the other customers. For instance consider the case of a radio or TV broadcast channel of capacity B. In this case, each customer consumes B individually without reducing the amount of capacity available to the other customers. Similar situations occur in the case of street lights, freeways and bridges and information that can be duplicated at zero cost. In fact, a customer benefits from the presence of other customers since they can share with him the cost of providing the public good. This produces a problem for the underlying game, in that a customer can reason that he need not contribute to the common cost of the good if others will pay for it anyway. This is known as the *free rider problem*, i.e. a customer expects other customers to pay for a good from which he also derives benefit. If all customers reason like this, the public good may never be provided, which is clearly a socially undesirable outcome. Such a situation is frequently encountered in communications when multicasting is involved, and hence deserves a more detailed discussion.

Suppose that two customers have utility functions of the form $u_i(B) + w_i$, where B is the total amount of the public good purchased (the bandwidth of the broadcast link), and w_i is the money available in the bank, $i = 1, 2$. Customer i pays for b_i units of the public good, hence $B = b_1 + b_2$. Assume that each starts with some initial budget w_i^0 and that the common good cost 1 per unit. Then, the optimal strategy of player i assuming that customer j will purchase a b_j amount of public good is

$$\max_{b_i} u_i(b_i + b_j) + w_i^0 - b_i$$

One can show by doing the complete analysis of this game that if one of the customers, say customer i, has a consistently higher marginal utility for the public good (i.e. $u_i'(C) > u_j'(C)$ for all $C \geq 0$), then the equilibrium strategy is for customer j to get a free ride from customer i. Customer i pays for the public good and customer j simply benefits without contributing. As a result, the public good will be available in a lesser quantity than the socially optimal one.

We can also construct a simple game in which the equilibrium is for the public good not to be purchased at all. Suppose the good is available in only two discrete quantities, b and $2b$, and each customer pays for either b or zero. For simplicity assume that the two customers are identical, and that their utility function satisfies $\phi_0 := u(b) + w^0 - b < \phi_1 := u(0) + w^0 < \phi_2 := u(2b) + w^0 - b < \phi_3 := u(b) + w^0$. The first inequality states that it is uneconomical for a single customer to provide b of the public good if the other customer provides 0. The last inequality motivates free riding as we will see. We can easily

Table 6.2 A game of purchasing a public good.
Here $\phi_0 < \phi_1 < \phi_2 < \phi_3$. Due to free-riding, the
equilibrium strategy is not to purchase the public
good

$i \setminus j$	contribute b	contribute 0
contribute b	ϕ_2, ϕ_2	ϕ_0, ϕ_3
contribute 0	ϕ_3, ϕ_0	ϕ_1, ϕ_1

write this as a prisoners' dilemma game, see Table 6.2, in which 'cooperate' and 'defect'
correspond to contributing a b or 0 respectively of the public One can easily check that
again 'defect-defect' is the equilibrium strategy, and so the public good is not purchased
at all.

There are many other examples of the prisoners' dilemma in real life. A multi-player
version arises when service providers compete over the price of a service that they all
provide. By forming a *cartel* they might all set a high price. But if they cannot bind
one another to the cartel then none can risk setting a high price, for fear another firm
will undercut it. However, if the game is a *repeated game*, rather than a *one-shot game*,
a cartel can be self-sustaining. The game is to be repeated many times and each player
tries to maximize his average reward over many identical games. In the cartel game it is
a Nash equilibrium strategy for all firms to adopt the strategy: 'set the high price until a
competitor sets the low price, then subsequently set the low price'. No firm can increase
its time-average reward by deviating from this strategy. Thus the cartel can persist and it
may require a regulator to break it.

In subsequent chapters we discuss related notions of cost-sharing and bargaining games
(Chapter 7), the principal-agent model in interconnection and regulation (Chapters 12
and 13), and auctions (Chapter 14). In these more general games there may be many
Nash equilibrium, or there may be none. If there are many, then it can be helpful to
introduce additional concepts to choose the equilibrium at which the game is actually
'solved'.

6.4.2 Cournot, Bertrand and Stackelberg Games

We next turn to games that model competition amongst a small number of service providers.
There are two cases to consider. In the so-called *Cournot model*: each supplier announces
as his strategic choice the quantities of services that he intends to supply. Prices adjust in
response to the aggregate supply, so that all the production can be sold, and each supplier
obtains a proportionate amount of the consumers' outlay.

In the second case, of the so-called *Bertrand model*, each supplier announces the prices
he intends to charge, and then customers buy services with preference for lower prices.
Both models are games that are played in a single round. Each player must decide what
to do without knowing what other players will do. This game has a simple solution in
the case of two suppliers with different marginal costs $c_1 < c_2$, which are known to both
suppliers. There is a continuum of Nash equilibria with $p_2 \in (c_1, c_2]$ and $p_1 = p_2 - \epsilon$,
for infinitesimally small ϵ. To see this, note that given that Player 1 chooses p_1, with
$p_1 > c_1$, Player 2 cannot undercut Player 1 on price without incurring a loss, so he has no
incentive to deviate from p_2. Given Player 2 chooses p_2, Player 1 maximizes his profit by
taking p_1 just less than p_2. Hence supplier 1 will always win, with a net benefit equal to

approximately $p_2 - c_1$ per unit sold. To choose amongst these equilibria we note that no player will wish to offer a price that is less than his marginal cost. For Player 2, $p_2 = c_2$ dominates the strategy $p_2 = p_2'$ for all $p_2' < c_2$, i.e., the first strategy is as good or better than the second, for all values of p_1. Thus, by imposing the constraint $p_2 \geq c_2$, we conclude that $(p_1, p_2) = (c_2 - \epsilon, c_2)$ is the equilibrium solution of the game.

We can also analyse the Cournot model. To illustrate some important properties of the resulting prices, we examine the simplest case of two competing firms who produce the same product. If firm i produces output at level x_i, then the total level of production is $x = x_1 + x_2$ and the resulting price in the market will be $p(x)$.

Modelling this as a one-shot game, each firm must choose an amount of output to be produced, and then, as a function of both choices, receive a pay-off (that is his net benefit). Clearly, the net benefit of firm i can be written

$$\pi_1(x_1, x_2) = p(x_1 + x_2)x_i - c_i(x_i)$$

where $c(x_i)$ is his cost for producing quantity x_i. A Nash equilibrium in this game is a pair of outputs x_1^*, x_2^* with the property that if firm i chooses x_i^* then there is no incentive for firm j to choose other than x_j^*, where $i, j \in \{1, 2\}, i \neq j$. This implies the first order conditions

$$\partial \pi_1(x_1, x_2)/\partial x_1 = p(x_1 + x_2) + p'(x_1 + x_2)x_1 - c_1'(x_1) = 0$$
$$\partial \pi_2(x_1, x_2)/\partial x_2 = p(x_1 + x_2) + p'(x_1 + x_2)x_2 - c_2'(x_2) = 0$$

These conditions define for each firm i its *reaction curve* $x_i(x_j)$, that is, its optimal choice of output as a function of its belief about the other firm's output x_j. The Nash equilibrium is the intersection of these curves. For example, suppose x_i is to be chosen within the interval $[0, 1]$, the inverse demand curve is $p(x_1 + x_2) = 1 - (x_1 + x_2)$, and $c_i(x_i) = 0$. Then $x_i(x_j) = \frac{1}{2}(1 - x_j)$ and the Nash equilibrium is at $(x_1, x_2) = (\frac{1}{3}, \frac{1}{3})$. One can show that under reasonable assumptions on the demand curve (such as concavity), the above equilibrium is always stable. That is, if the game is played in many rounds and players alternate in choosing their output based on the previous output of the other player, then their outputs will converge to the Nash equilibrium point. This is illustrated in Figure 6.8.

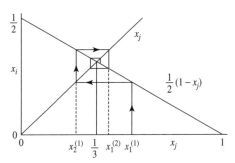

Figure 6.8 If the Cournot game is played in many rounds and players alternate in choosing their output based on the previous output of the other player, then their outputs will converge to the Nash equilibrium point. Here the inverse demand curve is $p(x) = 1 - x$ and both players have zero costs of production. Given that Player 1 produces $x_1^{(1)}$, Player 2 will produce $x_2^{(1)}$. Given that Player 2 produces $x_2^{(1)}$, Player 1 will produce $x_1^{(2)}$, and so on, with $x_1^{(n)}, x_2^{(n)} \to \frac{1}{3}$.

In the case of n competing firms, the first order conditions at equilibrium can be rewritten as

$$p(x)(1 + x_i/x\epsilon) = c_i'(x_i) \qquad (6.4)$$

where ϵ is the price elasticity. If all firms have same cost function the solution will be symmetric and

$$p(x)(1 + 1/n\epsilon) = c'(x/n) \qquad (6.5)$$

In this case the price is proportional to marginal cost. The markup depends upon the demand elasticity and converges to zero as the number of competing firms increases to infinity. This is consistent with what we know about the economic efficiency of perfect competition. By comparison, in the Bertrand model, with equal and constant marginal costs, the competitive equilibrium is independent of the number of competing firms and at a price equal to marginal cost.

In *duopoly* (i.e. a market with two firms), the *Stackelberg model* is interesting. This game is played in two steps. In the first step one player makes a move, and in the second step the other player moves, taking account of the first player's move. The game can be played with either *price leadership* or *quantity leadership*. Suppose firm 1 is the leader, who commits to price p. As above, suppose $x_i \in [0, 1]$, $x(p) = 1 - p$, but let $c_i(x_i) = x_i^2$. Firm 2 is the follower. He will take the leader's price as given, undercutting it by an infinitesimal amount and choosing his output level, x_2, to maximize $px_2 - x_2^2$, giving $x_2 = p/2$. Firm 1 sees residual demand of $1 - 3p/2$. One can check that he maximizes his profit by taking $p = 8/15$ and $x_1 = 3/15$, whence $x_2 = 4/15$. In this example, the follower does better than the leader.

In the game of quantity leadership firm 1 commits to supply a quantity x_1. Firm 2 observes this and then chooses to supply x_2. We continue with the model above, but take $c_i(x_i) = 0$. One can check that for any $x_1' \in [0, 1]$, a Nash equilibrium is

$$x_1 = x_1', \qquad x_2 = \begin{cases} \frac{1}{2}(1 - x_1'), & \text{if } x_1 = x_1' \\ 1 - x_1', & \text{if } x_1 \neq x_1' \end{cases}$$

Firm 2 threatens firm 1 with the threat: 'choose $x_1 = x_i'$ or I will flood the market and spoil it for us both'. Thus there is a continuum of Nash equilibria. However, not all threats are actually credible. For instance, for $x_1' = 1$, firm 2 is threatening firm 1 with the threat: 'if you don't flood the market, I will'. It is hard to see why this threat would be made. It would certainly not be carried out, since if firm 1 does not choose $x_1 = 1$, firm 2 has a better response than $x_2 = 1 - x_1$. To find a solution amongst the continuum of Nash equilibria we rule out incredible threats. To do this, we consider the subgame that is presented to firm 2 once firm 1 has chosen his output level x_1. The best response of firm 2 is to choose x_2 to maximize $x_2(1 - x_1 - x_2)$, i.e., $x_2(x_1) = \frac{1}{2}(1 - x_1)$. Thus firm 1 should choose x_1 to maximize $x_1(1 - x_1 - x_2(x_1))$, which gives $x_1 = 1/2$. Thus the strategy pair $(x_1, x_2) = (\frac{1}{2}, \frac{1}{4})$ is a Nash equilibrium for both the whole game, and the subgame that is presented to firm 2 once firm 1 has chosen his output level. In general, we say that a set of strategies is a *subgame perfect equilibrium* if it is a Nash equilibrium of the whole game and every subgame. In the above game, firm 1 does better than he does in the Cournot game and firm 2 does worse. Also, the leader does better than the follower. Depending on the particular circumstances, making the first move may or may not give an advantage.

6.5   A unifying social surplus formulation

Consider the problem of maximizing the social welfare, subject to the constraint that the supplier's profit is at least π. This can be formulated as the problem of maximizing a Lagrangian

$$L(x, \lambda) = [u(x) - c(x)] - \lambda \left[\pi + c(x) - \sum_j x_j p_j(x) \right]$$

where $u(x)$ is the maximum utility (summing over all customers) obtained by consuming in total x, and $c(x)$ is the minimum cost (shared by the producers) of producing x. Suppose both $u(x) - c(x)$ and $\sum_j x_j p_j(x) - c(x)$ are concave in x. Then there exists for each π, some nonnegative value of λ such that the solution to the constrained problem occurs where L is maximized. By taking $\gamma = \lambda/(1 + \lambda)$, this is at the same x that solves

$$\underset{x}{\text{maximize}} \left\{ \left[\sum_j x_j p_j(x) - c(x) \right] + (1 - \gamma) \left[u(x) - \sum_j x_j p_j(x) \right] \right\} \tag{6.6}$$

where $0 \leq \gamma \leq 1$. If $\gamma = 0$ we have the problem of maximizing social welfare. If $\gamma = 1$ we have the problem of maximizing producer profit. Thus γ can be interpreted as a measure of the supplier's market power. As in Section 5.5.1 we find that the maximum is where

$$\sum_j \frac{p_j - \frac{\partial c}{\partial x_j}}{p_j} \epsilon_{ij} = -\gamma \tag{6.7}$$

For independent goods (when cross-elasticities are zero), this equation is the same as that for Ramsey prices in (5.14),

$$\frac{p_i - \frac{\partial c}{\partial x_i}}{p_i} = -\frac{\gamma}{\epsilon_i} \tag{6.8}$$

Observe that (6.5) can be also obtained from (6.8) by setting $\gamma = 1/n$. This suggests that the form of prices resulting from oligopoly games can sometimes be obtained by some other formulation in which social welfare is maximized subjected to constraints on supplier profits. This shows the broad applicability of the form of prices in (6.7) and (6.8).

6.6   Further reading

A good text for much of the microeconomics presented in this chapter is Varian (1992). Game theory is a very rich subject and we have only touched upon some very basic ideas. More about game theory and models of competition can be found in Varian (1992) and Binmore (1992). The books by Karlin (1959) and Luce and Raiffa (1957) make good introductory reading. Eatwell *et al.* (1989) contains many interesting articles. More introductory material can be found in the course lecture notes of Weber (1998) and Weber (2001). Osborne and Rubenstein (1994) can be consulted for some advanced material. Slade (1994) considers the question at the end of Section 6.5: 'Do firms pursuing selfish objectives and behaving strategically act as if an agent were maximizing a fictitious-objective function?'

Part C

Pricing

7

Cost-based Pricing

This chapter is about prices that are directly related to cost. We begin with the problem of finding cost-based prices that are fair or stable under potential competition (Sections 7.1 –7.2). We look for types of prices that can protect an incumbent against entry by potential competitors, or against bypass by customers who might find it cheaper to supply themselves. We explain the notions of subsidy-free and sustainable prices. Such prices are robust against bypass. Similar notions are addressed by the idea of the second-best core. The aim now differs from that of maximizing economic efficiency. We see that Ramsey prices, which are efficient subject to the constraint of cost recovery, may fail sustainability tests.

In Section 7.3 we take a different approach and look at practical issues of constructing cost-based prices. Now we emphasize necessary and simplicity. Prices are to be computed from quantities that can be easily measured and for which accounting data is readily available. An approach that has found much favour with regulators is that of Fully Distributed Cost pricing (FDC). This is a top-down approach, in which costs are attributed to services using the firm's existing cost accounting records. It ignores economic efficiency, but has the great advantage of simplicity.

Section 7.3.5 concerns the Long-Run Incremental Cost approach (LRIC). This is a bottom-up approach, in which the costs of the services are computed using an optimized model for the network and the service production technologies. It can come close to implementing subsidy-free prices. We compare FDC and LRIC in Section 7.4, from the viewpoint of the regulator, who wishes to balance the aims of encouraging efficiency and competition, and of the monopolist who would like to set sustainable prices. The regulator may prefer the accounting-based approach of FDC pricing because it is 'automatic' and auditable. However, it may obscure old and inefficient production technology or the fact that the network has been wrongly dimensioned. These problems can be remedied by the LRIC approach, but it is more costly to implement.

Flat rate pricing is the subject of Section 7.5. In this type of pricing a customer's charge does not depend on the actual quantity of services he consumes. Rather, he is charged the average cost of other customers in the same customer group. We discuss the incentives that such a scheme provides and their effects on the market.

7.1 Foundations of cost-based pricing

In Chapters 5 and 6 we considered the problem of pricing in a context in which social welfare maximization is the overall aim. We posed optimization problems with unique

---

Pricing Communication Networks   C. Courcoubetis and R. Weber

solutions, each achieved by unique sets of prices. However, welfare maximization is not the only thing that matters. A firm's prices must ensure that it is profitable, or at least that it covers its costs. Cost-based pricing focuses on this consideration. Unfortunately, a fundamental difficulty in defining cost-based prices is that services are usually produced jointly. A large part of the total cost is a common cost, which can be difficult to apportion rationally amongst the different services. One can think of several ways to do it. So although cost-based prices may reasonably be expected to satisfy certain necessary conditions, they differ from welfare-maximizing prices in that they are usually not unique.

One necessary condition that cost-based prices ought reasonably to satisfy is that of fairness. Some customers should not find themselves subsidizing the cost of providing services to other customers. If so, these customers are likely to take their business elsewhere. This motivates the idea of *subsidy-free prices*. A second reasonable necessary condition is that prices should be defensive against competition, discouraging the entry of competitors who by posting lower prices could capture market share. This motivates the idea of *sustainable prices*. If prices do not reflect actual costs or they hide costs of inefficient production then they invite competition from other firms. Since customers will choose the provider from whom they believe they get the best deal, a game takes place amongst providers, as they seek to offer better deals to customers by deploying different cost functions and operating at different production levels. Prices must be subsidy-free and sustainable if they are to be *stable prices*, that is, if they are to survive the competition in this game.

Interestingly, the set of necessary conditions that we might like to impose on prices can be mutually incompatible. They can also be in conflict with the aim of maximizing social welfare maximization, since they restrict the feasible set of operating points, sometimes reducing it to a single point.

7.1.1 Fair Charges

Consider the problem of a single provider who wishes to price his services so that they cover their production cost and are fair in the sense that no customer feels he is subsidizing others. Unfair prices leave him susceptible to competition from another provider, who has the same costs, but charges fairly. Customers might even become producers of their own services.

Let $N = \{1, 2, \ldots, n\}$ denote a set of n customers, each of whom wishes to buy some services. For T that is a subset N, and let $c(T)$ denote the minimal cost that could by incurred by a facility that is optimized to provide precisely the services desired by the set of customers T. We call this the *stand-alone cost* of providing services to the customers in T. Assume that because of economies of scale and scope this cost function is *subadditive*. That is, for all disjoint sets T and U,

$$c(T \cup U) \leq c(T) + c(U) \tag{7.1}$$

In the terminology of cooperative games, $c(\cdot)$ is called a *characteristic function*.

The service provider wants to share the total cost of providing the services amongst the customers in a manner that they think is fair. Suppose he charges them amounts $c_1, \ldots, c_n$. Let us further suppose that he exactly covers his cost, and so $\sum_{i \in N} c_i = c(N)$. The charges are said to *subsidy free* if they satisfy the following two tests:

- The charge made to any subset of customers is no more than the stand-alone cost of providing services to those customers,

$$\sum_{i \in T} c_i \leq c(T), \quad \text{for all } T \subseteq N \tag{7.2}$$

- The charge made to any subset of customers is at least the incremental cost of providing services to those customers,

$$\sum_{i \in T} c_i \geq c(N) - c(N \setminus T), \quad \text{for all } T \subseteq N \tag{7.3}$$

The reason these conditions are interesting is that if either (7.2) or (7.3) is violated, then a new entrant can attract dissatisfied customers. If (7.2) is violated, then a firm producing only services for T and charging only $c(T)$ could lure away these customers. Similarly, if (7.3) is violated, then a firm producing only the services needed by $N \setminus T$ could charge less for these services than the incumbent firm. This happens because the incumbent uses part of the revenue obtained from selling services to $N \setminus T$ to pay for some of the cost of the services wanted by T. Next, we investigate certain variations and refinements of the above concepts.

7.1.2 Subsidy-free, Support and Sustainable Prices

Let reformulate the ideas of the previous section to circumstances in which charges are computed from prices. Suppose that a set of n services is $N = \{1, \ldots, n\}$ and an incumbent firm sells service i in quantity x_i, at price p_i, for a total charge of $p_i x_i$. Suppose that x_i is given and does not depend on $p = (p_1, \ldots, p_n)$. We call p a *subsidy-free price* if it satisfies the two tests

$$\sum_{i \in T} p_i x_i \leq c(T), \quad \text{for all } T \subseteq N \tag{7.4}$$

$$\sum_{i \in T} p_i x_i \geq c(N) - c(N \setminus T), \quad \text{for all } T \subseteq N \tag{7.5}$$

Inequalities (7.4) and (7.5) are respectively the *stand-alone test* and *incremental-cost test*. They have natural interpretation similar to (7.2) and (7.3). For instance, if (7.4) is violated then a new firm could set up to produce only the services in T and sell these at lower prices than the incumbent. Note that, by putting $T = N$, these tests imply $\sum_i p_i x_i = c(N)$. Thus the producer must operate with zero profit. Also, prices must be above marginal cost; to see this, consider the set $T = \{i\}$, imagine that x_i is small and apply the incremental cost test.

Example 7.1 (Subsidy-free prices may not exist) Consider a network offering voice and video services. The cost of the basic infrastructure that is common to both services is 10 units, while the incremental cost of supplying 100 units of video service is 2 units and the incremental cost of supplying 1000 units of voice is 1 unit. To be subsidy-free, the revenues $r_1(100)$ and $r_2(1000)$ that are obtained from the video and the voice services must satisfy

$$2 \leq r_1(100) \leq 12, \quad 1 \leq r_2(1000) \leq 11, \quad r_1(100) + r_2(1000) = 13$$

Thus, assuming that there is enough demand for services, possible prices are 0.006 units per voice service and 0.07 units per video service. Note that such prices are not unique and they may not even exist for general cost functions. Suppose three services are produced in unit quantities with a symmetric cost function that satisfies (7.1). Let $c(\{i\}) = 2.5$, $c(\{i, j\}) = 3.5$, and $c(\{i, j, k\}) = 5.5$, where i, j, k are distinct members of $\{1, 2, 3\}$. Then we must have $2 \leq p_i \leq 2.5$, for $i = 1, 2, 3$, but also $p_1 + p_2 + p_3 = 5.5$. So there are no subsidy-free prices. The problem is that economies of scope are not increasing, i.e. $c(\{i, j, k\}) - c(\{i, j\}) > c(\{i, j\}) - c(\{i\})$.

How can one determine if (7.4) and (7.5) are met in practice? Assume that a firm posts its prices and makes available its cost accounting records for the services. It may be possible to check (7.5) by computing and then summing the incremental costs of each service in T (though this only approximates the incremental cost of T because we neglect common cost that is directly attributable to services in T). Condition (7.4) is hard to check, as it imagines building from scratch a new facility that is specialized to produce the services in the set T. This cost cannot in general be derived from the cost accounting information of the firm which produces the larger set of services N. In practice, one tries to approximate $c(T)$, as well as possible given the available information.

There is another possible problem with the above tests. Although individual outputs may pass the incremental cost test, combinations of outputs may not. For example, suppose $N = \{1, 2, 3\}$. It is possible that the incremental cost test can be satisfied for every single good, i.e. for $T = \{i\}$, for all i, but not for $T = \{2, 3\}$. This could happen if there is a fixed common cost associated with services 2 and 3, in addition to their individual incremental costs, and each such service is priced at its incremental cost. Thus, the tests can be difficult to verify in practice.

In defining subsidy-free prices we assumed that services are sold in large known quantities (the x_is in (7.4) and (7.5)) using uniform prices, as happens when incumbent communications firms supply the market. In practice, individual customers consume small parts of each x_i and a coalition of customers may feel that it can 'self-produce' its service requirements at lower cost. In this case, it is reasonable to require (7.2) and (7.3). Clearly, such a 'consumer subsidy-free' price condition imposes restrictions on the cost function. For instance, imagine a single service has a cost function with increasing average cost. Selling the service at its average cost price violates (7.3) if individual customers request less than the total that is produced, although (7.4) and (7.5) are trivially satisfied for $N = \{1\}$. An appropriate definition is the following. Let us now write $c(x)$ as the cost of providing services in quantities $(x_1, \ldots, x_n)$. We say the vector p is a *support price* for c at x if it satisfies the two conditions

$$\sum_{i \in N} p_i y_i \leq c(y), \quad \text{for all } y \leq x \tag{7.6}$$

$$\sum_{i \in N} p_i z_i \geq c(x) - c(x - z), \quad \text{for all } z \leq x \tag{7.7}$$

Note these imply $\sum_{i \in N} p_i x_i = c(x)$. We can compare them to (7.2) and (7.3). For example, (7.6) implies that one cannot produce some of the demand for less than it is sold. They imply (7.4) and (7.5) (but are more general since they deal with arbitrary sub-quantities of the vector x, instead of looking just at subsets of service types), and hence a support price has all the nice fairness properties mentioned above. A last concern is whether such prices are achievable in the market, where demand is a function of price. Suppose p is the vector of support prices for x and, moreover, x is precisely the quantity vector that is demanded at price p. We call such prices *anonymously equitable prices*. Clearly, if they exist, these have a very good theoretical claim for being an intelligent choice of cost-based prices.

If prices affect demand

By allowing demand to depend upon price, we introduce subtle complications. Customers may feel badly treated even if the incremental cost test in (7.7) is passed. For example, if two services are substitutes then introducing one of them as a new service can reduce the demand for the other and the revenue it produces. Prices may have to increase if we are still to cover costs and this could mean that the price of the pre-existing service has to increase. This runs counter to what we expect: that adding a new service should allow prices of pre-existing services to decrease because of economies of scope in facility and equipment sharing. If the prices of pre-existing services increase then customers of these services will feel that they are subsidizing the cost of the new service.

To see this, let T be a subset of N, and define $p'_i = \infty$, $i \in T$, and $p'_i = p_i$, $i \notin T$. Thus, under price vector p' we do not sell any of the services in T (because their prices are infinite). If services in T are substitutes for those in $N \setminus T$, then we can have, (recalling $p'_i = p_i$ for $i \in N \setminus T$),

$$\sum_{i \in N \setminus T} p'_i x_i(p') > \sum_{i \in N \setminus T} p_i x_i(p)$$

i.e. when p' is replaced by p, the introduction of services in T reduces the demand for (and revenue earned from) services in $N \setminus T$. Noting that $\sum_{i \in N \setminus T} p_i x_i(p) = \sum_{i \in N} p_i x_i(p) - \sum_{i \in T} p_i x_i(p)$, we see that it is possible for $c(\cdot)$ to be such that

$$\sum_{i \in T} p_i x_i(p) > c(x(p)) - c(x(p')) > \sum_{i \in N} p_i x_i(p) - \sum_{i \in N \setminus T} p'_i x_i(p')$$

Here the incremental cost test (7.7) is passed (by the left hand inequality), but net additional revenue does not cover additional costs (the right hand inequality). Thus, the additional costs must be covered (at least in part) by increasing the charges levied on customers who were happy when only services in $N \setminus T$ were offered, rather than only making charges to customers who purchase services in T. These former set of customers may feel that they are subsidizing the later set of customers, and that these new services decrease the overall efficiency of the system. We conclude that, as a matter of fairness between customers, the second test condition (7.7) should take account of demand, and reason in terms of the net incremental revenue produced by an additional service, taking account of the reduction of revenue from other services. In other words, services are fairly priced if when service i is offered at price p_i the customers of the other services feel that they benefit from service i. They are happy because the prices of the services they want to buy decrease. This is called the *net incremental revenue test*. Let us look at an example.

Example 7.2 (Net incremental revenue test) Suppose a facility costs C and there is no variable cost. It initially produces a single service 1 in quantity $x_1 = a$ at price $p_1 = C/a$. Then, a new service is added, at no extra cost, and at a price p_2 that is just a little more than 0. As a result, demand for service 2 increases at the expense of demand for service 1. To cover the cost, p_1 must increase, making even more customers switch to service 2.

At the end, suppose that an equilibrium is reached where $p_1 = 10C/a$, $x_1 = 0.1a$ and $x_2 = 0.9a + b$. Note that, by our previous definition, these prices are subsidy-free, and (almost) all the revenue is collected by charging for service 1. These customers (the ones left using service 1) are right to complain that they subsidize service 2, since they see their prices increase after the addition of the new service. Indeed, choosing such a low price for service 2 results in an overall revenue reduction if prices of existing services are not allowed to increase. A fair price would be to choose p_2 in such a way that the overall net revenue (keeping the other prices, i.e. p_1, fixed) would increase. Then, the zero profit condition may be achieved by reducing the other prices and hence benefiting the customers of the other services. In our example, suppose that by setting $p_2 = p_1$ and keeping p_1 at its initial value, x_1 becomes $a/2$ and $x_2 = a/2 + b/2$. In other words, half the customers of service 1 find service 2 to suit them better at the same price, and so switch. There are also new customers that like to use service 2 at that price. Then the net revenue increase becomes $p_1 b/2 > 0$; so it is possible to decrease p_1 and allow customers of service 1 to benefit from the addition of service 2.

Finally, consider a model of potential competition. Imagine an incumbent firm sets prices to cover costs at the demanded quantities, i.e.

$$\sum_{i \in N} p_i x_i(p) \geq c(x(p)) \tag{7.8}$$

Suppose a competitor having the same cost function as the incumbent tries to take away part of the incumbent's market by posting prices p' which are less for at least one service. Suppose $x^E(p, p')$ is the demand for the services provided by the new entrant when he and the incumbent post prices p' and p respectively. Suppose that there is no p' and x' such that

$$\sum_{i \in N} p_i' x_i' \geq c(x'), \quad \text{and } p_i' < p_i \text{ for some } i, \quad \text{and } x' \leq x^E(p, p') \tag{7.9}$$

That is, there is no way that the potential entrant can post prices that are less than the incumbent's for some services and then serve all or part of the demand without incurring loss. Prices satisfying this condition are called *sustainable prices*. We have yet one more 'fairness test' by which to judge a set of prices.

The above model motivates the use of sustainable prices in contestable markets. A market is *contestable* when low cost 'hit-and-run' entry and exit are possible, without giving enough time to the incumbent to react and adjust his prices or quantities he sells. Such low barrier to entry is realized by using new technologies such as wireless, or when the regulator prescribes that network elements can be leased from incumbents at cost.

In the idea of sustainable prices we again see that price stability is related to efficiency. If prices are sustainable, a new entrant cannot take away market share if his cost function is greater than that of the incumbent. Hence sustainable prices discourage inefficient entry. However, if a new entrant is more efficient than the incumbent, and so has a smaller cost function, then he can always take away some of the incumbent's market share by posting lower prices. Thus an incumbent cannot post sustainable prices if he operates with inefficient technologies.

It can be shown that for his prices to be sustainable, an incumbent firm must fulfil a minimum of three necessary conditions:

1. He must operate with zero profits.

2. He must be a natural monopoly (exhibit economies of scale) and produce at minimum cost.

3. His prices for all subsets of his output must be subsidy free, i.e. fulfil the stand-alone and incremental cost tests.

The last remark provides one more motivation to use the subsidy-free price tests to detect potential problems with a given set of prices.

Ramsey prices

Unfortunately, there is no straightforward recipe for constructing sustainable prices. Constructing socially optimal prices that are sustainable is even harder. However, under conditions that are frequently encountered in communications, Ramsey prices can be sustainable. Recall that Ramsey prices maximize social welfare under the constraint of recovering cost. Again we see a connection between competition and social efficiency: in a contestable market, i.e. under *potential competition*, incumbents will be motivated to use prices that maximize social efficiency with no need of regulatory intervention.

However, Ramsey prices are not always sustainable. They are certainly not sustainable if any service, say service 1, is priced below its marginal cost and there are economies of scale. To see this, note that revenue from service 1 does not cover its own incremental cost since by concavity of the cost function $x_1 p_1 < x_1 \partial c / \partial x_1 < c(x) - c((0, x_2, \ldots, x_n))$. So a supplier who competes on the same set of services and with the same cost function can more than cover his costs by electing not to produce service 1. After doing this, he can slightly lower the prices of all the services that are priced above their marginal costs, so as to obtain all that demand for himself and yet still cover his costs.

Example 7.3 (Ramsey prices may not be sustainable) Whether or not Ramsey prices are sustainable can depend on how services share fixed costs, i.e., on the economies of scope. Consider a market in which there are customers for two services. The producer's cost function and demand functions for the services are

$$c(x_1, x_2) = 25x_1^{1/2} + 20x_2^{1/2} + F, \quad x_1(p) = x_2(p) = \frac{10^4}{(10 + p)^2}$$

The Ramsey prices are shown in Table 7.1. When the fixed cost F is 6 the Ramsey prices are not sustainable even though they exceed marginal cost. The revenue from service 2 is 169.45 and this is enough to cover the sum of its own variable cost and the entire fixed cost, a total of 162.76. This means that a provider can offer service 2 at a price less than the Ramsey price of 2.76 and still cover his costs. In fact, he can do this for any price greater than 2.62. However, if the fixed cost is 30 this is now great enough that it is impossible to cover costs by providing just one of the services alone at a lower price.[1] Hence, in this case, the Ramsey prices are sustainable. The lesson is that Ramsey prices may be sustainable if all services are priced above marginal cost and the economies of scope are great enough.

---

[1] The other possibility for a new entrant is to provide both services at lower prices. But it is impossible to lower both prices and still cover costs. If all prices are lower the consumer surplus must increase. Since we require the producer surplus to remain nonnegative, and it was zero at our Ramsey prices, this would imply that the social welfare — which is the sum of consumer and producer surpluses — would increase; this means we could not have been at the Ramsey solution.

Table 7.1   Ramsey prices may or may not be sustainable

	F = 6		F = 30	
	i = 1	i = 2	i = 1	i=2
Ramsey price, p_i	3.18	2.76	3.46	2.64
Demand, x_i	57.58	61.44	55.18	58.96
Marginal cost	1.65	1.28	1.68	1.30
Revenue, $x_i p_i$	183.02	169.45	191.02	178.26
Variable cost	189.70	156.76	185.71	153.57
Variable cost + F	195.70	162.76	215.71	183.57

To show how the existence of common cost plays a vital role in the sustainability of Ramsey prices, we can construct a simple example out of Figure 5.5.

Example 7.4 (Common cost and sustainability of Ramsey prices)   Suppose that two services are produced with same stand-alone cost function $A + bx$. First, consider the case in which there is no economy of scope, and hence the total cost is the sum of the stand-alone cost functions. Since both services are produced at equal quantities $x_i = x_j = x$ we have $x(p_i + p_j) = 2(A + bx)$ which implies $xp_i < A + bx < xp_j$. But $A + bx$ is the stand-alone cost for service j, which violates the sustainability conditions.

Now suppose that there are economies of scope and the fixed cost A is common to both services. Then $x(p_i + p_j) = A + 2bx$, and since $p_i > b$ we obtain $xp_j + bx < A + 2bx$. This implies $xp_j < A + bx$, which is the stand-alone cost for service j. Hence, the existence of common cost is vital for Ramsey prices to be sustainable. Observe that, in this particular case, any amount of common cost, A, will make Ramsey prices sustainable. In general, as suggested by Example 7.3, large values of A ensure sustainability.

7.1.3   Shapley Value

Let us now leave the subject of prices and return to the simple model at the start of the chapter, in which cost is to be fairly shared amongst n customers. The provider's charging algorithm could be coded in a vector function ϕ which divides $c(N)$ as $(c_1, \ldots, c_n) = (\phi_1(N), \ldots, \phi_n(N))$. Let us suppose that $\phi(T)$ is defined for an arbitrary subset $T \subseteq N$, and codes the way he would divide the cost of $c(T)$ amongst the members of the subset T if he were to provide services to only this subset of customers. Clearly, $\phi(\{i\}) = c(\{i\})$ being the stand-alone cost for serving only customer i.

Suppose that $T \subseteq N$ and i, j are distinct members of T. If $\phi_j(T) - \phi_j(T \setminus \{i\}) > 0$, then customer j pays more than he would pay if customer i were not being served. He might argue this was unfair, unless customer i can counter-argue that he is at least as disadvantaged because of customer j. But then if customer i is not to feel aggrieved then he must see similarly that customer j is at least as much disadvantaged. Putting this all together requires

$$\phi_i(T) - \phi_i(T \setminus \{j\}) = \phi_j(T) - \phi_j(T \setminus \{i\}) \tag{7.10}$$

On the other hand, if $\phi_j(T) - \phi_j(T \setminus \{i\}) < 0$, then customer j is better off because customer i is also being served. Customer i might feel aggrieved unless he benefits at least as much from the fact that customer j is present. But then customer j will feel aggrieved unless he benefits at least as much from customer i's presence. So again, we must have (7.10).

Surprisingly, there is only one function ϕ which satisfies (7.10) for all $T \subseteq N$ and $i, j \in T$. It is called the *Shapley value*, and its value for player i is the expected incremental cost of providing his service when provision of the services accumulates in random order. It is best to illustrate this with an example.

Example 7.5 (Sharing the cost of a runway) Suppose three airplanes A, B, C share a runway. These planes require 1, 2 and 3 km to land. So a runway of 3 km must be built. How much should each pay? We take their requirements in the six possible orders. Cost is measured in units per kilometer.

	Adds cost		
Order	A	B	C
A, B, C	1	1	1
A, C, B	1	0	2
B, A, C	0	2	1
B, C, A	0	2	1
C, A, B	0	0	3
C, B, A	0	0	3
Total	2	5	11

So they should pay for 2/6, 5/6 and 11/6 km, respectively.

Note that we would obtain the same answer by a calculation based on sharing common cost. The first kilometer is shared by all three and so its cost should be allocated as $(1/3, 1/3, 1/3)$. The second kilometer is shared by two, so its cost is allocated as $(0, 1/2, 1/2)$. The last kilometer is used only by one and so its cost is allocated as $(0, 0, 1)$. The sum of these vectors is $(2/6, 5/6, 11/6)$. This happens generally. Suppose each customer requires some subset of a set of resources. If a particular resource is required by k customers, then (under the Shapley value paradigm) each will pay one-kth of its cost.

The intuition behind the Shapley value is that each customer's charge depends on the incremental cost for which he is responsible. However, it is subtle, in that a customer is charged the expected extra cost of providing his service, incremental to the cost of first providing services to a random set of other customers in which each other customer is equally to appear or not appear.

The Shapley value is also the only cost sharing function that satisfies four axioms, namely, (1) all players are treated symmetrically, (2) those whose service costs nothing are charged nothing, (3) the cost allocation is Pareto optimal, and (4) the cost sharing of a sum of costs is the sum of the cost sharings of the individual costs. For example, the cost sharing of an airport runway and terminal is the cost sharing of the runway plus the cost sharing of the terminal. The Shapley value also gives answers that are consistent with other efficiency concepts such as Nash equilibrium.

The Shapley value need not satisfy the stand-alone and incremental cost tests, (7.2) and (7.3). However, one can show that it does so if c is *submodular*, i.e. if

$$c(T \cap U) + c(T \cup U) \le c(U) + c(T), \quad \text{for all } T, U \subseteq N \tag{7.11}$$

The reader can prove this by looking at the definition of the Shapley value and using an equivalent condition for submodularity, that taking the members of N in any order, say $i, j, k, \ldots, \ell$, we must have

$$c(\{i\}) \geq c(\{i, j\}) - c(\{j\}) \geq c(\{i, j, k\}) - c(\{j, k\}) \geq \cdots \geq c(N) - c(N - \{i\})$$

Note that choosing T and U disjoint shows that submodularity is consistent with $c(\cdot)$ being subadditive, i.e. (7.1).

7.1.4   The Nucleolus

The Shapley value has given us one way to allocate charges and it is motivated by a nice story of argument and counterargument. However, there are other stories we can tell. Let us call c an **imputation** of cost (i.e., an assignment of cost) if

$$\sum_{i \in N} c_i = c(N) \quad \text{and} \quad c_i \leq c(\{i\}), \text{ for all } i$$

That is, the provider exactly covers his costs and no customer is charged more than his stand-alone cost.

We now suggest a reasonable condition that the imputation c should satisfy. Suppose that for all imputations c' and subsets $T \subseteq N$ such that $\sum_{i \in T} c'_i < \sum_{i \in T} c_i$ there exists some $U \subseteq N$ (not necessarily disjoint from T) such that

$$\sum_{i \in U} c'_i > \sum_{i \in U} c_i \quad \text{and} \quad \sum_{i \in U} c'_i - c(U) > \sum_{i \in T} c_i - c(T)$$

So if a set of customers T prefers an imputation c' (because their total charge is less), then there is always some other set of customers U who can object because

- under c' the total charge they pay is more, i.e. $\sum_{i \in U} c'_i > \sum_{i \in U} c_i$, and
- they pay under c' a greater increment over their stand-alone cost, $c(U)$, than T pays under c over its stand-alone cost, $c(T)$.

so U argues that T should not have a cost-reduction at U's expense.

Then c is said to be in the *nucleolus* (of the coalitional game). It is a theorem that the nucleolus always exists and is a single point. Thus the nucleolus is a good candidate for being the solution to the cost-sharing problem. In the runway-sharing example, the nucleolus is $(1/2, 1, 3/2)$. Note that it is not the same as the Shapley cost allocation of $c = (2/6, 5/6, 11/6)$. The fact that c is not the nucleolus can be seen by taking $T = \{B, C\}$ and $c' = (3/6, 5/6, 10/6)$. There is no U that can object to this.

What would have happened if we had simultaneously tried to satisfy the conditions of both the nucleolus and Shapley 'stories'? The answer is that there would be no solution. The lesson in this is that 'fair' allocations of cost cannot be uniquely-defined. There are many definitions we might choose, and our choice should depend on the sort of unfairnesses that we are trying to avoid. We now end this section with a final story.

7.1.5   The Second-best Core

Thus far we have mostly been allocating cost without paying attention to the benefit that customers obtain. Surely, it is fair that a customer who benefits more should pay more. We end this section with a cost sharing problem that takes account of the benefit that customers obtain.

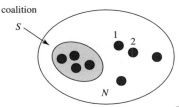

Figure 7.1   The second-best core. The monopolist fixes p s.t. $p^\top x - c(x) \geq 0$, where x is the aggregate demand, $x = \sum_{i=1}^{N} x^i(p)$, and $c(x)$ is the cost of producing x. The entrant targets a subset of customers S who he wishes to woo. He chooses p^S s.t. $(p^S)^\top x^S - c(x^S) \geq 0$, where $x^S = \sum_{i \in S} x^i(p^S)$, and such that the incentive compatibility condition holds, $CS_i(p_i^S) \geq CS_i(p)$, for all $i \in S$. We say p is in the second-best core if an entrant has no such possibility.

Suppose any subset of a set of customers N is free to bypass a monopolist by producing and supplying themselves with goods, at a cost specified by the sub-additive cost function c (which is the same as the monopolist's cost function). This subset must choose a price with which to allocate the jointly produced goods amongst its members. A price vector p is said to be in the *second-best core* if there is no strict subset of customers S who can choose prices p' so that they cover the costs of their demands at price p' and all members of S have at least the net benefit that they did under p. We express this as the requirement that

$$\sum_{i \in N} \sum_j p_j x_j^i(p) \geq c\left(\sum_{i \in N} x^i(p)\right)$$

and there is no $S \subset N$, and p' such that both

$$\sum_{i \in S} \sum_j p'_j x_j^i(p') \geq c\left(\sum_{i \in S} x^i(p')\right) \quad \text{and}$$

$$u_i(x^i(p')) - \sum_j p'_j x_j^i(p') \geq u_i(x^i(p)) - \sum_j p_j x_j^i(p), \quad \text{for all } i \in S$$

See also, Figure 7.1.

We can see that from the way that second-best core prices are constructed that they are also Ramsey prices. They maximize the net benefit of the customers in the set N subject to cost recovery, which is also what Ramsey prices do. However, although Ramsey prices always exist for the large coalition, they may be unstable, since smaller coalitions may be able to provide incentives for customers to leave the large coalition. Hence second-best core prices may not exist.

There is a subtle difference in the assumptions underlying sustainable prices and second-best core. In the second-best core model a customer who is a member of a coalition S must buy all his services from the coalition and nothing from the outside. So a successful entrant must be able to completely lure away a subset of customers, S. This is in contrast to the sustainable price model, where a customer may buy services from both the monopolist and the new entrant.

This difference means that sustainable prices are quite different to second-best core prices. Prices that are stable in the sense of the second-best core may not be stable if a customer is allowed to split his purchases. Also, prices that are not sustainable because a competitor may be able to price a particular service at a lesser price may be stable in the second-best core sense, since the net profit of customers that switch to the new entrant can be less. In the second-best core model customers must buy bundles of services and the price of the bundle offered by the entrant could be more.

In conclusion to this section, let us say that we have described a number of criteria by which to judge whether customers will see a proposed set of costs as fair, and presenting no incentive for bypass or self-supply. Anonymously equitable prices are attractive, but they may not exist. We would not like to claim that one of these many criteria is the most practical or useful in all circumstances. Rather, the reader should think of using these criteria as possible ways of checking what problems a proposed set of prices may or may not be present.

7.2   Bargaining games

Another approach to cost-sharing is to let the customers bargain their way to a solution.

7.2.1   Nash's Bargaining Game

Suppose that the cost of supplying x is $c(x)$, $x \in X$. Here, x is the matrix $x = (x_{ij})$, where x_{ij} is the quantity of service j supplied to customer i. Customer i is to pay a portion of the cost, c_i. Let us code all possible allocations of output and cost as $y \in Y$, where $y = (x, c_1, \ldots, c_n)$, with $x \in X$ and $\sum_i c_i = c(x)$. Suppose that, after taking into account the cost he pays, customer i has utility at y of $u_i(y)$. The customers are to bargain their way to a choice of point u in the set $U = \{(u_1(y), \ldots, u_n(y)) : y \in Y\}$, which we call the *bargaining set*. It is reasonable to suppose that U is a convex set, since if u and u' are in U then the utilities of any point on the line between them can be achieved (in expected value) by randomizing between u and u'.

To begin, suppose that there are just two players in the bargaining game. Infinite rounds of bargaining are to take place until a point in U is agreed. At the first round, player 1 proposes that they settle for $(u_1, u_2) \in U$. Player 2 can accept this, or make a counterproposal $(v_1, v_2) \in U$ at the second round. Now, player 1 can accept that proposal, or make a new proposal at the third round, and so on, until some proposal is accepted. We assume that both players know U. Note that only proposals corresponding to points on the northeast boundary of U need be considered, i.e. the players should restrict themselves to Pareto efficient points of U.

Rounds are s minutes apart. Let us penalize procrastination by saying that if bargaining concludes at the nth round, then the utility of player i is reduced by a multiplicative factor of $\exp(-(n-1)s\eta_i)$. If η_1 and η_2 differ then the players have different urgencies to settle. Note that this game is stationary with respect to time, in the sense that at every odd numbered round both players see the same game that they saw at round 1, and at every even numbered round they see the same game that they saw at round 2. Thus player 1 can decide at round 1 what proposal he will make at every odd numbered round and make exactly the same proposal every time, say (u_1, u_2). Similarly, player 2 can decide whether he will ever accept this proposal, and if not, what he would propose at the even numbered rounds, say (v_1, v_2).

Now there is no point in player 1 making a proposal that he knows will not be accepted. So, given v_2, he must choose $u_2 \geq e^{-s\eta_2}v_2$. But he need not offer more than necessary for his proposal to be accepted, and so he does best for himself taking a u such that $u_2 = e^{-s\eta_2}v_2$. Similar reasoning from the viewpoint of player 2 implies that $v_1 = e^{-s\eta_1}u_1$. In summary,

$$u_2 = e^{-s\eta_2}v_2 \quad \text{and} \quad v_1 = e^{-s\eta_1}u_1 \tag{7.12}$$

Let u and v be the two points on the boundary of U for which (7.12) holds. A possible strategy for player 2 is to propose (v_1, v_2) and accept player 1's proposal if and only if he would get at least u_2. A possible strategy for player 1 is to propose (u_1, u_2) and accept

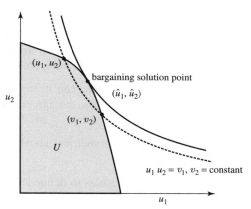

Figure 7.2 Nash's Bargaining game. Two players of equal bargaining power are to settle on a point in U. The Nash bargaining solution is at the point in U where the product $u_1 u_2$ is maximized.

player 2's proposal if and only if he would get at least v_1. The reader can check that this is a pair of equilibrium strategies, in the sense that player i can do no better if he changes his strategy while player j's strategy remains fixed, $j \neq i$. The equalities in (7.12) also imply that for all s,

$$u_1^{1/\eta_1} u_2^{1/\eta_2} = v_1^{1/\eta_1} v_2^{1/\eta_2}.$$

This means that the two points lie on a curve where $u_1^{1/\eta_1} u_2^{1/\eta_2}$ is constant. Recall that s is the number of minutes between rounds of bargaining. We see from (7.12) that as $s \to 0$, u_i and v_i tend to the same value, say $\hat{u}_i$. Assuming U is a closed and convex set, $\hat{u}$ must be the point on the boundary of U at which $u_1^{1/\eta_1} u_2^{1/\eta_2}$ is maximized. Figure 7.2 illustrates this for $\eta_1 = \eta_2 = 1$. Equivalently, writing $w_i = 1/\eta_i$, this is where $u \in U$ maximizes $w_1 \log u_1 + w_2 \log u_2$. Note that if player 1 has less urgency to settle, i.e., $\eta_1 < \eta_2$, then he has the stronger bargaining position, which is reflected in $\log u_1$ being multiplier by a greater weight than is $\log u_2$. There is a more subtle analysis that one can make of this game to prove that the solution we have found is also the unique subgame perfect equilibrium.

If there are more than two players, then it is reasonable to ask that at the solution point $\hat{u} = (\hat{u}_1, \ldots, \hat{u}_n)$, we should have that for each pair i and j the values of $\hat{u}_i, \hat{u}_j$ maximize $u_i^{1/\eta_i} u_j^{1/\eta_j}$ subject to $u \in U$ and $u_k = \hat{u}_k$, $k \neq i, j$. This condition is satisfied if we take $\hat{u}$ as the point in U where $\sum_i w_i \log u_i$ is maximized. We will meet this again, as 'weighted proportional fairness', in Section 10.1.

The *Nash bargaining solution* is usually defined with η_i the same for all i. Additionally, we suppose that if bargaining breaks down then the players obtain utilities $d_1, \ldots, d_N$. The solution to the Nash bargaining game (d, U) says that

$$u \text{ should be chosen in } U \text{ to maximize } \prod_{i=1}^{N} (u_i - d_i) \qquad (7.13)$$

The generalization in which u should maximize $\prod_i (u_i - d_i)^{w_i}$ comes from imagining that if u is chosen then there are actually w_i players who accrue benefit u_i. Thus the choice of u_i affects w_i players and the choice of u_j affects w_j players. If $w_i > w_j$ there is more 'bargaining power' influencing the choice of u_i than u_j. There are several other ways to motivate the solution (7.13), including the following axiomatic approach.

Let $f(d, U)$ be a function that determines the agreement point of the bargaining game (d, U). That is, $u = f(d, U)$. It is defined as $f(d, U) = d$ if they cannot agree. The players might at least agree that f should be consistent with the following 'rules'. Rather surprisingly, if they do, then one can prove that (7.13) must characterize f:

1. *Pareto optimality.* If $f(d, U) = u$, then there can be no $v \in U$ such that $v \geq u$ and $v_i > u_i$ for at least one i. In other words, the agreement point must be on the boundary of U.

2. *Symmetry.* If $d_1 = \cdots = d_N$ and U is symmetrical about the line $u_1 = \cdots = u_N$, then $f_1(d, U) = \cdots = f_N(d, U)$.

3. *Linear invariance.* If any player, say 1, decides to define a different point as his point of 0 utility, and/or to linearly rescale the units in which he measures his utility, then the bargaining solution is essentially unchanged. It becomes transformed in the natural way. That is, if $d' = (a+bd_1, d_2, \ldots, d_N)$ and $U' = \{(a+bu_1, u_2, \ldots, u_N) : u \in U\}$, then $f_1(d', U') = a + bf_1(d, U)$, $f_j(d', U') = f_j(d, U)$, $j = 2, \ldots, N$.

4. *Independence of irrelevant alternatives.* If $U \subset U'$, $f(d, U') = u$ and $u \in U$, then $f(d, U) = u$. This says that if the set U is increased to U' and u is the solution within U', but u happens to lie in U, then it must also be the solution for the bargaining game (d, U).

Example 7.6 (A merger of two firms) Suppose firm 1 is a cable operator who provides both cable local access and cable TV content. Suppose firm 2 is a provider of an Internet portal service. Both have customers and they intend to merge, since they expect the merger of the two businesses to be worth more than they are separately. Suppose that separately they are worth d_1 and d_2, and together they will be worth d_3, where $d_3 > d_1 + d_2$. How much should the value of the new firm be distributed fairly amongst the owners of the two firms at merger? The Nash bargaining paradigm suggests that they should receive u_1, u_2, where these maximize $(u_1 - d_1)(u_2 - d_2)$, subject to $u_1 + u_2 = d_3$ (assuming both owners have linear utilities). This gives $u_i = d_i + (d_3 - d_1 - d_2)/2$, $i = 1, 2$.

For example, suppose $d_1 = 10$, $d_2 = 20$ and $d_3 = 40$. The Nash solution is $u = (15, 25)$. Each gets half of the added-value. Note that this is the same as in the Shapley allocation. Firm 1 would be considered to bring d_1 or $d_3 - d_2$ depending on whether he adds value first or second. The average of these is $d_1 + (d_3 - d_1 - d_2)/2$.

7.2.2   Kalai and Smorodinsky's Bargaining Game

Of course, there are other reasonable axioms that could be agreed. Suppose rule 4 of the axioms specifying the solution of the Nash's bargaining game is replaced by a monotonicity condition which says that if U is increased then no one must be worse off. More precisely, use instead the rule

5. *Monotonicity.* Suppose $U \subset U'$, and for all i

$$\sup\{u_i : u \in U'\} = \sup\{u_i : u \in U\}$$

and for $j \neq i$

$$\sup\{u_j : u_i \geq t, u \in U'\} \geq \sup\{u_j : u_i \geq t, u \in U\}, \quad \text{for all } t$$

Then $f_j(d, U') \geq f_j(d, U)$ for all $j \neq i$.

This is the *Kalai and Smorodinsky bargaining game*. It turns out that there is precisely one way to satisfy axioms 1–3 and 5. Let $m_i = \sup\{u_i : u \in U\}$ and $m = (m_1, \ldots, m_N)$. Then $f(d, U)$ must be the point in U on the line joining d to m whose Euclidean distance to m is least.

Consider again Example 7.6. The Nash solution is $u = (15, 25)$. The Kalai and Smorodinsky solution is $u = (16, 24)$. Just as we saw in our discussion of Shapley value and nucleolus solutions, there can be more than one solution concept. Note that there is no solution concept that obeys all the 'reasonable' axioms 1–5.

7.3 Pricing in practice

Many methodologies have been proposed for assigning costs to services. Most of them follow basic common principles and are motivated by the requirements of fairness and stability that have been mentioned in Section 7.1. They differ in the details of how they define and assign costs. We start with a brief overview of the practical problems and methodologies. We examine various types of cost, the accounting bases for defining costs, and methods for mapping the costs of input factors to costs of services.

7.3.1 Overview

In the previous sections we have characterized the properties that prices should possess if they are to be stable under competition. However, this has not provided us with a recipe for constructing prices. In practice, we do not know the complete cost function. That is, we do not know the cost of producing any arbitrary bundle of services. We know only the current cost of producing the bundle of services that is presently being sold. Another practical difficulty is that most of the cost may be common cost, which cannot be attributed to any particular service so far as the accounting records show. For example, accounting records may not show part of a maintenance crew's cost as attributed to providing a video-conferencing service. Usually only a small part of the total cost is comprised of factors that can be attributed to a single service. This is a major problem when trying to construct cost-based prices.

In practice, we can identify some key principles that are closely related to concepts of fairness. These include the principles of *cost causation* (the cost of a service should be related as much as possible to the cost of the factors that are consumed by the service), *objectivity* (the cost of the service should be related to the cost factors in an objective way), and *transparency* (the cost of a service should be related to the cost factors in a clear and formulaic manner, and so that it can be easily checked for possible inconsistencies).

The first two of these principles are difficult to implement since, as we have commented above, the accounting records usually attribute only a small part of the total cost to individual services, and so the greatest part of the cost, i.e., the common cost, may be unattributed. One solution is to make each service pay for part of the common cost. This is the Fully Distributed Cost (FDC) approach that we investigate in Section 7.3.3. Unfortunately, the division of the common cost amongst the services is rather *ad hoc*. Since common cost accounts for a large proportion of the cost, prices can be 'cooked' in many ways, making certain prices artificially large or small.

The definition of subsidy-free prices suggests that a reasonable way to construct the price of a service (actually a lower bound on the price) is to calculate the incremental cost of the service. This clearly includes the directly attributable cost from the accounting records. Although the sum of the incremental costs of the services still leaves some common cost unaccounted for, this part of the common cost is much smaller than that which is left over after considering only the directly attributed costs. This restricts the range that possible prices may take if they are to avoid cross-subsidization. Let us see this through an example.

Suppose that a factory produces two tourist souvenirs, one of wood and one of bronze. The only factors that are directly attributed to the production of the souvenirs are the quantities of wood and bronze consumed, say W and B, with respective costs $c(W)$ and $c(B)$. Other factors that are used in producing the souvenirs are considered to be common cost. These are quantities of labour and electricity, say L and E, with costs $c(L)$ and $c(E)$. There is a single accounting record for each, and no information on how to attribute these costs to the production of the souvenirs. How should we split the overall cost so as to define the cost of each product?

If we use FDC, we must find a way to split the common cost, i.e, we must define the coefficients γ_l and γ_e, which in turn define the cost of production of wooden and bronze souvenirs to be

$$c_w^{FDC}(x) = c(W) + \gamma_l c(L) + \gamma_e c(E)$$
$$c_b^{FDC}(y) = c(B) + (1 - \gamma_l)c(L) + (1 - \gamma_e)c(E)$$

where x and y are the quantities of wooden and bronze artifacts produced. Note that if the cost of labour and electricity are substantial compared to the cost of wood and bronze, then γ_l and γ_e play a significant role in determining the items' prices (which are found for each product by dividing the cost of production by the number of items produced). This approach can produce prices that are not subsidy-free. For instance, suppose we take $\gamma_l = \gamma_e = 0$. Then the cost of the bronze souvenirs that must be recovered is $c(B) + c(L) + c(E)$, and this probably exceeds than the stand-alone cost of producing the same quantity of bronze souvenirs on their own.

There are various approaches to reducing the amount of unattributed common cost. They are of varying difficulty and cost of implementation. The incremental cost approach needs to calculate the difference between the cost of the facility that produces both types and the cost of the facility that produces a single type. Suppose that $c_{w,b}(x, y)$ is the cost of a facility that can produce both types and it operates at production levels x, y. Similarly, $c_w(x)$ and $c_b(y)$ are the costs of facilities that are optimized to produce only wooden or bronze artifacts at production levels, x and y, respectively. Then the incremental costs are

$$c_w^{incr}(x) = c_{w,b}(x, y) - c_b(y), \quad c_b^{incr}(y) = c_{w,b}(x, y) - c_w(x) \qquad (7.14)$$

The problem is that the accounting records hold only the actual cost $c_{w,b}(x, y)$. Evaluating $c_w(x)$ or $c_b(y)$ requires creative thinking. Could we use $c_{w,b}(x, 0)$ instead of $c_w(x)$? The answer is probably not. The factory was built to produce the products simultaneously, and many design decisions were taken to optimize the joint production. This implies that $c_{w,b}(x, 0)$ is greater than $c_w(x)$. Moreover, calculations of $c_{w,b}(x, 0)$ from the accounting records may be very inaccurate. This is because the only quantity that is related to producing wooden souvenirs and is easy to compute from accounting records is $c_{w,b}(x, y) - c(B)$. However, $c_{w,b}(x, y) - c(B)$ is greater than $c_{w,b}(x, 0)$, because it includes all the common cost (electricity and labour) as if both souvenirs were produced in quantities x, y (since we've only subtracted the directly attributable cost). Using $c_{w,b}(x, y) - c(B)$ as a proxy for $c_w(x)$ in (7.14) will lead to an underestimate of the incremental cost of bronze souvenirs and to an overestimate of the stand-alone cost of wooden souvenirs.

There are two solutions to this problem. The first is the so-called *bottom-up approach*, in which each stand-alone cost is computed from a model of the most efficient facility that specializes in the production of that one product, using current technology. Thus, we construct $c_w(x)$ and $c_b(y)$ from scratch, by building models of fictitious facilities that produce just one or the other of these products. This contrasts with the approach of FDC, which is a *top-down approach*, in that it starts from the given cost structure of the existing facility and attempts to allocate the cost that has actually incurred to the various products. The methodology of LRIC+, which we mention later, has been traditionally associated with a bottom-up approach.

The second solution is to adopt a *top-down approach*, but attempt to reduce the unaccounted-for common cost. One way to do this is to refine the accounting records, keeping more information on how the common cost is generated. There are several ways to do this. The *activity-based costing* approach defines several intermediate activities that contribute to the production of the end products.

Examples of activities related to communication networks are repair, operation, network management, consumer support, and so on. The cost of each such activity can be computed from accounting information about the amounts of the input factors that are consumed by each activity, usually gathered through questionnaires. For example, we would keep track of how many man-hours of labour are used for repair. This allows for a large part of the cost of labour to be attributed to specific activities and so be subtracted from the common cost (though some common cost will always remain).

Now, since each activity could be contributing to the production of a number of end products, we need to say for each product what percentage of each activity this product consumes. This can be done fairly accurately by monitoring the operation of the facility, and logging appropriate information. This activity-based approach is a refinement of the FDC approach. By drastically reducing the unaccounted-for common cost, it reduces the inaccuracy that stems from the ad hoc splitting of that cost.

In the following sections we refine some of the above concepts. We start by providing useful definitions concerning the various types of cost viewed from different perspectives.

7.3.2 Definitions Related to the Cost Function

In this section we remind the reader of several important definitions and concepts concerning the cost function that we use in our pricing approach.

The cost of a particular service can be divided into *direct cost* and *indirect cost*. Direct cost is the part of the cost that is solely attributed to the particular service and will cease to exist if the service is not produced. Indirect cost is other cost that is related to the provision of the service. The following should be noted:

- A cost may have a direct relation with a service, but no accounting information is kept to quantify it. Such a cost can become direct cost by refining the accounting system.
- A cost may arise from the provision of a group of services and there may be a logical way to specify the percentage of the cost that is related to the provision of each service. This is called an *indirectly attributable cost*. For example, consider a telephone switch that is used by both local and long-distance calls. One could measure the numbers of calls of each type that use the switch, and divide the cost of the switch proportionally between them.
- An *unattributable cost* is one that cannot be straightforwardly divided amongst the services. An example is the cost of the company's management.

The quantity (and hence cost) of a specific factor may be fixed, or it may vary with the amount of service produced. The *fixed cost* of a service is the sum of all factor costs that remain constant when the quantity of the service changes. For example, the cost of the buildings may be a fixed cost in providing long-distance calls. *Variable cost* is the cost of those factors whose quantities depend on the amount of the service produced. For example, in producing wooden souvenirs, the cost of wood is a variable cost. Note that direct and indirect cost factors can contribute to either the fixed or the variable cost of a service.

Is the cost of the building really a fixed cost? If a firm reduces its output, then it might rent a smaller building and reduce its costs. Thus whether a cost is fixed or variable depends upon the time frame over which the firm is allowed to re-optimize its production capabilities. In the short run, reducing output will not allow for re-optimization of facilities and so the cost of the present facilities is a fixed cost. In the long run it is variable cost. This suggests that we define the *short-run incremental cost* for an increase in output of Δx as the increase in total production cost required for this increase in output when the firm may not re-optimize its production procedures. Similarly, the *long-run incremental cost* is the increase in cost required when the firm may re-optimize its production procedures. Clearly the long-run incremental cost is always less than the short-run incremental cost. The (short-run or long-run) marginal cost is the incremental cost when Δx is very small.

In the same way, consider a service that is produced in an amount x. We can define the short and the long-run incremental cost of this service as the difference between the cost of producing it in an amount x and not at all. Note that when defining subsidy-free prices we did not mention the type of the incremental cost involved. Using long-run incremental cost restricts the possible prices to a smaller interval and reflects more accurately the situation in a competitive market in which firms can reorganize and re-optimize their production facilities to become more efficient. This is the rational for using long-run incremental cost in (7.14).

Another useful notion is *average cost*; it is obtained by dividing the cost by the quantity of service produced. In general, the short-run average cost decreases due to economies of scale, but beyond a certain level of production it increases because factors that cannot be changed in the short run produce inefficiencies (e.g. lack of space at the production facility produces congestion).

In many cases, when we talk about cost recovery in the context of a regulated firm, we do not imply that the firm must make zero accounting profits but that it must make zero economic profits. The cost which must be recovered through prices refers to the economic cost of the firm. That means that the total cost that should be recovered includes, apart from the cost registered in the accounting books of the firm, also a reasonable rate of return on the capital employed. The inclusion of this reasonable profit margin make it possible for the company to improve, to make new investments and also to compensate its shareholders. However, this return is not as such to permit the company to create super profits, but only profits that equal to the economic cost of capital or, in other words, the opportunity cost of capital under the specific risk conditions.

Historic and current costs

An accounting system can use *historic cost* or *current cost* to assign costs to factors. Historic cost is easier to use since it is the actual amount paid to purchase the various factors (equipment, etc). Such information is readily available in the accounting records of the firm. Together with depreciation information it can be used to compute the yearly cost of the equipment.

Current cost (the exact terminology being 'current cost for modern equivalent assets') is a completely different notion; it reflects the cost of the equipment if it were bought today. Current cost can be hard to define since technological innovation makes older equipment obsolete. New equipment would be more capable and efficient. One should take this into account and rescale cost appropriately. For example, if new switches have double the capacity of installed switches then the current cost of an installed switch would be half that of a new one.

Using current costs for computing the depreciation of network equipment leads to lower prices than using historic costs, due to technology improvements. This is a main reason they are favoured by regulators. Obliging a firm to use prices based on current costs motivates the firm to maintain and operate an efficient network, using state-of-the-art technology, since otherwise cost recovery may not be possible. A problem with current costs is that they are not directly available in the accounting system of the firm and must be constructed by specialists. In that sense, accounting systems based on current costs are not as objective and can be audited only by experts (and hence are less 'auditable').

Interestingly enough, in some cases, historic costs may lead to lower prices! We see this when computing prices for renting the use of the local loop to competitors that would like to sell high-bandwidth access services over the local loop (the copper wire pair that connects the premises of customers to the telephone network). (See also the discussion about unbundling in Section 7.3.5). In most locations the local loop network belongs to existing incumbent operators (telephone companies), is many years old and is already largely depreciated. On a historic cost basis the price of renting the local loop would be nearly zero. This contrasts with the price that should be charged if a current cost basis is used. Although new technology, such as fibre and wireless, can help to reduce costs and improve performance, the cost of the access service in such a network may be substantially greater than the incumbent operator's historic cost. Should the regulator be prefer lower prices and hence use of historic costs? Although these can increase competition and lead to lower prices to consumers, they have some serious drawbacks. They do not provide incentives for alternative access networks of newer technologies to be built by other operators, since such networks will have to charge higher prices (based on current costs), and so be less competitive. Also such low prices may not provide enough incentives to the incumbent operator to improve and maintain the access network, and the quality of the services sold will probably deteriorate. For these reasons, regulators prefer the use of current costs for pricing such services, even if these lead to higher prices.

Note that bottom-up models are naturally combined with current costs (since the network model is built from scratch), while top-down models such as FDC are naturally combined with historic costs found in the accounting records. In the next sections we investigate methodologies for assigning costs to services.

7.3.3 The Fully Distributed Cost Approach

We have already mentioned that a virtue of the Fully Distributed Cost (FDC) approach is its simplicity in directly relating prices to information that is available in the accounting and billing system of the firm. Such information can be easily checked for its accuracy, which makes an FDC costing model *auditable*. The idea of the FDC approach is to simply divide the total cost that the firm incurs amongst the services that it sells. This can be made a mechanical process: a program takes the values of the actual costs of the various operating factors and computes for each service its corresponding portion of the total cost. The parameters of the program are the coefficients used to divide the costs of the input factors amongst the services produced.

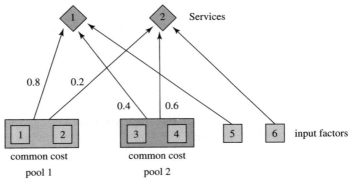

Figure 7.3 In the FDC approach the cost of input factors are assigned to services. For example, service 1 is assigned 0.8 of the cost factors in the first cost pool and 0.4 of the cost factors in the second cost pool. The different common cost pools and the coefficients for sharing the cost of the common factors are defined by the designer of the system.

The FDC approach is illustrated in Figure 7.3. The idea is to put all the cost of factors that are not uniquely identified with a single service into a number of common cost pools. Since only a small part of the cost is directly attributable to a single service most of the cost will be common cost. Next, one defines coefficients to apportion the common cost among the services, in a way that may depend on the particular common cost pool. Since there is no other information available in the accounting system (as in the case of an activity based model investigated next), such a function mapping cost factors to cost of services is constructed in a rather ad hoc way.

Formally, suppose service i is produced in quantity y_i and has a variable cost $VC_i(y_i)$ that is directly attributable to that service. There is a shared cost $SC(y)$, $y = (y_1, \ldots, y_n)$, that is attributable to all services, and which for simplicity we assume is assigned to a single cost pool. The price for the quantity y_i of service i is defined to be its cost, i.e.

$$p_i(y_i) = VC_i(y_i) + \gamma_i SC(y)$$

where $\sum_i \gamma_i = 1$. The price per unit is defined as $p_i = p_i(y_i)/y_i$. The γ_is may be chosen in various ways: as proportions of variable costs, quantities supplied, or revenue, i.e. proportional to $VC_i(y_i)$, y_i or $y_i p_i$.

Clearly, once the coefficients γ_i are defined, then the construction of the prices is trivial and can be done automatically using accounting data. This avoids building a model of the facility from scratch as required by the bottom-up models. Due to its simplicity and the ability to audit the price constructing procedure, FDC pricing has been popular with network operators and regulators, at least in the early days of the price regulation process in the communications market.

However, there are a number of problems with FDC pricing. First, there is no reason that the prices constructed are in any sense optimal or stable. A major reason is that the coefficients for apportioning the common cost factors are constructed in some arbitrary way, without taking into account important information about the operation of the facility. Second, these prices hide potential inefficiencies of the network such as excess capacity, out-of-date equipment, inefficient operation, bad routing and resource allocation. This is because there is no way to track down the actual reasons that certain prices for services are exceedingly high.

The refinement of the FDC model through the definition of activities helps to link more accurately a larger part of the common cost to particular services, and so improves the subsidy-free properties of the resulting pricing scheme. Also when a price is higher than anticipated, one can, in principle, trace the activities involved and find potential inefficiencies. Of course, there is always the possibility that prices may be artificially high due to the particular choice of the apportioning coefficients in splitting the common activity cost. The following example helps to clarify this issue and motivate the activity-based model described in more detail in Section 7.3.4.

Consider, as above, a facility that produces wooden and bronze souvenirs, with the cost function

$$c(y_w, y_b) = s^f x^f + s^l (x_0^l + a_w y_w + a_b y_b) + s^w \theta_w y_w + s^b \theta_b y_b \qquad (7.15)$$

where s^f is the per unit cost of the fixed factor x^f (e.g., the cost of the building), s^l is the per unit cost of the labour factor, s^w, s^b are the per unit costs of wood and bronze respectively, x_0^l is the fixed amount of labour that is consumed independently of the production (e.g. secretarial support), a_w and a_b are coefficients that relate the levels of production of the souvenirs to the amount of consumed labour that is directly attributed to the production, and θ_w and θ_b relate these levels of production to the amount of raw materials consumed. Note that $s^f x^f + s^l x_0^l$ is a fixed cost, whereas $(s^l a_w + s^w \theta_w) y_w + (s^l a_b + s^b \theta_b) y_b$ is a variable cost.

Consider first the case of simple FDC pricing without activity definitions and no explicit accounting information on how labour effort is spent. In this case, $\text{VC}(y_w) = s^w \theta_w y_w$, $\text{VC}(y_b) = s^b \theta_b y_b$, the common cost is the remaining part $\text{SC}(y_w, y_b) = s^f x^f + s^l (x_0^l + a_w y_w + a_b y_b)$, and the FDC prices are of the form

$$p_w(y_w) = s^w \theta_w y_w + \gamma_w [s^f x^f + s^l (x_0^l + a_w y_w + a_b y_b)] \qquad (7.16)$$

$$p_b(y_b) = s^b \theta_b y_b + (1 - \gamma_w)[s^f x^f + s^l (x_0^l + a_w y_w + a_b y_b)] \qquad (7.17)$$

Now suppose two activities are defined, related to the production of the souvenirs. In each activity, there is exact accounting of the labour effort required for the production of each souvenir. Now $\text{VC}(y_w) = (s^l a_w + s^w \theta_w) y_w$, $\text{VC}(y_b) = (s^l a_b + s^b \theta_b) y_b$, and the common cost is reduced to $\text{SC}(y_w, y_b) = s^f x^f + s^l x_0^l$. The resulting FDC prices are

$$p_w(y_w) = (s^l a_w + s^w \theta_w) y_w + \gamma_w (s^f x^f + s^l x_0^l) \qquad (7.18)$$

$$p_b(y_b) = (s^l a_b + s^b \theta_b) y_b + (1 - \gamma_w)(s^f x^f + s^l x_0^l) \qquad (7.19)$$

We can make the following observations:

1. The prices in the simple FDC approach less accurately relate prices to actual costs. Suppose, $0 \approx a_w << a_b$. That is, wooden souvenirs are extremely easy to construct and the greater part of labour effort is spent on bronze souvenirs. Let there be equal sharing of the common cost, so $\gamma = 1/2$. Then the price of wooden souvenirs in (7.16) subsidizes the production of bronze souvenirs as it pays for a substantial part of the labour for making them. This cross-subsidization disappears in (7.18).

2. Suppose that the facility is built inefficiently and that the amount of building space is larger than would be required if new technologies were used. This fact is hidden in both (7.16) and (7.18). However, if one develops a bottom-up model for the facility, the corresponding factor in this model will be less, say $x^f/2$. This will reduce the corresponding prices $p_w(y_w)$ and $p_b(y_b)$. This discrepancy between the prices obtained by the top-down and the bottom-up model indicates the existence of

inefficiencies in the way the firm operates. In general, it is not straightforward to track down the exact reason for such inefficiencies. This is because running these models produces sets of numbers instead of nicely shaped functions that one can easily understand and compare. If we refine the cost model by introducing more activities this can help us to better understand the relation between the cost of services and input factors.

3. Consider the price of wooden souvenirs. The variable part of the price in (7.18) is a better approximation of the long-run incremental cost of producing the amount y_w of wooden souvenirs than the variable part in (7.16). The reason that it may not be equal to the long-run incremental cost is that if only one souvenir is produced, then the common cost could be reduced (perhaps a smaller facility is needed, or one secretary will suffice rather than two). Unfortunately, this reduction cannot be extracted from the accounting data. Once again, one must construct a 'virtual' model of a facility that is specialized in constructing only bronze souvenirs, so that one can subtract the appropriate cost. This again shows the weakness of the top-down models that are the basis of FDC pricing.

As we have already remarked, FDC is naturally combined with historic costs. This is because accounting data concerns the firm's actual costs. It is not impossible to use current costs, but this requires modifications to the accounting system as discussed earlier.

7.3.4 Activity-based Costing

A top-down approach for assigning actual costs to network services that is well-accepted in practice is shown in Figure 7.4. It is based upon a hierarchy of four levels and is a refinement of the traditional FDC approach. The bottom level consists of the input factors that are consumed by the network operator, such as salaries of personnel, depreciation]of network elements, cost of capital, depreciation of buildings and vehicles, marketing cost, overhead, power consumption, and the cost of renting raw bandwidth. Depreciation is the yearly estimate of the cost of asset usage and corresponds to the decrease of asset value

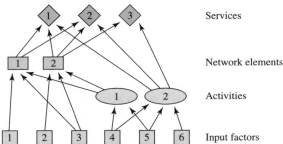

Figure 7.4   The cost of input factors can be assigned to services in a hierarchical fashion. The lowest level are input factors that are consumed by the network operator, such as labour and depreciation of network elements. The next level consists of labour-intensive activities that are required for the network to operate and produce services. The next level consists of the network elements such as the routers, switches and links. The last level consists of services. Input cost factors are allocated to network elements and activities. Activities (activity costs) are allocated to network elements or directly to services. The cost of network elements is allocated to each service in proportion to its use by that service. Usually the cost of a service also includes the cost of capital it employs. A crucial decision, besides the definition of the activities, is the definition of the coefficients to apportion the costs of one level to the next level up.

during a year's operation. The goal is to apportion these cost elements to the services that the network provides.

The next level is the activity level. Activities are labour-intensive processes that are required for the network to operate and produce services. Usually, an activity has a well-defined purpose, such as maintaining certain equipment, managing network elements, operating the communications links, supporting certain customer services, or operating the business. Using information about time spent that staff spend on each activity, one defines coefficients that allow the cost of the input factors (mostly labour) to be shared amongst the activities. By this procedure the accounting system is enriched with a specification of higher-level activities and their relations with the basic cost factors.

The next level up consists of network elements such as routers, switches and transmission links. The cost of each network element is computed by apportioning the input factors that are related to the particular element (equipment depreciation, power consumption, space rent, etc.), and the activities that are concerned with the operation and management of the network element. These include input factors and activities that have a broader scope and are of the common cost type, such as general expenses and company management cost. The cost of these must be apportioned in an ad hoc way. An example is the salaries of the members of the board of the company. This overhead cost might allocated in proportion to the other costs that have been assigned more rationally. Sometimes such overhead costs can be directly attributed to services. The gathering of both activity costs and all other costs (such as depreciation to network elements) is usually called network costing.

The idea so far is that the complete cost of the network should be allocated to the various elements of the network and to activities that only deal with the provision of services. For instance, a particular router will be assigned a cost that sums its depreciation cost, power consumption, space rent, and the cost of all the activities (and hence indirectly the input factors) that contribute to its operation, such as management and maintenance. Customer support is an activity that purely relates to services rather than network elements. Although some of the common cost will always be allocated in a rather *ad hoc* manner, the use of activity-based costing can greatly reduce the need for it. In computing the cost of an activity, one carefully accounts for the amounts of the factors that it consumes, and so reduces the unaccounted-for common cost. Traditional models for computing the cost of services might consider the complete labour cost as common cost, and so allocate it to the network services in a completely *ad hoc* way. The definition of activities provides the relevant information to allocate such cost more accurately.

The last level is the service level. Services (such as local and long-distance calling, leased lines, interconnection, IP connectivity, and so on) are sold to customers. They make use of the network elements and the service related activities. Once again, we must define coefficients to apportion the costs of network elements and (non network element related) activities amongst the services they support. For example, the access service to a customer uses the copper wires that connect the customer to a concentrator (which is located at the street level and serves customers on the same street). It also uses the cable (or fibre) that connects the concentrator to the switch at the premises of the network provider. Hence, the network element related cost of the service to the customer includes the cost of the copper wires (which are not used by others), and of part of the cost of the concentrator, the cable and the input port of the switch, and the management and support activities related to these network elements (which are used by others). To compute the total cost of the access service one must also apportion the cost of activities such as customer support, marketing and company management. Some of these activities are easier to apportion than others (e.g.

one may be able to estimate the proportion of time that the customer support staff give to the access service). Clearly, the cost of the service depends on the number of customers that share the common cost and so is customer-driven.

Similarly, local telephone service uses many network elements that are also used for other services. For example, links and switches are also used for long-distance and international calls. One would like to say what percentage of the use of these elements is due to local telephone service. This can be done by defining a measure of usage, such as call-minutes. One measures the total number of call-minutes that each network element provides, and then computes a call-minute cost for that network element by dividing the cost of the network element by this number of call-minutes. The cost of providing a telephone call is computed by summing the call-minute costs that it consumes at all network elements it uses. Because call routing is not fixed and it would be very expensive to account for each call's particular route through the network, this is often done on an average basis. This motivates the usual tariff for telephone service: the constant part (the monthly rental) reflects the cost of the access service, and the variable part, computed from call-minutes, reflects the cost of carrying the telephone calls through the network of the service provider.

We make two comments about the above approach. First, it hides potential inefficiencies of the network provider. Even if a network element is underutilized, its cost is completely shared by the services that use it and there is no incentive for the provider to improve its efficiency. If the provider were only allowed to recover the cost of a network element in proportion to its actual utilization, he would have a clear incentive to improve the efficiency of his network (by better routing, resolving potential bottlenecks, reselling spare capacity, and so on). This highlights a key difference in the bottom-up and top-down approaches. In the top-down approach, the cost of the existing facility is allocated amongst the services sold. In the bottom-up approach, a model of the facility is constructed. This model uses state-of-the art technologies and optimally dimensions the facility, in some cases taking into account the topology and structure of the existing network. It is used to derive the cost of the network elements and of the corresponding services. Prices that arise from a bottom-up approach provide incentives for improving the efficiency of the network. They also enhance competitiveness by preventing entry by inefficient competitors. This is one reason that regulators suggest bottom-up models for pricing network services.

The second comment concerns the inadequacy of using activity-based costing (or any FDC costing type of model) for determining the incremental cost of a service. The problem is that a top-down approach provides for an one-way function that takes as inputs the incurred input factor costs and the coefficients for apportioning the costs in the various level of the model, and computes the costs of the various services. It does not provide the means to answer the question 'if service a were not produced, what would the cost of producing the rest of the services be'? To answer this question, one must go backwards and check all the cost factors that contribute to the cost of the given service. To estimate the reduction of each such cost factor if service a is not produced one must be able to determine the fraction of the cost that was allocated to the service which is variable and the fraction that is fixed. Clearly, one should reduce the cost factor only by the variable amount, and reapportion the fixed part among the services that continue to be provided. This information is not available in traditional accounting systems and is rather complex to obtain. Recent trends show that many large communication companies are in favour of constructing such advanced top-down costing models that allow the accurate calculation of incremental costs, mainly due to auditability requirements imposed by regulators and for comparing the resulting prices. Most of these models employ current costs instead of the traditionally used historic costs.

7.3.5 LRIC+

We have already discussed important properties of fairness and stability that prices should possess. Prices based on LRIC+ share the nice properties of subsidy-free prices, are close to the prices that would prevail in an actual contestable market and send economic signals that promote efficient forward-looking investment decisions.

The key notions we will use in this section are the long-run incremental cost (LRIC) of a service, and the stand-alone cost (SAC) of a service. We remind the reader the definition of the long-run incremental cost through an example, in which we also introduce LRIC+. We also assume, as typical in communications, that the cost functions of the firms that produce services exhibit economies of scope (by the existence of common cost).

Consider a firm that offers quantities y_1 and y_2 of services 1 and 2. Let the cost for this mix be $c(y_1, y_2)$. The LRIC for service 1 is defined as $LRIC(y_1) = c(y_1, y_2) - c(y_2)$, where $c(y_2)$ is defined for a facility whose production plan is optimized to produce only type 2 service (i.e. if we stop production of service 1, then we have enough time to optimize the production plan to produce only y_2 of service 2). Similarly, the SAC of a service is the cost for building and operating a facility that produces only that service. Since in the definition of the SAC we do not take account of economies of scope in producing a larger number of services, we have that $LRIC(y_1) \leq SAC(y_1)$.

The motivation for using LRIC as a basis for constructing prices is the idea of subsidy-free prices of Section 7.1.2. The methodology for implementing LRIC is based on constructing bottom-up models from which to compute $c(y_1, y_2)$, $c(y_1)$ and $c(y_2)$. We have already mentioned that in a bottom-up model the network is designed from scratch using the most cost-effective technologies. The cost of the services is computed by apportioning the cost of the network elements (similarly as in the activity-based approach), and by adding the cost of labour and the rest of the overheads as a simple markup on the cost of the infrastructure. Such a markup follows the trends observed in actual networks.

A problem with using $LRIC(y_1)$ as a price for the quantity y_1 of service 1 is that the sum of the prices constructed according to LRIC will not in general cover the production cost. For example,

$$LRIC(y_1) + LRIC(y_2) = c(y_1, y_2) + [c(y_1, y_2) - c(y_1) - c(y_2)]$$
$$\leq c(y_1, y_2)$$

as the term in square brackets is in general negative. Some amount of common fixed cost is not recovered. A way to remedy this is to distribute this amount of common cost amongst the prices of the services. Since there are many ways to distribute this common cost, we impose the further constraint that the resulting price, $p(y_i)$ should satisfy

$$LRIC(y_i) \leq p(y_i) \leq SAC(y_i) \tag{7.20}$$

and the sum of the prices equal the total cost. This approach is known as LRIC+ , and is a practical application of (7.4) and (7.5).

It is natural to use current cost with LRIC+ because the aim is to construct prices that would prevail in a competitive market. The use of current rather than historic costs does not pass on the inefficiencies of the operator due to high historical costs and inefficient out-of-date technologies. Furthermore, it provides incentives for improving efficiency, since this is the only way the operator can make some profit at these prices. Also, the bottom-up nature of the model makes such costs a natural candidate to be used as inputs.

The use of LRIC+ has also disadvantages. Traditional accounting systems do not provide any information that can be used by an LRIC+ model, and hence such models must be built from scratch. Also, prices based on LRIC+ are hard to audit as we have discussed earlier. This motivated the recent development of LRIC+ systems based on top-down models using current costs (see the discussion in Section 7.3.4).

There is another fairness perspective on the use of LRIC+. Consider an incumbent local exchange carrier (ILEC) who is forced to unbundle part of his network, e.g. the part that accesses his customers. The concepts of ILECs and unbundling are discussed in more detail in Section 13.3 and Section 13.4.1.

Unbundling requires the incumbent to sell the services of the access network in a stand-alone manner, instead of selling them in combination with other services (such as local and long distance telephony). Examples of such services are the physical layer consisting of the copper wires that form the local loop between the premises of the customers and the local exchange of the carrier, and the raw bandwidth that can be provided by xDSL modems operating over the copper local loop wires.

The price of the unbundled access service makes a big difference to competition from other providers. If it is very high, then competitors will prefer to build their own access network, which is very costly and risky. If the price is low, then there will be fierce competition in providing higher level services over the access network, but the incumbent will have no incentive to upgrade the access network or improve its quality, and other operators will have no incentive to build access networks using alternative new technology. How should such prices be defined? Regulators have proposed the use of LRIC+ (see discussion on historic vs. current costs in Section 7.3.2). But is this fair to the incumbent operator?

If the incumbent is required to rent elements of the access network to his competitors then he will be unable to use these elements to provide the services he presently sells to his customers (which he is probably doing at a high profit). Thus, unbundling has an opportunity cost for the incumbent. LRIC+ penalize the incumbent for inefficiencies, but does not take account of his historic costs, or this opportunity cost. To be fair, should not the price include this opportunity cost? In the next section we describe a pricing scheme that has been proposed as an alternative to LRIC+ , and which does take account of the incumbent's opportunity cost. This is known as the Efficient Component Pricing Rule (ECPR) for network elements. As we will see, such a pricing scheme has serious inefficiencies compared to LRIC+.

7.3.6   The Efficient Component Pricing Rule

The Efficient Component Pricing Rule (ECPR) is an alternative to LRIC+ that does take account of the incumbent's opportunity cost. Unfortunately, ECPR has severe incentive problems and must be used with care. This section should be seen as a case-study that demonstrates that a pricing rule that looks plausible at the first sight may be very inadequate in particular situations.

Consider a service AB offered by an incumbent by means of two network elements, A and B, as shown in Figure 7.5.

New entrants in the above market possess only element B and so require element A in order to provide service AB. From their viewpoint, element A is the bottleneck to providing service AB. For example, A might be the local-loop part of a telephone connection and B

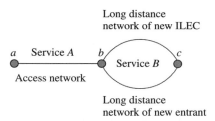

Figure 7.5   The Efficient Component Pricing Rule for pricing network services. Service A connects a to b; service B connects b to c and service AB connects a to c. According to the ECPR an incumbent should charge for A a rental price of $p_A = p_{AB} - c_B = c_A + (p_{AB} - c_A - c_B)$, where p_i, c_i are the price and cost of providing a unit of service i. Note that p_A is the cost of service A plus the private opportunity cost to the incumbent of not offering a unit of service AB.

might be the wide area network part. Let c_A and c_B be the costs of elements A and B in providing a single unit of AB. Let p_{AB} be the price the incumbent charges its customers for a unit of AB.

After unbundling, the new entrant should be able to offer service AB, by renting element A from the incumbent. According to the Efficient Component Pricing Rule the incumbent should change a rental price of

$$p_A = p_{AB} - c_B = c_A + (p_{AB} - c_A - c_B)$$

Notice that, in this case, the incumbent receives c_A plus his profit for providing a unit of AB. That is, he receives both the cost of providing element A and the opportunity cost he incurs through being unable to offer a unit of AB because he has given up a unit of A. In ECPR, this opportunity cost is defined for equivalent services: a long-distance customer of the incumbent will become a long-distance customer of the new entrant.

For example, if $c_A = \$0.50$ and $c_B = \$0.40$ and $p_{AB} = \$1.20$ (i.e. there is a profit of $\$0.30$), then $p_A = \$0.80$. Notice that this equals the sum of c_A and the profit of the incumbent.

The main motivation for the ECPR is that a new entrant can only survive in the market if he is efficient in producing B, i.e. if he produces B at a cost c'_B that is no more than c_B. Otherwise, he will have to price AB as $p_A + c'_B = p_{AB} - c_B + c'_B > p_{AB}$, and be uncompetitive against the existing price. Thus ECPR deters inefficient entrants. However, this may be the only good thing that can be said of EPCR. Against it can be said the following:

1. It reduces the profit of efficient entrants by including the private opportunity cost of the incumbent in the rental price of A. If the incumbent is inefficient, then this amounts to a tax on the entrants. Let us see this in detail. Assume that the new entrant is efficient with $c'_B < c_B$. Then by undercutting the incumbent's prices by ϵ he makes a profit of $c_B - c'_B - \epsilon$ instead of $p_{AB} - \epsilon - c_A - c'_B$, if he were just paying for the actual cost of A. The difference of these two terms is the private opportunity cost of the incumbent $p_{AB} - c_A - c_B$, and has a form of a tax paid by the new entrant to the incumbent.

2. It guarantees the incumbent's profit margin through inclusion of his private opportunity cost. Even when the incumbent is inefficient, he is guaranteed the same income. Repeating the previous argument, an efficient entrant posts a price $p_{AB} - \epsilon$ and gets all the market AB. For each unit of service he pays the incumbent $p_{AB} - c_B$.

So the incumbent continues to make the same profit as before on every unit of service that the new entrant sells. Since the new entrant is forced to lower prices for AB in order to obtain market share, demand will increase, which translates into increased profits for the incumbent. An inefficient incumbent sees his profits increasing instead of being pushed out of the market!

3. It does not provide any motivation for estimating accurately the cost of the element A. Hence ECPR perpetuates the historic costs of the incumbent in this part of the network.

4. The incumbent has no incentive to be more efficient and reduce c_A. However, a reduction in c_B will increases the rental price of A.

5. Since ECPR deters certain entrants, it can increase the incumbent's market share of provision of the AB service. This may lead to a reduction in the marginal cost of B, through economies of scale, and thus increase the rental price of A, thereby further reducing competition.

6. If the true costs are not easily obtained then the incumbent may claim that the cost of A is greater than c_A (without affecting his total cost $c_A + c_B$); this increases the rental price of A. He essentially achieves double cost recovery.

7. There are administrative problems, because the same network element may have to be rented at different prices, depending on the service for which it is employed. This means the ECPR can produce price discrimination and that it can be very complicated to determine the price p_A. For example, if the new entrant wants to use A for providing local service over the access network, the opportunity cost will be different.

To conclude, the ECPR is inconsistent with the typical goals of a regulator because it does not lead to price decrease, is not cost-based, and can lead to price discrimination.

7.4   Comparing the various models

Let us summarize some advantages and disadvantages of the pricing schemes introduced in the last sections. The advantages of FDC based prices that use historic costs are

- they are easier to develop since they are based on linear relations with the actual cost information and are easier to understand by accountants;
- they are based on accounting data that are retrieved and kept any way in the information system of the firm;
- they are easy to audit by regulators.

The disadvantages of such prices are that:

- they do not provide incentives for improving the efficiency of the provider and deploying newer technologies since they cover his full historic cost;
- they are not always based on causal relations but depend on arbitrarily chosen coefficients for sharing the non directly attributable cost; hence these do not reflect the actual cost of services. This problem can be reduced if one uses the activity-based costing scheme.

LRIC+ combined with bottom-up models using current costs has the advantages that:

- it generates prices that are subsidy-free, hence stable and in many cases economically efficient;
- since it is not based on historic costs, it does not include inefficiencies that are due to decisions made in the past, and provides the right competitive signals to the market;

and the disadvantages that

- it is hard to develop due to the complexity of the bottom-up models and the large amounts of information needed to input the right parameters to the models;
- since these prices are not based on traditional accounting procedures, accountants find them hard to understand.

It must be clear to the reader by now that top-down models are based on actual costs while bottom-up models deal with hypothetical systems and hence serve for different purposes. LRIC+ prices that are based on top-down models using current costs do not have auditability disadvantages but still hide potential inefficiencies in the network. They are presently favoured by regulators as the way that incumbent operators should construct prices for wholesale services that are sold to competitors. Such systems tend to displace the older generation of costing systems based on FDC. Traditional LRIC+ based on bottom-up models results in even lower prices, and is mainly used to detect network inefficiencies. This is the case if the prices constructed by the top-down and bottom up models differ significantly.

7.5 Flat rate pricing

In *flat rate pricing* the total charge that a customer pays for a service contract is fixed at the time the contract is purchased. That is, it is determined *a priori*, even though the actual cost of the contract to the service provider can only be known *a posteriori*, i.e. at a later time, which might be only at the end of the contract. For example, if a connection consumes resources which cannot be predicted at the time its contract is made then the cost of the contract cannot be know *a priori*. In most cases, prices reflect some average concept of cost taken over all the contracts in the past history of the system. Examples of flat rate contracts include flat-fee Internet access with a monthly fee that depends upon access speed, dining at an 'all-you-can-eat' restaurant, and charging a VBR ATM service according to its traffic contract parameters without taking account of the actual bandwidth used.

There are two advantages that have led to the wide use of flat fees. First, a flat fee is simple to implement. Second, customers often prefer the predictability of a flat fee. However there are also serious drawbacks. Flat rate pricing tends to produce high social cost because of the waste of resources. It is unstable under competition because if light users subsidize heavy users, the light users are likely to switch to a competitor who offers them a fairer pricing scheme. It is easy to see that flat rate charging may produce prices that are not subsidy-free in the senses of Section 7.1. However, since flat rate is widely used in practice, it deserves a detailed discussion and clarification of its inefficiencies. To make things more concrete, we use the example of flat-fee Internet access and use a simple economic model to capture most of the essential aspects.

Consider the case of a service provider who provides Internet access by selling customers contracts for a fixed access speed at a flat monthly fee of p_{flat}. The resource consumed by a customer during one month is approximated by x, the total monthly volume of bytes

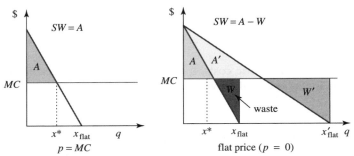

Figure 7.6 Social waste under flat rate pricing. If a user is charged a price $p = MC$ then he consumes x^* and the social welfare is the area A. However, if he is charged a flat price, say $p = 0$, then he has no incentive to reduce his consumption and so consumes x_{flat}. This makes the social welfare $A - W$ where W is social waste. Thus charging only a flat fee encourages social waste. For a demand function with a greater demand, so the demand at $p = 0$ is x'_{flat}, the social waste of W' is even greater.

of inbound traffic that he receives from the Internet.[2] The monthly cost to the provider is $a + bv$, where v is the total volume of inbound Internet traffic. Assume customers have individual demand functions, which are linear, start from the same point on the price axis and end at the point $(x^i_{\text{flat}}, 0)$, on the monthly volume axis, where x^i_{flat} is the maximum amount a customer of type i can consume at price zero.

First we show that flat rate pricing encourages social waste. Consider a customer of type i, but omit for simplicity the superscript i in the notation that follows. The optimal price under which social welfare is maximized is the marginal cost $MC = b$, which results in a social surplus equal to the area of A, the upper triangle in the left part of Figure 7.6. If the customer is faced with a flat-fee charge, he has no incentive to reduce his resource consumption and consumes up to his maximum possible level, i.e., x_{flat}. In this case the consumer's utility is the area of the triangle bounded by the demand line, and the social cost of its provision bx_{flat}. This reduces the optimal social surplus value A by the amount W; we call W the social waste that is induced by the flat rate price. This is shown in the right part of Figure 7.6, which also suggests that social waste increases with the value of the maximum resource consumption x_{flat}. Note that x_{flat} may vary for different customer types due to their different maximum network access speeds or time available for Internet access.

Consider now a flat fee that is computed to cover the variable cost of the network. Suppose we compute the average amount $\bar{x}_{\text{flat}}$ consumed under flat fee pricing and then charge all customers for that amount. This is illustrated in Figure 7.7, where it is assumed that customer types are uniformly distributed in terms of maximum resource consumption.

In this case, since the charge is equal to $b\bar{x}_{\text{flat}}$. Those who consume less than $\bar{x}_{\text{flat}}$ find it unprofitable to participate and choose to end their contracts. This is because their utility for using the service is less than what they must pay. Suppose we continue this argument in many rounds. In each round the flat fee is updated to reflect the new average consumption cost after some customer types have left. Eventually, only the heaviest users of the service will remain. Clearly, the customer base is greatly reduced. If profits are a mark-up of total value of contracts this will also greatly reduce profits.

It is interesting to analyze the Internet flat rate pricing example from the perspectives of fairness and price stability introduced previously. Assume that the fixed cost of the provider

---

[2] As present practice suggests, typical users of the Internet mostly download information instead of uploading.

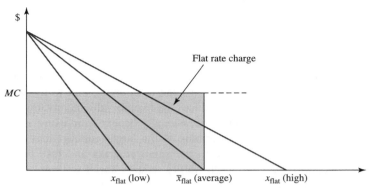

Figure 7.7 Cross subsidization with a flat fee. Suppose a flat fee is charged, sufficient to cover the cost of average usage, i.e. fee= $x_{\text{flat}}MC$. However, having paid that fee, a low users will consume $x_{\text{flat}}(low)$ and find that he has negative net benefit. Given this, he will chose not to buy from this service provider.

is shared equally among its N customers, i.e. each one pays a flat fee of $a/N + b\bar{x}_{\text{flat}}$. We can now observe the following:

- Flat rate pricing is not subsidy-free, since high consuming customers are charged below their incremental cost. This is the case if the fixed cost is shared among many customers and hence $b\bar{x}_{\text{flat}}(high) > a/N + b\bar{x}_{\text{flat}}$.
- If the fixed cost a is small enough then flat rate prices are neither sustainable or in the second-best core. For example, suppose $a = 0$. If a new provider offers contracts priced at $b\bar{x}_{\text{flat}} - \epsilon$ to users that consume less than $\bar{x}_{\text{flat}}$, then he will attract those users and cause the prices of the incumbent to collapse, since under these price the incumbent can no longer recover his costs. The service differentiation can be effected by the new entrant using simple policing mechanisms which restrict users that consume above their contract levels. In Chapter 8 we investigate several related issues and discuss why subsidy-free prices are desirable. A similar result will be obtained if the new entrant charges its customers proportionally to the actual amounts of service they consume.

One way to reduce the bad effects of flat rate pricing is for the contract to restrict the range allowed to the resource usage, v. This produces prices that are fairer to individual customers. For example, instead of defining a contract for which v can be anywhere in the interval $[0, M]$, one could define m contracts, such that the ith contract limits v to the interval $[0, k_i]$, where $0 < k_1 < \cdots < k_m < M$. Each of these contracts is priced on a flat rate basis. To enforce the contract constraints, the service provider must police the users' traffic: if a customer exceeds the maximum resource usage allowed by his contract, then he must be blocked from using more of the service. Clearly, a customer has an incentive to predict as accurately as possible his resource usage and purchase the cheapest contract that accommodates his needs. For customers with low variance, and hence predictable resource usage, this scheme can work well, and it clearly reduces social waste and cross-subsidization between heavy and light users. However, if a customer finds usage too difficult to predict because of large variance of v, and blocking is costly (he cannot afford to be precluded from using the service when he needs to), then he may have to purchase the largest contract. Consider a customer with a low average, say $0.1M$, but with a non-negligible probability to consume M, and who cannot afford blocking. This customer may need to buy the largest contract $[0, M]$, and will feel that he is not charged fairly, since he pays the same amount as a customer who always uses M.

However, if v has a large variance, measured within a large period $T = \sum_i T_i$, but its contribution in a smaller sub-period T_i is predictable given the available information of the customer at that time, then it may be preferable for the customer to purchase a new contract in every sub-period. This may be a good strategy if the overhead of purchasing and activating contracts is negligible for the network and its customers. In general, forcing the customer to choose a single contract type, instead of allowing him to dynamically switch contracts according to his actual needs, reduces the overall value he obtains from using the network. For example, a light user might occasionally benefit from using a larger contract, with greater service quality. Since subscribing to the larger contract may be exceedingly expensive, this customer will subscribe to the small contract and lose the capability of higher quality. Hence denying contract flexibility reduces social welfare.

Note, also, that policing a contract may require some measurements, and this increases the implementation cost. However, this cost is generally less than the costs of the measurements, accounting and billing involved in usage charging. Often, policing is done automatically by the line access speed that connects the customer to the network.

The previous discussion suggests that the policing that may be required to reduce the bad effects of flat rate pricing can greatly reduce customer flexibility, especially for those customers with bursty resource usage requirements. This deficiency can be alleviated if the charge for a contract has a usage component. For example, instead of performing 'hard policing', by dropping packets and blocking usage, we might perform 'soft policing', by simply measuring usage and making the customer pay extra if he exceeds what his contract allows. If the customer's cost of blocking is substantial, then he may prefer to pay this extra charge occasionally. He might even buy insurance against such an event from a third party. Such 'soft' mechanisms increase the value of the service to customers, and so increase social welfare. They are stable because customers end up paying more closely for the resources consumed.

7.6 Further reading

The book of Mitchell and Vogelsang (1991) contains comprehensive treatment of theoretical aspects of cost-based pricing. The net incremental revenue test is due to Baumol (1986). Activity based costing is discussed in the book of Hilton, Maher and Selto (2003). A thorough discussion of the merits of LRIC for pricing access network services is in Economides (2000). For definitions of the ECPR see Willig (1979) and Baumol (1983). The inefficiency of ECPR is discussed by Economides (1997). The discussion on flat rate pricing is based on Edell and Varaiya (1999).

Introductory material on bargaining and cooperative games can be found in the course lecture notes of Weber (2001). The bargaining game is due to Nash (1950), and its treatment as a competitive game is due to Zeuthen (1930) and Rubinstein (1982). The variation of Section 7.2.2 is that of Kalai and Smorodinsky (1975). For further details we recommend the books of Luce and Raiffa (1957), Karlin (1959), Eatwell *et al.* (1989), Binmore (1992) and Osborne and Rubenstein (1994).

8

Charging Guaranteed Services

In Section 2.1.5 we defined a guaranteed service as one for which there is a contract between the service provider and the customer. This contract specifies obligations for both parties. The service provider agrees to provide a service with certain quality parameters so long as the customer's traffic satisfies certain constraints.

In general, a contract for a guaranteed service may allow some flexibility. Certain contract parameters, such as maximum peak rate, may be renegotiated and allowed to change their values during the life of the service. For example, the contact might specify that the network guarantees no information loss so long as the user sends at no more than a maximum rate of h Mbps. The value of h may be renegotiated at the beginning of every minute to be some value between 1 and 2. Thus there is a part of the contract which guarantees no cell loss at a rate of 1. Any extra rate above this must be negotiated. One possibility is that the extra rate must be bought in a bandwidth auction. This auction is run by the network operator so as to better utilize spare capacity. A second possibility is that the operator posts a price $p(t)$ and lets the user choose how much bandwidth in excess of 1 he wishes to buy. He sets $p(t)$ to reflect the present level of congestion in the network. Seeing $p(t)$, the user must choose the amount of bandwidth in excess of 1 he would like.

Chapter 10 is about charging flexible contracts and pricing methodology that gives users incentives to make such choices optimally. However, in this chapter we restrict attention to guaranteed services whose contracts do not allow the users such flexibility. We suppose that all contract parameters are statically defined at the time the contract is established. Equivalently, we restrict attention to that portion of the contract which has no flexibility and for which the network is bound to provide some minimal requirements, known at the time the contract is established and persisting throughout its life. In the example above, this portion of the contract is the obligation to provide a 1 Mbps rate at no cell loss. We use ideas of previous chapters to develop a theory of charging for such contracts. We do this in various economic contexts, such as the maximization of the social welfare or the supplier's profit.

Most interesting guaranteed services have contracts that specify minimum qualities of service that the network must provide, such as minimum throughput rate, maximum packet delay or maximum packet loss rate. This means that the network must reserve resources to meet the requirements of the active service contracts, and if network resources are finite, the network must operate within its technology set. Recall from Chapter 4 that the technology defines the set of services and their quantities that it is within the network's capability to provide at one time. In this chapter we analyse, in different economic contexts, the

---

Pricing Communication Networks   C. Courcoubetis and R. Weber
© 2003 John Wiley & Sons, Ltd   ISBN 0-470-85130-9 (HB)

form of prices that result from considering the particular structure of the constraints of technology sets.

An important distinction between service contracts for communications services and some other economic commodities is that the former do not specify fully the resources that are required to produce a unit of output. For example, the resources that are required to produce a particular model of personal computer are fixed before its manufacturing starts, whereas a connection whose service contract specifies only an upper bound on the connection's maximum rate may use buffer and bandwidth in a way that can only be known to the network once the connection ends. The fact that some information is known only 'a posteriori', rather than 'a priori', makes the problem of pricing service contracts quite complex. We will see that by including component of usage in the tariff we can produce a charge that more accurately reflects the actual resource consumption. This type of charge can provide a customer with the incentive to change his prospective network usage in a way that benefits overall system efficiency. Perhaps he might smooth his traffic and make it less bursty, or use some sort of compression scheme to reduce its total volume. If there is no usage component in the charge then customers have no incentive to conserve resources; instead, they may be wasteful of resources and behave in ways that reduce the overall efficiency and capacity of the network. We argue that flat rate pricing can lead to exactly this sort of waste, and that pricing methods which include a usage charge are to be preferred.

Chapter 4 presented the concept of an effective bandwidth as a proxy for the quantity of network resources consumed by a bursty connection. In Section 8.1 we discuss market models for which it is or is not appropriate to use effective bandwidths as the basis for pricing network connections. In Section 8.2 we investigate the more complex problem of constructing tariffs for service contracts. We discuss the pros and cons of flat rate pricing and give justifications for using tariffs that take account of actual network resource usage and charge proportionally to effective bandwidths.

As we see in Section 8.3, it is important that the tariffs for service contracts be incentive compatible. A network can be more competitive and fairer to its users if it presents them with a range of tariffs, each of which is intended for a specific user type. In the simplest case, a network might offer two different tariffs: one for heavy users and one for light users (as we did in Example 5.5.3). The network cannot prevent a heavy user from choosing the tariff that is intended for light users, but it can construct the tariffs so that heavy users pay less on average if they choose the tariff that is intended for them, rather than the tariff that is intended for light users. This gives users the incentive to make choices that are informative to the operator, who can tell whether the a customer's consumption of network resource is more likely to be heavy or light, before any resources are actually consumed. This information can help the operator to dimension and operate his network more efficiently, for the benefit of all his customers. At the end of Section 8.3 we explain the competitive advantage of such tariffs, and consider some related problems of arbitrage and splitting.

Section 8.4 describes three simple pricing models that make use of this type of pricing. Section 8.5 presents a simple example to illustrate the long-term interaction between tariffing and the load on the network.

8.1 Pricing and effective bandwidths

A simple example will illuminate the relationship between the prices for services and their effective bandwidths. Suppose a network operator offers two contract types to his customers

that the optimal price and the operating point are completely determined by the degree of competition and the price elasticity of demand, as summarized by θ and ϵ (recalling that ϵ, the price elasticity of demand, is a function of p). In other words, the revenue maximizing price of the service does not depend on the amount of resources it consumes in the network, but only on its demand. The marketing department should construct the tariff for the service from market research. There is no need to consult the engineering department and to better understand what use the service contract actually makes of network resources.

If the constraint of the technology set is active, then $\mu > 0$ in (8.2). In this case the price depends also on the shadow cost and the derivative of the constraint. However, if the market is highly competitive (so θ is approximately 1) and there is a negligible marginal variable cost, then we obtain (approximately)

$$p(x^*) = \mu g_1'(x^*) \tag{8.4}$$

Thus the price has a simple form, which we can exploit further. Since $g_1(x)$ is a constraint of the technology set, we can use the results from Section 4.5 to approximate $g_1(x) = b_1$ locally at x^* by $x^*\alpha(x^*) = C$, where C is the effective capacity of the link, and obtain

$$p(x^*) = \lambda\alpha(x^*)$$

Note that if there are multiple contract types the same analysis holds. Then $g_1(x) = b_1$ is approximated by $\sum_i x_i \alpha_i(x^*) = C$ and

$$p_i(x^*) = \mu \frac{\partial g_1(x^*)}{\partial x_i} = \mu\alpha_i(x^*) \tag{8.5}$$

where α_i is the effective bandwidth of contract type i. Thus

$$\frac{p_i(x^*)}{p_j(x^*)} = \frac{\alpha_i(x^*)}{\alpha_j(x^*)} \tag{8.6}$$

That is, optimal prices are proportional to the effective bandwidths of the corresponding contracts. They also depend upon the shadow price of the resource that is constrained as $g_1(x) = b_1$. In this case the marketing department must surely consult the engineering department to obtain some reasonable approximations for the effective bandwidths of the services. Marketing research should help in determining the value of μ.

Another practical approach is to use tatonnement to find the appropriate prices. This requires no *a priori* knowledge of the value of μ. We only need to know the relative values of the effective bandwidths. The tatonnement proceeds in an iterative fashion as follows. Pick a set of prices in proportion to the effective bandwidths. This corresponds to choosing a value of μ. Determine whether for these prices the demand lies inside or outside the technology set and then respectively inflate or deflate all prices by the same small percentage. Repeat this step, until the demand lies just inside the technology set.

In practical terms, given that the network operator wishes to solve (8.1), the value of the shadow price μ is the amount he would be willing to pay to increase by one unit the constant b_1 of the binding constraint. In our case, this corresponds to increasing C, the effective capacity of the link. If the price for increasing C in the actual market is less than μ then there is an incentive is to expand the network. Observe that μ depends upon demand. The greater the demand for services, the greater μ will be.

In general, if there are multiple contract types, then contract types can be substitutes and complements for one another. If the price for one contract type increases, the demands for other contract types can increase and decrease. In the general case, maximizing L gives in place of (8.2), and generalizing (6.7),

$$\sum_j \frac{p_j - \frac{\partial c}{\partial x_j} - \mu \frac{\partial g_1}{\partial x_j}}{p_j} \epsilon_{ij} = -(1 - \theta) \tag{8.7}$$

Example 8.1 (Pricing minimum throughput guarantees) Consider a single link that can carry Q bytes in total within a period of length T. The contract of a transport service is defined in terms of the maximum number of bytes, say q, that the network will transport on behalf of the contract during this period. In other words, the network guarantees a throughput rate of q/T over the time window of length T. Such a contract does not specify any other performance guarantee. Let us suppose that each contract that is accepted by the network is required to make all the bytes that it wishes to have transported available at the beginning of the period T (since it would clearly be very troublesome if the data were available only towards the end of the period). How should the network price this contract? Should prices be in proportion to q?

Based on our previous discussion, the answer depends on competition aspects. In a social welfare optimization context, prices of contracts should be proportional to effective bandwidths. Let us discretize the size of the possible contracts and enumerate them so that q_i is the size of a contract of type i, $i = 1, \ldots, k$. Now the technology set of the network is $\sum_i x_i q_i \leq Q$, where x_i is the number of contracts of type i. Hence the effective bandwidth is $\alpha_i = q_i$, and the optimal prices are of the form $p_i = \lambda q_i$. In other words, λ is the price per byte, and is the same for all contracts irrespectively of their size. Clearly, such a simple charging scheme is not optimal when the network operator has market power. He may use volume discounts to effect price discrimination in selling his service and so obtain larger revenues from his customers. If the operator can use personalized pricing, then he will wish to make each user a take-it-or-leave-it offer.

In concluding this section, we observe that we have not yet spoken about one further important aspect of the pricing problem that is special to the nature of transport services and makes pricing decisions even more complex. This concerns arbitrage. By their nature, transport service contracts can be combined and re-sold in smaller units. For instance, one may buy a contract with a large effective bandwidth and resell it to other customers in terms of a number of different contracts with smaller effective bandwidths. The traffic of these customers must be multiplexed and then demultiplexed at the end, at some cost. However, if there is little competition and marginal variable cost is near 0, the implication of (8.3) is that prices should be computed solely on demand assumptions. But these prices can be impractical. This is because high prices for certain services provide the incentives for customers to buy cheaper service types and then disguise them as the expensive service types, i.e. to use them to transport the data of the applications which would otherwise buy the expensive services. Such an incentive is reduced if prices reflect actual resource consumption. Alternatively, network operators may avoid such commoditization of their transport services by combining them with other offerings such as security, reliability and global availability. Personalizing a service according to the customer's needs is an important tool for achieving greater revenues. Hence in practice, revenue maximizing operators will choose prices that are related to effective bandwidths to provide for a stable environment in which to offer services. Such choices must also take account of demand, personalization capabilities, and the cost of service resale by third parties. We return to these issues in Section 8.3.5.

Finally, we extend our results to the general case of pricing contracts for connections over a network instead of single link.

8.1.1 The Network Case

We let L be a set of links and R be a set of routes, a route being a set of links. Connections are made over routes, and use contracts from a finite set of contract types, K. Suppose that a connection using route r has contract type k. Then, as in Section 4.13, we can assume for simplicity that the effective bandwidth α_k that is consumed by a contract is the same on each link of the route, and so depends only upon the type of the contract. Denote by C_j the effective capacity of link j.

Let x_{rk} be the demand for contracts of type k over route r, and assume that this demand arises from the users' aggregate utility function $u(\{x_{rk}\})$. In this case, taking account of (4.28), the social welfare maximization problem becomes

$$\underset{\{x_{rk}\}}{\text{maximize}\, u(\{x_{rk}\})}, \quad \text{subject to} \sum_{r:j\in r} \sum_k \alpha_k x_{rk} \le C_j, \quad \text{for all } j \in L \qquad (8.8)$$

where $\{r : j \in r\}$ is the set of routes that use link j. The Lagrangian is now $L = u(\{x_{rk}\}) + \sum_j \lambda_j (C_j - \sum_{r:j\in r} \sum_k \alpha_k x_{rk})$. As in the single link case, we take the derivative with respect to x_{rk} and find that the optimal price for contracts of type k on route r is given by

$$p_{rk} = \alpha_k \sum_{j:j\in r} \lambda_j \qquad (8.9)$$

If λ_j is the shadow price of effective capacity on link j, then the quantity $\sum_{j:j\in r} \lambda_j$ is the charge per unit of time of a unit of effective bandwidth along route r. This again suggests that optimal prices should be proportional to effective bandwidths. The price for a contract over route r is equal to the product of the effective bandwidth of the contract and the price of a unit of effective capacity along route r.

Such prices can be computed by a tatonnement. Each link of the network posts its price for effective capacity. These lead to prices for contracts along all routes. The demand for contracts adjusts itself to these prices. Each link now increases or decreases its price depending on whether or not there is excess demand for effective capacity at that link. Iterating this procedure, prices eventually converge to ones that achieve the optimum in (8.8).

As a simple application, consider the following approach for pricing guaranteed quality services using the Integrated Services architecture described in Section 3.3.7. To establish the contract, the originating node declares, in addition to its quality of service requirements, its maximum willingness to pay (per unit time) for the connection. In the process of establishing the connection, bandwidth is reserved at each link, and the available budget is decremented by the cost of the bandwidth at each link. If it is found that the budget is sufficient, then the connection is established and the price is set to the sum of these costs. Otherwise, the connection is rejected, or it is allowed to renegotiate a reduced bandwidth requirement. The links constantly update prices to reflect available capacity. Prices should rise if the available capacity becomes small, say less than 10% of the total link capacity.

8.2   Incentive issues in pricing service contracts

In practice, service contracts specify constraints which restrict the *maximum* amount of resource usage. This contrasts with other economic goods for which the resource use is specified exactly. For example, a traffic contract might specify a maximum access rate or a leaky bucket constraint. The fact that a traffic contract only constrains the maximum resource consumption creates a number of interesting incentive issues. In this section, we discuss the impact of the structure of tariffs on actual resource usage. This motivates the construction of tariffs that combine *a priori* and *a posteriori* contract information.[1] Such tariffs include an element of usage charge and make sense from the viewpoint of both the network and users. Let us consider the user's viewpoint first.

Consider a simple model for a user application that needs a contract to transport data with a constant rate x through the network. (In general, x may be an effective bandwidth.) If all network applications were of this type, differing only in the value x, and this were a known parameter, things would be simple. Each user would request a contract that exactly fits the needs of his application, and pay appropriately. Unfortunately, in practice, x is not known and so we must model it as a random variable. For instance, the application may be known to produce data at a rate, x, which randomly takes a value in the range $[x_1, x_2]$, independently chosen each time the user starts the application. What contract should the user select? One possibility would be for him to play safe and buy a contract for x_2. But this contract may be very expensive. A second possibility is for him to purchase a contract for a rate y between x_1 and x_2, which would be sufficient most of the time. However, the downside it that when x exceeds y, the policing mechanisms of the network will trim the rate and the application will experience unacceptable performance. This will reduce the value of the service to the customer. The user may also feel that he is charged unfairly every time x is less than y, since he pays for y even though he does not use it. Such a user would benefit from a contract that allows his applications to use the range of rates up to x_2 (to reflect *a priori* information, that x_2 is known), but charges him something that reflects the actual rate he uses (the *a posteriori* information about x).

Consider now the network's perspective. We have argued in Section 8.1 that charging in proportion to the effective bandwidths may be the optimal approach under appropriate market conditions. However, there are subtleties in the conversion of an effective bandwidth into a charge. As we have already discussed, these subtleties arise because contracts specify a range of possible effective bandwidths, rather than a unique one. An additional complexity is that users may alter their traffic generating applications in response to the incentives that are provided by whatever effective bandwidth definition is used to price the contract.

Let us investigate two extreme possibilities. Consider first the problem of designing an effective bandwidth pricing scheme that is based only on *a priori* information. That is, it does not take account of the actual traffic that is carried under the contract. For simplicity, suppress the coefficient μ from the effective bandwidth charge, and assume that the network has all the information it needs to compute the effective bandwidths.

The *a priori* information that might be available for all connections of type j, could include the fact that all connections of this type are subject to the same traffic contract. Perhaps this contract is defined in terms of leaky bucket parameters. The *a priori*

---

[1] *A priori* information consists of the contract's static parameters and knowledge of the amount of resources that connections using this type of contract have consumed in the past. The *a posteriori* information includes the amount of resources that the connection actually consumed; it may include statistics about the traffic that was generated during the connection's life.

information might also include data on past connections of type j. For example, one might estimate the effective bandwidth of connections of type j in the following way. Suppose that we have seen n_j connections of type j. We take the kth connection that we have seen of type j, divide its duration T_k into intervals of length t, and then compute

$$\frac{1}{n_j} \sum_{k=1}^{n_j} \left[\frac{1}{T_k/t} \sum_{i=1}^{T_k/t} e^{sX_{jk}[(i-1)t,it]} \right] \tag{8.10}$$

where $X_{jk}[(i-1)t, it]$ is the number of bytes of traffic that was measured from connection k in the interval $[(i-1)t, it]$ (with $i = 1$ denoting the start of the connection). This would give us an empirical estimate of the expectation

$$E e^{sX_j[0,t]}$$

which appears in the effective bandwidth definition (4.5). By taking the logarithm of this and multiplying by $1/st$, we could make an estimate of the effective bandwidth of a connection of type j, say $\tilde{\alpha}_j(s, t)$. (Note that we must average over many connections of type j. Because we have not assumed ergodicity of sources of type j, the evaluation of (8.10) may differ significantly between two connections of this type.) We can now simply charge each newly admitted connection of type j an amount per unit time equal to the empirical estimate $\tilde{\alpha}_j(s, t)$. That is, each connection of type j is charged proportionally to the average effective bandwidth of past connections of the same type.

This is really the same as flat rate pricing, in which all users pay an identical rate of charge, calculated from the average resource usage of previous similar users. It is also the charging method of an all-you-can-eat restaurant. In such a restaurant, each customer is charged not for what he eats, but for the average amount that similar customers have eaten in the past; (we say 'similar customer', because some restaurants have a lower price for children or different prices depending on the time of day). The existence of all-you-can-eat restaurants demonstrates that this charging scheme is viable. It is analogous to the charging scheme used when local telephone calls are unmetered, or when the only cost a student pays to browse the WWW is the cost of waiting for a free seat in the computer room. However, all-you-can-eat restaurants are not for everyone. They encourage diners to overeat; they tend to serve only the lower quality part of the market. Customers with small appetites may feel that they are overcharged. Others are put off by the bare-bones, help-yourself, no-frills ambiance.

We can identify two problems with a flat charging scheme. The first concerns a user who has connections of type j but whose traffic usually has an effective bandwidth that is less than the average for this type. Such a user may feel that he is being overcharged, and subsidizing other users of connection type j whose traffic usually has a greater effective bandwidth than his. Consequently, he may defect to a service provider who uses a charging method that is more favourable to him. The second problem is that customers have an incentive to overconsume. Since the charge does not depend on usage, customers have no incentive to use applications in ways that conserve resources. Network resources will be wasted, and probably congestion will increase. The result is that the typical contract will have a larger effective bandwidth, and this must eventually be reflected in a greater contract price. As before, customers with light usage may change providers, and ultimately the network will be left with only the heaviest users. This is known as the *adverse selection problem*. Thus, it is clear that a flat pricing scheme has severe problems. Similar problems occur with a form of peak rate pricing, in which the operator defines the effective bandwidth as the greatest effective bandwidth that can result under the given contract.

Having examined one extreme, let us examine the other: a charge based completely on *a posteriori* measurements. For example, one might charge the kth connection of type j proportionally to

$$\hat{\alpha}_{jk}(s, t) = \frac{1}{st} \log \left(\frac{1}{T_k/t} \sum_{i=1}^{T_k/t} e^{s X_{jk}[(i-1)t, it]} \right) \tag{8.11}$$

This is the effective bandwidth of this connection measured *a posteriori*. Apart from the difficulty of interpreting this complicated tariff to users, there is the following conceptual flaw. Suppose that a user requests a connection policed by a high peak rate, but then actually transmits very little traffic over the connection. Then the *a posteriori* estimate of the effective bandwidth given by (8.11) will be near zero, and hence the charge near zero, even though the a priori expectation may be much larger, as assessed by either the user or the network. The network bears too much of the risk inherent in uncertainty about the user's traffic, since the network may have to allocate at least some resources on the basis of *a priori* information about the connection.

Our discussion of the two extreme cases above has highlighted the flaws in two possible approaches to charging. A third approach, which we believe to be the most reasonable, attempts to circumvent these flaws. It creates a charge that is close to the actual effective bandwidth of the connection. Like the first approach, it takes account of *a priori* information in the contract. This ensures that some charge is made for resources that must be reserved even if they are not used. Like the second approach, it also takes into account actual usage. This ensures users have an incentive not to overconsume.

The key idea of the approach is that the charging scheme is framed in terms of a menu of several tariffs. The user chooses in advance the tariff from which he would like his charge to be computed. Clearly, he will choose the tariff under which he expects the smallest charge. This is the one for which he would expect the smallest average charge, given what he knows about his likely use under the contract. The network can use the information about the tariff selection to better estimate the effective bandwidth of the particular contract. Hence, the network can do a better job of call acceptance, utilize its resources better, and in principle provide more services. This alignment of incentives between the individual choices made by users and the network's goal of optimizing its performance is what we call the incentive compatibility property of the charging scheme. We explain more details in the next section.

8.3   Constructing incentive compatible tariffs from effective bandwidths

In this section we present an incentive compatible charging scheme. It is based on the effective bandwidth concept. It avoids the problems of a charge that is based only on *a priori*, or only on *a posteriori*, information. The key idea is to approximate the effective bandwidth by an upper bound that depends on both *a priori* and *a posteriori* information, i.e. upon both the static parameters of the contract and actual measurements. This gives a good approximation of the actual effective bandwidth of the traffic stream produced by the contract. We bound the effective bandwidth by a set of linear functions of parameters that are measured *a posteriori*, with coefficients that depend on the static parameters known *a priori*. These linear functions become the basis for simple charging mechanisms. In particular, users are offered the set of linear functions as tariffs. If the user knows the expected value of the parameters that are to be measured, he can choose the tariff that minimizes his expected charge. Even if he does not know these expected values precisely, better estimates of them can help him to select a better tariff. Although this method can be used for arbitrary measurements, we illustrate it by considering simple measurements of the contract's duration and the volume of bytes carried under it.

There are other important issues that we also discuss. We show by an example that providers who use such effective bandwidth schemes have a competitive advantage over those who use flat rate schemes. We also discuss, as at the end of Section 8.1, issues of contract arbitrage, resale and splitting. By their nature, transport contracts do not specify the ownership of the bytes carried. Hence, a customer may himself become a transport service provider by selling parts of his transport capability to other customers. Pricing schemes that leave open this possibility are usually not desirable, since they are vulnerable to competitive entry.

8.3.1   The Time-volume Charging Scheme

We illustrate our approach by describing how one class of traffic might be charged. Suppose this class of traffic uses a traffic contract under which the user must send at no more than a maximum rate h (the *a priori* information). Imagine that a connection uses this contract and sends data at a mean rate m (the *a posteriori* information). It can be shown that amongst possible traffic of mean rate m and peak rate no more than h the traffic with the greatest effective bandwidth is one that is periodically on and off, and has on and off phases of long duration. As we have seen in Example 4.5, this type of traffic has an effective bandwidth given by

$$\alpha_{\text{on-off}}(s, t) = \frac{1}{st} \log \left[1 + \frac{m}{h} \left(s^{sth} - 1 \right) \right] \tag{8.12}$$

Here, s and t are defined by the operating point of the multiplexer.

Think of $\alpha_{\text{on-off}}$ as a function of m, where without confusion we can write it as $\alpha_{\text{on-off}}(m)$, a concave function of m. Note that the network does not know the value of m when the contract is established. Now for our traffic contract, parameterized by the peak rate h, we define a family of tariff lines, parameterized by the parameter m, each of which takes the form

$$f_m(M) = a(m) + b(m)M$$

which as a function of M lies above the curve $\alpha_{\text{on-off}}(M)$ and is tangent to it at $m = M$. Note that $a(m)$ and $b(m)$ also depend upon h, s and t through the definition of $\alpha_{\text{on-off}}$ in (8.12), but because these are fixed we do not indicate the dependence on them explicitly. The user chooses a tariff, or, equivalently *states a value of m*. The final charge is $T[a(m)+b(m)M]$, where M is the *measured* mean rate of the user's traffic. Equivalently, the charge is $a(m)T + b(m)V$, where $V = TM$ is the volume of traffic carried (measured in cells or bytes) (see Figure 8.1).

Does such a scheme really charge for effective bandwidths? Can we make the user reveal his mean rate, m, through his choice of tariff? If he does not know m, can we give him an incentive to estimate it at the time the connection is set up?

One can easily see from Figure 8.1 that a user's expected charge is minimized when he chooses the tariff with $m = E[M]$, i.e. when the parameter of the tariff equals the expected value of the measured mean rate of the connection. In the figure, we suppose $E[M] = 1$. The choice of the tariff $f_m(M)$, with $m = E[M] = 1$, produces an average charge of $2T$. The choice of $f_{m'}(M)$ produces an average charge of $2.4T$. This is the notion of 'incentive

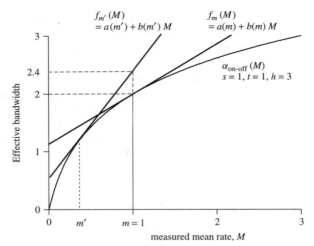

Figure 8.1 Implicit pricing of an effective bandwidth. The effective bandwidth is plotted against the mean rate, M, for a fixed peak rate h. The user is free to choose any tangent to this curve, and is then charged $a(m)$ per unit time and $b(m)$ per unit volume. He minimizes his average charge rate to 2 by selecting $m = E[M] = 1$. If he chooses the tariff indexed by some other value, say m', the average charge will be greater, here $Ef_{m'}(M) = 2.4$.

compatibility': the tariffs are designed so that if a user knows his M and chooses amongst the tariffs in a self-interested way, he will choose the tariff that reveals the true value of his M.

In practice, the user may not know his M in advance, but he has the incentive to make a good estimate of it. If he can make a good estimate, he will be rewarded by being charged less than he would be otherwise. The network operator is provided with some information about the likely mean rate of the connection. He can use this information to help reserve resources appropriately. Thus, the risk that the network reserves the wrong amount of resources is more evenly shared between the provider and the user.

By giving the user a set of tariff choices as above, we obtain several desirable consequences:

- The total charge takes the very simple form $a(m)T + b(m)V$. To charge for time and volume is perhaps the simplest usage-based scheme one could imagine, yet it is firmly based in the theory of effective bandwidths.
- The tariff coefficients depend upon known traffic contract parameters, h, s and t, and so can be easily computed.
- The charge accounts both for resource reservation (which is charged by the time component) and actual usage (which is charged by the volume component).
- The charge requires only simple accounting. It should be simple to measure T and V.
- By allowing a user to specify m to be used in his tariff, a new dimension is added to the traffic contract. Thus, users obtain added-value from the fact that they can choose to be charged in a way that fairly reflects their actual resource usage.

Note that, in this example, the tariff coefficients $a(m)$ and $b(m)$ depend upon the traffic contract through the single parameter h and on the operating point through s and t. One can repeat the analysis for contracts involving more than one static parameter. For example, if a contract is policed by K leaky buckets with static parameters $\{h_k = (\rho_k, \beta_k), k = 1, \ldots, K\}$, then we can take these into account using an effective bandwidth approximation such as (4.20), or even better (4.22). Viewed as functions of the parameter m, these define curves whose tangents have coefficients that depend upon all the static parameters of the contract. These tangents are to be used as tariffs.

8.3.2 Using General Measurements

There are reasons for devising charging schemes based on more refined measurements. Consider two connections. One sends constantly at rate m. The other has the same mean rate m, but sends as an on-off source with peak rate h. Observe that, under the time and volume charging scheme, the expected charges for these two connections are the same, even though the latter one is more bursty, has a greater actual effective bandwidth and consumes more network resources. The problem is that if the only observation available to the network is the amount of data sent then these connections look identical.

The solution is to allow for more detailed measurements. Just as we can allow for more than one static parameter, we can also allow for more than one measured parameter. Let us discretize the duration of the contract into a large number time intervals, T, and let $X_1, \ldots, X_T$ denote the values of the source rate in these intervals. Let $\{g_i(\cdots), i = 1, \ldots, L\}$ be a set of measurement functions. For example, $g_1(X)$ might measure the mean rate $(1/T) \sum_t X_t$. Perhaps $g_2(X)$ measures the fraction of time the actual rate is within 10% of the peak rate h, i.e. $(1/T) \sum_t 1\{X_t > 0.9h\}$. Let us find the greatest effective bandwidth subject to some constraints on the value of the measurements $g(X) = (g_1(X), \ldots, g_L(X))$. Define the largest effective bandwidth possible, given a vector of *a priori* parameters $\mathbf{h}$, and the vector a posteriori measurements $\mathbf{m}$, as

$$\alpha(\mathbf{m}) = \underset{\{X_t\}}{\text{maximize}} \; \frac{1}{st} \log E e^{sX[0,t]}, \quad \text{subject to } g(X) = \mathbf{m}, \quad \{X_t\}_{t=1}^T \in \Xi(\mathbf{h})$$

where $\Xi(\mathbf{h})$ is the set of traffic processes consistent with the traffic contract parameters $\mathbf{h}$. One can show that $\alpha(\mathbf{m})$ is a concave function of $\mathbf{m}$. We can construct linear tariffs as tangent hyperplanes to $\alpha(\mathbf{m})$ at points corresponding to different values of $\mathbf{m}$. These are of the form

$$f_{\mathbf{m}}(\mathbf{M}) = a_0 + a_1 M_1 + \cdots + a_L M_L$$

where $\mathbf{M} = (M_1, \ldots, M_L)$ represent a vector of measurement values and $\mathbf{m}$ is the point that specifies the above tangent. The coefficients $a_0, \ldots, a_L$ are functions of the particular $\mathbf{m}$ and the rest of the parameters $\mathbf{h}$ that define the function $\alpha(\mathbf{h}; \cdot)$. The user chooses a tariff, indexed by $\mathbf{m}$, and has an expected charge of $E f_{\mathbf{m}}(\mathbf{M})$, which is minimized if he chooses the index $\mathbf{m}$ so that $m_i = E g_i(X)$.

There are many measurement schemes of the above type that one could use. For example, we propose the following as a refinement of the time-volume scheme. The rate interval $[0, h]$ is divided into two bands, $[0, A]$ and $[A, h]$. The measurement function divides the total volume V into two parts V_1 and V_2, where V_1 is the total volume of data transmitted during unit time intervals in which less than A bytes were transmitted, and V_2 is the total volume of data transmitted during unit time intervals in which more than A bytes were transmitted. One can solve the optimization problem and construct the tariffs. These are of the form $f(V_1, V_2) = a_0 + a_1 V_1 + a_2 V_2$, where $a_1 < a_2$. Note that such a scheme produces higher prices for the on-off source than for the source transmitting constantly at the mean rate, and these prices approximate better the actual effective bandwidth if the coefficients are rightly chosen. On the other hand, it cannot distinguish between sources producing the same values of V_1 and V_2, and may be further refined.

What are the pros and cons of schemes using finer measurements? On the one hand, the charge may more accurately reflect effective usage and in that sense be fairer. On the other hand, there are increased costs for measurement and accounting. The obvious question is whether such costs are justified by the accuracy of the resulting tariff, this accuracy being reflected into providing better price stability and fairness. Experimental results suggest that a substantial improvement is achieved by the simple time-volume tariffs compared to the flat rate tariffs, and that further refinements may not be worth the added complexity. Many network operators even consider the measurement of time and volume a burden they would rather avoid.

8.3.3   An Example of an Actual Tariff Construction

We consider in more detail the construction of tariffs of the form $aT + bV$ for simple contracts that specify only the peak rate h of the source.

Let $\alpha_{\text{on-off}}(m)$ be the value of the effective bandwidth computed using (8.12) and showing the dependence on m, where for simplicity we take $t = 1$. We provide formulas for the coefficients $a(m)$ and $b(m)$. Since these are coefficients of the tangent to $\alpha_{\text{on-off}}(\cdot)$ at m, simple algebra gives

$$b(m) = \frac{e^{sh} - 1}{s\left[h + m\left(e^{sh} - 1\right)\right]}, \qquad a(m) = \alpha_{\text{on-off}}(m) - mb(m) \tag{8.13}$$

The tariff $f_m(M) = a(m) + b(m)M$ is shown in Figure 8.1. $M = V/T$ is the measured average rate of the connection. As we have said, $f_m(M)$ is a rate of charge that depends upon the rate M. Hence, over a period T the charge is $T \times [a(m) + b(m)M] = a(m)T + b(m)V$.

It is not essential to provide the user with a continuum of tariff choices: the function $\alpha_{\text{on-off}}(m)$ may be well approximated by a small number of tangents, say q, especially if the capacity C is large compared to the peak rate h. Observe that m simply labels a linear function, and that the presentation of tariff choices may be entirely couched in terms of $a(m_i, h, s, t)$ and $b(m_i, h, s, t))$, $i = 1, \ldots, q$, where we now also make explicit the dependence of the tariff coefficients on the fixed parameters h, s and t. These specify a charge per unit time and a charge per unit volume, respectively, with no mention of the word 'mean'.

Thus, under the tariff f, the user has no incentive to 'cheat' by choosing a tangent other than the tangent that corresponds to his expected mean rate. The property that the expected cost per unit time under the best declaration is equal to the effective bandwidth has several further incentive compatible properties. If a user shapes his traffic to have a different mean or peak, or does a better job of characterizing traffic by better prediction of M, then he gains through a reduction of charge that exactly equals the reduction in the expected effective bandwidth of his traffic. Thus, users are discouraged from doing more work to determine the statistical characteristics of their connections than is justified by the benefit the network obtains from better characterization.

In practice, a constant coefficient is added to the tariff. The tariff takes the form $aT + bV + c$, where c is chosen to discourage traffic splitting, as we discuss in Section 8.3.5.

Example 8.2 (A numerical example)   We now illustrate the ideas of this section with a numerical example. Suppose that the predominant traffic offered to a link of capacity 100 Mbps is of three types, with peak and mean rates as shown in Table 8.1. Calculations show that for mixes of this traffic it is reasonable to take $s = 0.333$ in (8.12). Note that

Table 8.1 Typical charges for traffic with low mean rate. The various charges are expressed in the same units (of resource usage per second or per megabit) and are directly comparable with one another

Service type	Rate (Mbps)		Charge	
	Peak h	Mean m	Fixed (s^{-1}) $a(h, m)$	Variable (Mbit^{-1}) $b(h, m)$
1	0.1	0.04	2.7×10^{-4}	1.0
2	2.0	0.02	1.3×10^{-4}	1.4
3	10.0	0.01	1.1×10^{-3}	7.9

almost all of the charge for these three service types arises from the variable charge $b(h, m)$. The charging rates are 0.040, 0.028 and 0.080, respectively.

While the predominant traffic may be of types 1, 2 and 3, connections are not constrained to just these types. For example, a connection with a known peak rate of 2 Mbps could select any pair $(a(2, m), b(2, m))$ from Figure 8.2, or a connection with a known peak rate of 10 Mbps could select any pair $(a(10, m), b(10, m))$ from Figure 8.3. We can calculate similar tariffs for sources with other peak rates.

For a peak rate of 0.1 Mbps the bandwidth $\alpha_{\text{on-off}}(M)$ is almost linear in M, producing a variable charge $b(m)$ per unit of traffic that is almost constant in m. Since statistical multiplexing is efficient for sources with such low peak rates, very little incentive need be given to determine mean rates accurately. Peak rates above 2 Mbps produce more concave effective bandwidths, and hence more incentive to accurately estimate the mean.

Observe that the total charge for service type 1 is greater than that for service type 2: for these service types statistical sharing is relatively easy, and the advantage of a lower mean rate outweighs the disadvantage of a greater peak rate. The total charge for service type 3 is, however, more than twice as great as that for service types 1 and 2: statistical sharing becomes more difficult with a peak rate as high as 10% of the capacity of the resource. Observe that for the three service types shown in Table 8.1, almost all of the user's total cost arises from the variable charge.

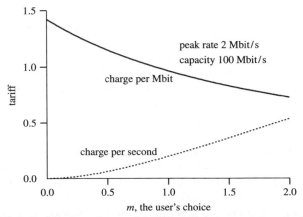

Figure 8.2 Tariff choices for a peak rate of 2 Mbps. The user can choose a lower charge per megabit, with a higher charge per second.

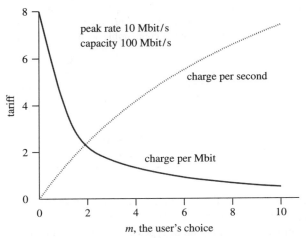

Figure 8.3 Tariff choices for a peak rate of 10 Mbps. The charge per second is typically greater with a peak rate of 10 Mbps per second than with a peak rate of 2 Mbps, since statistical sharing of the resource is more difficult.

Table 8.2 Typical charges for traffic with high mean rate. The charging rates are 1.2, 3.9 and 5.6, respectively

Service type	Rate (Mbps)		Charge	
	Peak h	Mean m	Fixed (s^{-1}) $a(h, m)$	Variable $(Mbit^{-1})$ $b(h, m)$
4	2.0	1.0	0.2	1.0
5	10.0	1.0	1.7	2.2
6	10.0	2.0	3.0	1.3

For the service types shown in Table 8.2 much more of the total cost arises from the fixed charge, more than half in the case of service type 6.

8.3.4   Competition

An operator who uses tariffs of the type described in the previous section has a competitive advantage. Let us illustrate by example why this is.

Suppose there are two identical service providers. There are also two classes of customer, and so two types of contract. The peak rates within both classes are h, but mean rates are m_1 and m_2, with $m_1 < m_2$. Suppose, initially that both service providers adopt flat rate pricing schemes based on peak rate and charge p per minute for each contract independently of the actual mean rate. Suppose also, for simplicity, that each contract lasts just one minute.

The two types of traffic have demand functions $x_1(p)$ and $x_2(p)$ of new contracts per minute. Since each contract lasts for one minute, $x_1(p)$ and $x_2(p)$ are also the number of contracts of each type that are in the system at each moment in time (recalling Little's Law from the end of Section 4.4). Initially, the customers are shared equally between the two networks. Assume also that the networks of the two providers are not fixed and hence the costs of the networks are not sunk. Instead, each provider must rent network resources from a wholesale market at a price of $\$c$ per unit effective bandwidth per minute. Suppose p is chosen such that revenue just covers costs, i.e.

$$\sum_{i=1,2} x_i(p)p - 2c \sum_{i=1,2} \frac{1}{2}x_i(p)\alpha_i(s,t) = 0 \qquad (8.14)$$

At this point, each network has capacity $C = (1/2)\sum_i x_i(p)\alpha_i(s,t)$, and earns revenue $R = cC$. In this case, p is also the cost of the effective bandwidth of the typical (average) customer.

Imagine that supplier 1 now adopts a pricing scheme in which he charges the two contracts prices $p_1 = \alpha_1 c$ and $p_2 = \alpha_2 c$. (We omit for notational convenience the dependence on s, t.) Note that if no customers are allowed to change supplier then his revenue is unchanged and customers pay for the cost of the effective bandwidth they consume.

Now suppose that customers do change suppliers, seeking the lowest price, and the networks adjust their capacities in response to demand. Since $\alpha_1 < \alpha_2$ and $p_1 < p < p_2$, what happens now is that supplier 1 attracts all the customers of type 1 to his network. Since he is charging according to effective bandwidth he is indifferent between customers of the two types, both in terms of resource usage and in revenue generation. However, as we have explained in Section 4.6, there is an interaction between the traffic mix and the operating point. Because type 1 customers have a smaller mean rate, such customers are easier to multiplex, and so by filling his network mostly with type 1 customers, his operating point will change to one for which s is smaller. He can operate more efficiently and his profit increases to above 0. Meanwhile, the second supplier is left with all the type 2 customers, and once he buys the bandwidth required to maintain the service contract his profit falls below 0. He might try to increase his profit by raising p. At the end of the day he will have to raise p to at least p_2', where $p_2' = \alpha_2' c > p_2$ and α_2' is the effective bandwidth of type 2 customers when a network has only customers of this type. At this point, even the type 2 customers will prefer to choose supplier 1, where because of more efficient multiplexing (i.e. a lower value of s) they occupy a smaller amount of effective bandwidth. Thus, in this simple model of competition, supplier 2, who insists on charging all contracts the same price, is completely driven out of business by supplier 1.

8.3.5   Discouraging Arbitrage and Splitting

We have provided a methodology that charges services proportionally to their effective usage. However, there are a number of criteria by which we should check whether a pricing scheme is sound. One of these has to do with the fact that prices should, if possible, eliminate the possibility that a customer might profit from *arbitrage* or *splitting*. Arbitrage occurs when a customer can make a profit by buying a service of a certain type and then repackaging and reselling it as a different service at market prices. Splitting takes place when a user splits a service into smaller services, and pays less this way than if he had bought the smaller services at market prices.

If prices are proportional to effective bandwidths then arbitrage opportunities are eliminated. This is because the total effective bandwidth of the new services that are created cannot exceed the effective bandwidth of the service that was purchased. Hence, the total revenue cannot exceed the cost. Unfortunately, splitting is encouraged. This is because the network treats each subcontract as a smaller independent source of traffic, and due to the resulting multiplexing gain charges a less total effective bandwidth.

As an example of splitting, consider a source with peak rate h, mean rate m and effective bandwidth a_{total}. Suppose it is split into two traffic streams, each with peak rate $h/2$, mean rate $m/2$ and effective bandwidth α_{split}. Splitting will be beneficial to the user if it will result in a less total charge, i.e. if

$$2\alpha_{split} < \alpha_{total}$$

Unfortunately, it is easy to check (using, for example, (8.12)) that such an inequality can hold. This is because correlated traffic streams are erroneously charged as if they were independent. Of course, in reality the user must take account of the fact that splitting and then reassembling his traffic at the destination is costly in terms of equipment and delay. If this cost is substantial, then splitting may not be profitable.

Traffic splitting is undesirable to the service provider, because it reduces his revenue, generates large amounts of correlated traffic on the same route, exhausts the set of available connections, and increases the signalling overhead for setting up more connections. However, splitting can be beneficial to the provider, if each sub-contract is routed along a disjoint path. In that case, each link will carry uncorrelated traffic generated by smaller sources, and so multiplexing will be easier.

A simple way to discourage splitting is to add a fixed charge to the tariff. This results in what we call an *abc-scheme*, of the form $aT + bV + c$, in which a and b are as before, but c is large enough to discourage splitting. Note that c should be greater for connections that last longer, since given any value of c, if a connection lasts sufficiently long, there will be always an incentive to split.

Another possibility is to use a *homothetic tariff*, satisfying $\alpha(h, m) = k\alpha(h/k, m/k)$.[2] Such tariffs are computed from a function $\alpha(x, y)$, which concave in y and increasing in x. This can serves as the basis for constructing a whole family of tariffs. The convexity in m creates an incentive for users to reveal their true mean rates and the fact that α is homogeneous of degree one means that nothing can be gained by splitting. However, a disadvantage is that the charge is not proportional to the effective bandwidth, although it can be close.

8.4   Some simple pricing models

In this section we discuss three examples for pricing simple models of services using the ideas of this chapter.

8.4.1   Time-of-day Pricing

Consider a transport service that is sold in peak and off-peak periods, $t = 1, 2$, respectively. Let $u_i(x_1^i, x_2^i)$ denote the utility to user i of sending flows of mean rates x_1^i and x_2^i during periods 1 and 2, respectively. Let C_t be the capacity available during period t. A global planner has the problem

$$\underset{\{x_1^i, x_2^i\}}{\text{maximize}} \sum_{i=1}^{N} u_i(x_1^i, x_2^i), \quad \text{subject to} \sum_{i=1}^{N} x_t^i \leq C_t, \; t = 1, 2$$

---

[2] A function f is homothetic if $f(x) = g(h(x))$, where g is strictly increasing and h is homogeneous of degree 1, i.e. $h(tx) = th(x)$ for all $t > 0$.

As in Section 5.4.2, the maximum is achieved by setting prices p_1, p_2, and then posing to user i the problem

$$\underset{\{x_1^i,x_2^i\}}{\text{maximize}} \left[u_i(x_1^i, x_2^i) - p_1 x_1^i - p_2 x_2^i \right]$$

If at every i and (x_1^i, x_2^i) user i has a greater marginal utility for sending data in the peak period than in the off-peak period, then p_1 will be greater than p_2. Note that peak and off-peak usage may be near substitutes for one another. In practice, it is likely that the demand for off-peak usage will increases with p_1, as users substitute off-peak usage for peak usage.

The interpretation of this simple model is that the network sets its prices so that its capacity is fully used at all times. (In practice, prices are chosen to keep the load just below C_t, to leave room for some burstiness in the traffic.) These prices can be determined by a market mechanism such as a tatonnement. The network increases or decreases each p_t depending on whether the demand $\sum_i x_t^i$ is greater or less than C_t. The customers purchase the capability to sustain an amount of throughput that varies during the periods. In that sense, this is a model of a service that guarantees some minimum throughput. The model implicitly assumes that each customer is small (the parameter s in the effective bandwidth formula is nearly zero), and hence his effective bandwidth can be approximated by his mean rate.

The above can also be viewed as a model of regulating a best-effort service. There is no strict guarantee of performance in terms of throughput, delay or packet loss. Customers see the posted prices and decide on the amount to send. The network uses prices to avoid the performance degradation that occurs when the total input rate exceeds capacity. Such prices are computed based on the past history of demand during the different time periods. Hence, the optimization problem is solved by considering some estimate of the actual demand. For this reason, there is no guarantee that demand will always be less than the available capacity and the network may become temporarily overloaded. This is an example of a 'better-than-best-effort' service, since most of the time performance will be acceptable. For a purely best-effort service the network would set the prices to zero, and so have no feedback loop with its customers.

8.4.2 Combining Guaranteed with Best-effort

In this example we price a single link that operates a priority service. Type 1 traffic receives priority service in the sense that type 2 traffic is served only if there is no traffic of type 1. To keep the model simple, assume that there is a single type of applications that may need either type 1 or type 2 service. In both cases, it has an effective bandwidth of 1 kbps and a mean rate of $1/2$ kbps. Let x denote the sum of the effective bandwidths of all type 1 traffic. Let y denote the sum of the mean rates of type 2 traffic. Then the constraints of the system are

$$x \leq C \text{ and } y + x/2 \leq C \tag{8.15}$$

where C is the capacity of the link (which, for simplicity, we assume is also equal to the effective capacity). The first constraint is the quality of service constraint, and the second is the stability constraint. The latter provides for a 'better-than-best-effort' service as discussed in Section 8.4.1.

Suppose that the user population has a utility of $u(x, y)$ for x and y amounts of type 1 and type 2 service. Assuming that the demand for best-effort traffic of type 2 can always exceed C, the operating point is always on the boundary of the acceptance region defined by (8.15), and there are two possible cases: either $x = C$ and $y + x/2 = C$, or $x < C$ and $y + x/2 = C$. Let p_i be the optimal price for type i, and let p_q, p_m be the shadow prices of the quality constraint and the mean rate constraint, respectively. In the first case, $p_1 = p_q + 0.5 p_m$ and $p_2 = p_m$. Here, type 1 traffic is charged for its mean rate on equal terms as type 2 traffic and additionally pays a premium that reflects its demand for quality. Note that to restrict demand of type 1 to the technology set it is not enough to price only the mean rate. However, it is enough in the second case, where we take $p_1 = 0.5 p_m$ and $p_2 = p_m$. Here demand for type 1 is very elastic, and even a small price such as $0.5 p_m$ is enough to keep it within C.

Observe that, given the prices for type 1 and type 2 services, customers can self-select and choose which of the transport services they wish to use. The fact that services are substitutes is captured in the definition of u. The magnitude of the cross elasticity depends upon the quality of the type 2 service. The better is the quality level ensured by keeping the utilization of the link low, through a high p_m, the better is the chance that customers of type 1 will switch and use the type 2 service. Such service cannibalization may be annoying to a profit maximizing network operator. He may prefer to keep the quality of the best-effort service as low as possible, and so he may even wish to degrade the best-effort service by adding extra delays or purposely losing packets. Clearly, such practices reduce social welfare. Of course, he must balance the any revenue he could gain this way against the revenue that would he would loss because some of the best-effort customers find a degraded service unacceptable. Service and price personalization may reduce the incentives for customers to switch services (see Section 6.2.2).

As a last comment, note that there is social benefit in keeping p_m small since this increases the number of users that use and benefit from the network. Thus enough capacity must be provided to meet the best-effort traffic demand at this price. If accounting and billing costs are high, then one may decide to take $p_0 = 0$. However, as we see in today's Internet, free service usually becomes very congested and is of little value.

8.4.3 Contracts with Minimum Guarantees and Uncertainty

Finally, we consider a model in which a link of bandwidth C is shared by n users of an 'Available Bit Rate' (ABR) service and some other users. A user of the ABR service, say i, can request a minimum guaranteed bandwidth, say x_i. He obtains bandwidth of $x_i + Z x_i$, where Z depends upon the loading of the link and so is not guaranteed in advance. Let us suppose that this extra bandwidth is obtained by dividing the leftover bandwidth in proportion to the minimum rates requested, so that, taking Y as the total bandwidth used by the non-ABR users,

$$Z = \frac{C - Y - \sum_j x_j}{\sum_j x_j}$$

Now suppose that $\sum_i x_i = C_1 < C$, and Y has some distribution over the interval $[0, C - C_1]$, which depends upon the allocated bandwidth $C - C_1$. This is a model for what happens when ABR services are provided under ATM, where the minimum rate is the parameter MCR (Minimum Cell Rate). It is also what happens in a frame relay service, where the minimum rate is the customer's request for CIR (Committed Information Rate).

Note that the utility of a user depends upon the decisions of the other users. This is an example of a congestion effect, in which there are negative externalities. The larger the number of users that contend, the smaller is the share of the left-over bandwidth that each user obtains.

Suppose that capacity of $C - C_1$ is reserved for the non-ABR traffic, and it obtains total utility $u_0(C - C_1)$. If any part of this capacity is not used fully, then the unused part is allocated to ABR traffic. Also, suppose that the ABR users do not differentiate between the values of guaranteed and extra bandwidth. Then the expected social welfare takes the form

$$SW(x, C_1) = \sum_i Eu_i \left(\frac{x_i}{C_1}(C - Y) \right) + u_0(C - C_1)$$

The expectation is taken over Y, and we have substituted $\sum_j x_j = C_1$ (which we imagine now is known to the ABR users). Note that $SW(x, C_1)$ is a concave function of x (since the average of concave functions is concave). Thus, there is some p such that SW is maximized under the constraint $\sum_j x_j = C_1$ by presenting user i with the problem of maximizing $Eu_i(x_i(C - Y)/C_1) - px_i$. That is, given that the link provider has already chosen C_1, an economically efficient choice of x_i can be induced with linear pricing of MCR.

In practice, however, not all the ABR users will make use of the bandwidth simultaneously. Let us suppose that user i is present only with probability α_i. Let X_i be a random variable that is equal to x_i or 0 with probabilities α_i and $1 - \alpha_i$, respectively. The expected social welfare is now

$$SW(x, p, C_1) = \sum_i \alpha_i Eu_i \left(\frac{x_i}{x_i + \sum_{j \neq i} X_j}(C - Y) \right) + u_0(C - C_1)$$

Consider a problem in which this is to be maximized subject to two constraints. The first constraint is that the expected total quantity of MCR request should be equal to some C_2, i.e. $\sum_j \alpha_j x_j = C_2$, where C_1 and C_2 have been chosen optimally, with $C_2 < C_1$. Thus, they are to be treated as given constants in what follows. Assuming that n is large so that the variance of $\sum_j X_j$ is small relative to its mean of C_2, we can make the approximation

$$SW(x, p, C_1) \approx \sum_i \alpha_i Eu_i \left(\frac{x_i}{(1 - \alpha_i)x_i + C_2}(C - Y) \right) + v(C - C_1)$$

Again, one can check that this is a concave function of $x_1, \ldots, x_n$.

The second constraint is a probabilistic version of the constraint that the total requested bandwidth of the ABR traffic should not exceed C_1. It should be very unlikely that a request for an ABR connection cannot be met, e.g. the logarithm of the probability of the event $\sum_j X_j > C_1$ should be less than $-\gamma$, for some $\gamma > 0$. Using the fact that $\sum_j X_j$ is approximately Gaussian and making a large deviations estimate (cf. the effective bandwidth formula for a Gaussian source is Section 4.7), this constraint is approximately equivalent to the set of constraints

$$\sum_i \left[\alpha_i x_i + s\alpha_i(1 - \alpha_i)x_i^2/2 \right] \leq C_1 - \gamma/s, \quad \text{for all } s > 0 \qquad (8.16)$$

After solving the welfare maximization through inclusion of these constraints in a Lagrangian, we see that social welfare maximum can be achieved in a decentralized way by facing user i with a charge of the form $x_i[p_1 + p_2 + p_2 s^*(1 - \alpha_i)x_i/2]$, where p_1 is the shadow price of the first constraint, s^* is the value of s for which (8.16) is active, and p_2 is the shadow price of this constraint. To see this, we imagine that by past experience user i knows C, C_2 and the distribution of Y, and thus that $\sum_{j \neq i} X_j \approx C_2 - \alpha_i x_i$. His problem is

$$\text{maximize}_{x_i} \, u_i \left(\frac{x_i}{(1 - \alpha_i)x_i + C_2}(C - Y) \right) - x_i[p_1 + p_2 + p_2 s^*(1 - \alpha_i)x_i/2]$$

Here p_2 is non-zero only if some constraint of (8.16) is active. The charge depends to some extent on x_i^2, but only if $\alpha_i < 1$, reflecting the fact that some reservation of resource must be made even if user i is not present. We can interpret the term $(1 - \alpha_i)x_i$ as peak minus mean rate. Note, however, that since s^* depends upon all the values $\alpha_1, \ldots, \alpha_n$, one should not be misled into thinking that the part of the charge $p_2 s^*(1 - \alpha_i)x_i^2/2$ decreases with α_i. More detailed analysis shows, as one would expect, that if all else is fixed, then the charge $x_i[p_2 + p_2 s^*(1 - \alpha_i)x_i/2]$ increases in α_i.

We can conceive of other charging schemes. For example, an ABR user could be charged a fixed charge, F, plus p times his requested MCR. The fixed charge would make it unprofitable for some users to use the service at all, and could be chosen so that (8.16) is satisfied. Then p could be chosen as the shadow price of the constraint $\sum_j \alpha_j x_j = C_2$.

8.5   Long-term interaction of tariffs and network load

As we have explained in Section 8.3, there is an interdependence between whatever tariffs are posted and the operating point of the link. The network operator posts tariffs that have been computed for the current operating point of the link, expressed through the parameters s and t. These tariffs provide the customers with incentives to change some parameters in their contracts to minimize their anticipated costs. Once customers do this, the demand for the various contract types changes, and the operating point of system moves on the boundary of the technology set. This changes s and t, therefore the network operator must calculate new tariffs, for the new operating point. This long term interaction between the network and the customers continues until an equilibrium is reached. We illustrate now, by a very simple example, that if the network operator uses the effective bandwidth charging approach, then an equilibrium point is reached at which social welfare is maximized, as measured by the number of customers admitted to the system.

We consider the customer's problem of picking a leaky bucket (ρ, β) for the traffic contract of a particular application. Clearly, there are many values of the parameters ρ, β for which the traffic of the application is conforming. This suggests the idea of an *indifference curve*, $\beta = G(\rho)$, such that for a given ρ, the minimum value of β for which the traffic is conforming is $G(\rho)$. (In some variations of this the traffic is conforming with some high probability.) As shown in Figure 8.4, some key properties of the indifference curve are that it is convex, tends to infinity as ρ tends to the mean rate m, and is zero for $\rho = h$. For simplicity, we assume that all customers' connections are statistically identical and are policed with the same leaky bucket (ρ, β). Assuming also that they have identical contracts, their indifference curves are the same. The network consists of a shared link with capacity C and buffer B, and uses deterministic multiplexing to load the link. It will be filled just to capacity if prices are set appropriately.

As we have seen in Section 4.12, the value of s that is relevant to deterministic multiplexing (i.e. zero cell loss), is $s = \infty$, and the effective bandwidth of a connection

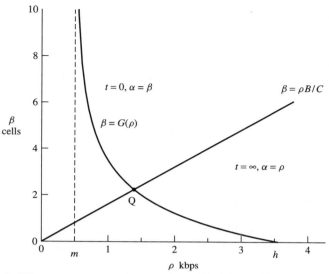

Figure 8.4 The indifference curve of a leaky bucket and the effective bandwidth for deterministic multiplexing. For points on the curve $\beta = G(\rho)$ lying above Q, users will wish to decrease their bucket size β, since their charge is proportional to the effective bandwidth, which is β. For points below Q, they will wish to decrease their leak rate ρ, since their effective bandwidth is now ρ.

policed with (ρ, β) is $\alpha_j(\infty, t) = \bar{X}[0, t]/t = \rho_j + \beta_j/t$, for $t > 0$ and $\alpha_j(\infty, 0) = \beta_j$. The acceptance region is defined by the two linear constraints

$$\sum_j \rho_j \leq C \quad \text{and} \quad \sum_j \beta_j \leq B \qquad (8.17)$$

where the left-hand side of each inequality is the sum of the effective bandwidths at $t = \infty$ and $t = 0$, respectively.

Suppose that the system proceeds in lock-step. Given that the operating point is where all customers choose the leaky bucket policer $(\rho, \beta) = (\rho, G(\rho))$, a price will be set of $p = \lambda\alpha$, for an effective bandwidth of $\alpha = \max\{C/\rho, B/\beta\}$, and where λ is chosen so that the demand n is maximal, subject to $n\rho \leq C$ and $n\beta \leq B$.

Consider the point Q at the intersection of $\beta = G(\rho)$ and the line $\beta = \rho B/C$ (see Figure 8.4). It is easy to show that this point maximizes the number of customers who obtain service while satisfying the constraints of the indifference curve and the technology set. Consider a point (ρ, β) on G (i.e. initial choice of a contract (ρ, β) by customers). One can easily see that if this point is below Q, the system will fill so that the active constraint is the one corresponding to $t = \infty$ and $\alpha = \rho$, whereas if the point lies above Q, then the active constraint is the one corresponding to $t = 0$ and $\alpha = \beta$.

Assume now that the network uses our charging approach. If the customers choose $(\rho, G(\rho))$ below Q, then the first constraint will be active ($t = \infty$) and the charge will be proportional to ρ; this will guide customers to reduce ρ and move towards Q. If the customers choose $(\rho, G(\rho))$ above Q, then the second constraint will be active and the charge will be proportional to β; this will guide customers to reduce β and move towards Q. Assuming that, to avoid oscillations, only small changes to their traffic contracts are allowed, the point Q will be reached eventually.

Since both constraints are active at Q, the charge will be proportional to a linear combination of the effective bandwidths corresponding to the active constraints at Q, i.e.

it will be of the form $\lambda_1 \rho + \lambda_2 \beta$, where λ_1, λ_2 are the shadow prices of the technology set constraints (8.17) in the problem of maximizing n. These prices, for bandwidth and buffer, are proportional to the coefficients of the tangent to G at Q, and Q is an equilibrium, since the users minimize their charges by choosing the $(\rho, G(\rho))$ of Q.

We conclude by reminding the reader of the simplicity of our model. In practice, users do not renegotiate contracts in lock-step. Also effective bandwidth charges are only approximations of the actual effective bandwidth. Since these approximations are translated into user incentives, poor effective bandwidth approximations may lead to inefficient equilibria.

8.6   Further reading

The idea of using simple tariffs based on effective bandwidths was first introduced by Kelly (1994a,b), and was subsequently refined in the work of Courcoubetis, Kelly and Weber (2000), from which much of the material in this chapter is borrowed. Courcoubetis, Kelly, Siris and Weber (2000) have conducted experiments with actual and synthetic traffic to evaluate the accuracy of the pricing scheme based on (8.12) compared to ones that use other more accurate effective bandwidth approximations.

Important motivation for studying guaranteed services comes from ATM technology. The early paper of Low and Varaiya (1993) on charging ATM services has been influential in economic modelling.

Most of the ideas in this chapter were conceived when the authors collaborated in a project called CA\$hMAN, which was a collaborative project within the European ACTS (Advanced Communication Technologies and Services) program. The project ran for three years from September 1995 and aimed to develop and evaluate potential charging schemes for ATM and possibly for the young Internet. It was a multidisciplinary project, integrating mathematical models for usage-sensitive charging schemes with implementation of an experimental platform using specially developed measurement hardware and management software. In CA\$hMAN we developed mathematical models that allowed us to find simple functional approximations for resource usage, and we studied charging schemes derived from these models. Participants in the project included network operators and equipment manufacturers, and user trials were conducted in three European countries. Songhurst (1999) edited a comprehensive book on the CA\$hMAN project.

9

Congestion

This chapter deals with the phenomenon of congestion, and the ways in which it can affect pricing decisions. Internet users are especially familiar with the phenomenon of congestion and the decline in service quality that occurs as the number of simultaneous users increases. This is similar to the congestion effects experienced by car drivers when journey times and accident numbers increase because more cars are on the road. In this chapter, we show how pricing can be used to control congestion and increase the value of services to users.

In previous chapters we have neglected the effects of congestion by supposing that the benefit that a user obtains from a communications service depends only upon parameters of that service and the amount of the service he obtains. We have imagined that if user i buys a quantity of a service x_i at a price p then his net benefit takes the form

$$u_i(x_i) - px_i \tag{9.1}$$

where for user i's demand we write x_i rather than x^i, since his demand is one-dimensional. Once we know the users' demand functions, the suppliers' cost functions and their technology sets, the problems of maximizing the suppliers' profit or the social welfare can be solved by using prices to allocate services to the users who value them most and to match the demand for services to supply. This is all true for a service that has statically defined guarantees that may not vary during the life of the service. In this case, congestion is expressed in terms of packet loss rate or packet delay, and a maximum tolerable level of congestion is part of the service specification. Call admission control is used to maintain congestion below this level. Hence $u_i(x_i)$ in (9.1) denotes the utility of using a quantity of service x_i that has this level of congestion.

When services have contracts with dynamic parameters (e.g. the maximum sending rate may vary during the life of the service), and there is no strict guarantee on minimum performance levels, users will be tempted to demand the most that they can from the network. But a decision by the network to grant such requests to all its customers may make performance intolerable.

It is clear that (9.1) fails to capture the effects of the arbitrary levels of congestion that can occur if the network does not use controls such as call admission to restrict the maximum congestion level. In modelling congestion, we suppose that when a user receives more of a service the value of the service deteriorates, as it is experienced by him and all other users.

Pricing Communication Networks   C. Courcoubetis and R. Weber
© 2003 John Wiley & Sons, Ltd   ISBN 0-470-85130-9 (HB)

The models in this chapter are concerned with the effects of congestion and pricing that take congestion into account. In the case of services sharing a common resource pool, we model congestion by supposing that user i has a net benefit that depends upon the amounts of service demanded by other users. That is, he enjoys net benefit of a form such as

$$u_i(x_i, y) - px_i \tag{9.2}$$

where $y = \sum_i x_i/k$, for some constant k. Here k parameterizes the resource capacity of the system. The intuition is that congestion depends upon the load of the system, as measured by y. Full load may correspond to $\sum_i x_i = k$.

If user i requests a quantity of service that is small compared with the total requests of all users, then y does not vary much with different choices of x_i, and so the problem of maximizing (9.2) reduces to that of maximizing (9.1), with y taken as fixed. In this chapter we suppose y is not fixed, and consider the problem of determining p so that when the market is in equilibrium we maximize some measure such as social welfare or the service provider's profit.

When a participant in a market can, without suffering penalty, make choices of variables that adversely affect the utilities of other participants, we say there is a *negative market externality*. Congestion is a good example of a negative market externality. Positive market externalities are also possible. For example, when a consumer purchases a particular model of mobile phone he increases the popularity of that phone; its increased popularity encourages the manufacturer to provide spare parts and accessories, making it more valuable to all its owners.

Returning to our model of congestion: how can users be posed problems of maximizing their individual net benefits so that social welfare is maximized when they do so? The answer, which we give in Section 9.1, is to price congestion. Economists say that we 'internalize the externality'. The user's final charge has two parts: a charge for the cost for providing the service and a charge for congestion. The congestion charge is used to manage congestion and to determine how the available resources are shared amongst users. In general, moderate levels of congestion are usually desirable. This is because zero congestion may require very inefficient use of resources. A high level of congestion uses resources more efficiently, but services are degraded too much. Ideally, a mechanism for controlling congestion by pricing should be self-tuning, and automatically find a good compromise between service degradation and effective resource usage.

In Section 9.2 we make the connection between congestion pricing and sharing finite resources under a capacity constraint. In Section 9.3 we give examples in which users share congested resources. We look at models with blocking, loss and delay. Section 9.3.2 looks at the problem of pricing more than one service, when the services are substitutes and both subject to congestion. An important notion, described in Section 9.4, is the realization of a congestion price as a sample path shadow price. Finally, in Section 9.5 we present a general model for congestion pricing.

9.1   Defining a congestion price

The following simple model illustrates the basic ideas of congestion pricing. Consider the case of a single service quantified in terms of a single dynamic parameter. Suppose n users make demands for quantities of this service. The producer can supply capacity k at cost $c(k)$. For the moment, we take k as fixed and pose the problem of maximizing social welfare as

$$\underset{\{x_1,\ldots,x_n \geq 0\}}{\text{maximize}} \sum_{j=1}^{n} u_j(x_j, y) - c(k) \tag{9.3}$$

where $y = \sum_i x_i / k$. Here u_i is assumed to be strictly increasing and concave in x_i and strictly decreasing and convex in y. Note that the only constraint on x_i is $x_i \geq 0$. The maximum occurs at the stationary point where

$$\frac{\partial u_i(x_i, y)}{\partial x_i} + \frac{1}{k} \sum_{j=1}^{n} \frac{\partial u_j(x_j, y)}{\partial y} = 0, \quad i = 1, \ldots, n \tag{9.4}$$

These equations give the socially optimal demands, say $x_1^*, \ldots, x_n^*$.

Now suppose user i is presented with a price

$$p_E = -\frac{1}{k} \sum_{j=1}^{n} \frac{\partial u_j(x_j^*, y)}{\partial y} \tag{9.5}$$

where because of the assumption that u_j is decreasing in y, we have $p_E > 0$. This price represents the marginal decrease in social welfare due to congestion by a marginal increase in usage. The suffix E reminds us that we are pricing the externality. So user i is faced with the problem

$$\underset{x_i \geq 0}{\text{maximize}} \left[u_i(x_i, y) - p_E x_i \right] \tag{9.6}$$

He solves this where

$$\frac{\partial u_i}{\partial x_i} + \frac{1}{k} \frac{\partial u_i}{\partial y} - p_E = \frac{\partial u_i}{\partial x_i} + \frac{1}{k} \frac{\partial u_i}{\partial y} + \frac{1}{k} \sum_{j=1}^{n} \frac{\partial u_j(x_j^*, y)}{\partial y} = 0 \tag{9.7}$$

Now if n is large, it is reasonable to suppose that

$$\left| \frac{1}{k} \frac{\partial u_i(x_i, y)}{\partial y} \right| \ll \left| \frac{1}{k} \sum_{j=1}^{n} \frac{\partial u_j(x_j, y)}{\partial y} \right| \tag{9.8}$$

Then (9.7) is approximately the same as (9.4), and so is solved by $x_i = x_i^*$. Thus, individual maximization of net benefit induces the socially optimal solution.

Note that in solving (9.7), user i regards all other users' demands as fixed. Thus $x_1^*, \ldots, x_n^*$ is a Nash equilibrium. That is, user i has no incentive to do other than choose $x_i = x_i^*$, provided $x_j = x_j^*$, for all $j \neq i$.

Suppose the demand of each individual user is small. Then we might expect congestion prices to converge to optimal price under a tatonnement procedure having the following repeated steps: (a) the network determines p_E from (9.5) and communicates it to the users; (b) each user i solves (9.6) to determine a new x_i^*, assuming y is insensitive of his choice of x_i^*.

Note that if the network is to solve (9.5) then it needs to know the users' utility functions. This may be difficult in practice. What the network might actually do is to vary p_E until it finds an equilibrium at which no users wish to increase or decrease their demands. Similarly, we might wonder how user i can solve (9.6) without knowing the value of y. The simple answer is that $u_i(x_i, y)$ is actually a function of the form $u_i(x_i, D(y))$, where D is the value of the congestion for a given y. So user i needs only observe the value of the congestion caused by y, rather than y itself. This value of D is used when solving locally (9.6). In the following sections, we discuss some important properties of congestion prices.

9.1.1 A Condition for Capacity Expansion

Thus far, we have imagined that the capacity is fixed. Let us suppose that $c(k)$ is increasing and convex in k. The maximized social welfare is

$$w(k) = \sum_{j=1}^{n} u_j(x_j^*, y^*) - c(k)$$

Investment is worthwhile if $w'(k) > 0$. We can compute,

$$w'(k) = \sum_{i=1}^{n} \frac{\partial w}{\partial x_i^*} \frac{dx_i^*}{dk} + \frac{\partial w}{\partial y} \frac{dy}{dk} - c'(k)$$

Now since social welfare is maximized, $\partial w / \partial x_i^* = 0$. Thus, with $x^* = \sum_j x_j^*$, $y = x^*/k$, we have

$$w'(k) = \sum_{i=1}^{n} \frac{\partial u_i}{\partial y} \frac{\partial y}{\partial k} - c'(k) = -\frac{x^*}{k^2} \sum_{i=1}^{n} \frac{\partial u_i}{\partial y} - c'(k)$$

Capacity expansion is worthwhile if $w'(k) > 0$, and from the above this is equivalent to

$$p_E x^* > k c'(k)$$

Thus, congestion prices may be used by the service provider as signals for deciding whether to expand or reduce the capacity of the network.

9.1.2 Incentive Compatibility

As noted above, the network operator must know the utilities of the users if he is to compute the congestion price from (9.4). In particular, he must know the derivatives of their utility function, i.e. their sensitivities to degradation in performance due to congestion. Is there any reason to expect that users will be truthful in declaring these sensitivities? Clearly not. Users may adopt complex strategies in which their declarations are far from the truth. We investigate such incentive compatibility issues in Sections 9.4.3 and 9.4.4.

9.1.3 Extensions

The definition of the load $y = \sum_i x_i / k$ is natural for a single link network in which x_i is an average flow and k is the bandwidth of the link. In principle, congestion measures, such as delay and packet loss, can be directly determined given the statistics of the traffic and the service discipline of the link. But is our model useful for more general situations, in which we desire to price dynamic parameters of the contract that are not average flows, but quantities such as leaky bucket parameters?

Take as an example traffic contracts in which each source is policed by its peak rate h and the leaky bucket (ρ, β), where the values of h and ρ are fixed, and β is dynamic. How should the network charge for β? One could now let x_i denote the amount of β in contract i. Then the congestion price would be the marginal total decrease in the utilities of all users due to congestion when the β in some contract is marginally increased. One may consider a contract as producing a worst-case traffic or traffic corresponding to a typical source policed with the contract parameters.

A similar analysis holds for a network in which the traffic of user i passes through a number of links, say $\ell_1, \ldots, \ell_m$. In this case, u_i depends upon the congestion at each of the links, and hence is a function $u_i(x_i, y_{\ell_1}, \ldots, y_{\ell_m})$, where, summing over j such that user j uses link ℓ, the load on link ℓ is $y_\ell = \sum_{j:\, j \in \ell} x_j / k_\ell$. Hence, if we define the congestion price p_E^ℓ at link ℓ to be

$$p_E^\ell = -\frac{1}{k_\ell} \sum_{i=1}^{n} \frac{\partial u_i(x_i^*, y_{\ell_1}, \ldots, y_{\ell_m})}{\partial y_\ell} \tag{9.9}$$

then the same global optimization problem is solved by charging each user the sum of the congestion prices on the links that his traffic uses.

9.2 Connection with finite capacity constraints

Thus far, we have placed no constraint on the total amount of services provided. In previous chapters, we have considered problems in which there is a capacity constraint, and the problem of maximizing social welfare takes the form

$$\underset{\{x_1, \ldots, x_n \geq 0\}}{\text{maximize}} \sum_{i=1}^{n} u_i(x_i), \quad \text{subject to} \sum_{i=1}^{n} x_i \leq C \tag{9.10}$$

Recall that the problem of maximizing social welfare can be decentralized by presenting user i with a price λ_C, where λ_C is the shadow price of the constraint.

Should one think of λ_C as a congestion price? It is certainly true that when x_i increases there is a reduction in the amount of resource that remains for the other users, and so an increase in x_i negatively impacts the benefits they obtain. Nonetheless, we believe it is best to reserve the terminology of 'congestion price' for circumstances in which x_i appears explicitly in the utility of the other users, and there is no hard capacity constraint.

It is still interesting to observe a close connection between the models. For purposes of illustration, consider a series of problems indexed by $m = 1, 2, \ldots$, for which the utility of user i is $u_i(x_i, y) = a_i \log x_i - D^{(m)}(y)x_i$. The interpretation of $D^{(m)}(y)$ is of the average amount of congestion cost incurred by user i for every unit of flow he sends to the network. For instance, it may be average delay suffered by every packet sent. Thus, $D^{(m)}(y)x_i$ represents the (average) congestion cost suffered this user when transmitting at total rate x_i. The net utility of the user is the utility of transmitting at rate x_i reduced by the cost due to performance degradation. Note that both terms must be measured in the same units. Now the problem of maximizing social welfare is

$$\underset{\{x_1, \ldots, x_n \geq 0\}}{\text{maximize}} \sum_{i=1}^{n} \left[a_i \log x_i - D^{(m)}(y)x_i \right]$$

where $y = \sum_i x_i$. Suppose the congestion cost is $D^{(m)}(y) = 1/[m(C - y)]$. Note that $D^{(m)}(C) = \infty$, but for any $y < C$, $D^{(m)}(y) \to 0$ as $m \to \infty$. The idea here is that the congestion price is negligible when $y < C$, but tends to infinity as y approaches C. Each curve becomes steeper as m grows. A physical interpretation is that as m increases the traffic in each flow becomes less random, tending to a constant flow with the same rate, and congestion phenomena tend to appear rather suddenly as y approaches C.

For this model the congestion price is $p_E = 1/m(C - y^*)^2$ where y^* depends on m. Some algebra shows that as $m \to \infty$, $y^* \to C$ and $p_E \to \sum_i a_i / C = \lambda_C$, where λ_C is simply the shadow price of the constraint in (9.10) when we put $u_i(x_i) = a_i \log x_i$. Hence, one can think of the congestion cost as penalizing violation of a flow feasibility constraint. Note that for any finite m we have $y^* < C$, so the capacity constraint is not active. This is what often happens in problems with both congestion costs and capacity constraints. If the congestion cost grows very large before the capacity constraint in reached, then the capacity constraint is not active at the solution point.

9.3 Models in which users share congested resources

In this section we present several models of congestion, and illustrate some ways that the social welfare maximization problem can be solved using congestion prices.

9.3.1 A Delay Model for a $M/M/1$ Queue

An important version of the model in Section 9.1 is one in which there is explicit congestion cost, and the utility function of user i takes the form

$$u_i(x_i, y) = v_i(x_i) - \gamma_i D(y)x_i \qquad (9.11)$$

Here, $\gamma_i D(y)x_i$ is a congestion cost and γ_i parameterizes the sensitivity of user i to congestion. For example, this congestion cost might arise as the product of x_i and the average delay experienced by a packet belonging to user i when packets are served at a $M/M/1$ queue. Assuming service rate 1 and Poisson arrivals at rates $x_1, \ldots, x_n$, the average delay in the queue is $1/(1 - y)$, so we have

$$\gamma_i D(y)x_i = \gamma_i \frac{1}{1 - y}x_i, \quad y \leq 1$$

Note that $v_i(x_i)$ increases in x_i, and $D(y)$ increases in x_j, for all j. The social welfare is

$$\sum_{i=1}^{n} \left[v_i(x_i) - \gamma_i D(y)x_i \right]$$

and this is maximized by choosing $x_1, \ldots, x_n$ to satisfy

$$\frac{\partial v_i}{\partial x_i} - \gamma_i D(y) - \frac{\partial D(y)}{\partial y} \sum_{j=1}^{n} \gamma_j x_j = 0 \qquad (9.12)$$

Denote the solution point as $x_1^*, \ldots, x_n^*$. The same point can be reached if user i is charged the congestion price given in (9.5), namely,

$$p_E = \left. \frac{\partial D(y)}{\partial y} \right|_{y=y^*} \sum_{j=1}^{n} \gamma_j x_j^*$$

Note that p_E is just the extra cost suffered by all users due to a marginal increase in user i's demand. Customer i seeks to maximize his net benefit of

$$v_i(x_i) - \gamma_i D(y)x_i - p_E x_i$$

under the assumption that he is a small participant and so his choice of x_i does not affect p_E or D. Taking the partial derivative of the above with respect to x_i once again leads to (9.12), and we see that use of this congestion price maximizes social welfare.

9.3.2   Services Differentiated by Congestion Level

Thus far, we have simplified our discussion by supposing that users buy only one service. However, the ideas extend to models with more than one service. The following is an interesting example. Suppose a service is sold in two versions. The services are perfect substitutes except that they differ in the levels of congestion present. The problem of maximizing social welfare takes the form

$$\underset{\{x_1^i, x_2^i \geq 0, \; i=1,\dots,n\}}{\text{maximize}} \sum_{i=1}^{n} \left[v_i(x_1^i + x_2^i) - \gamma_i D_1(y_1)x_1^i - \gamma_i D_2(y_2)x_2^i \right] \tag{9.13}$$

where $y_t = \sum_i x_t^i$ is the total load carried in version t and the congestion cost is of the same form as in (9.11). This has similarities with the model for 'time-of-day pricing' discussed in Section 8.4.1. In that model, the versions of the service correspond to peak and off-peak periods and one has constraints $y_t \leq C_t$, $t = 1, 2$. As in the previous section, the problem can be decentralized by pricing congestion. User i should be faced with the problem

$$\underset{\{x_1^i, x_2^i \geq 0\}}{\text{maximize}} \left[v_i(x_1^i + x_2^i) - \gamma_i D_1(y_1)x_1^i - \gamma_i D_2(y_2)x_2^i - p_1 x_1^i - p_2 x_2^i \right] \tag{9.14}$$

where $p_t = \partial D_t(y)/\partial y \sum_i \gamma_i x_t^i$ is evaluated for the optimal $\{x_1^i, x_2^i : i = 1, \dots, n\}$. As above, the users treat $D_1(y_1)$ and $D_2(y_2)$ as fixed. It is easy to see that the user problem is solved by taking either $x_1^i = 0$ or $x_2^i = 0$. As we see in our discussion of Paris Metro Pricing in Section 10.8.1 it can sometimes be economically efficient to divide capacity in the manner above, so that users buy versions of the service that are best matched to their sensitivities to congestion.

In the differentiated services IP network architecture, described in Section 3.3.7, different versions of the service (say, bronze, silver and gold) are given different priorities by the network. We can model that idea here, by imagining that version 1 receives higher priority service than version 2, and so D_2 depends upon y_1 and y_2, but D_1 depends only upon y_1. The congestion prices are now

$$p_1 = \frac{\partial D_1(y_1)}{\partial y_1} \sum_i \gamma_i x_1^i + \frac{\partial D_2(y_1, y_2)}{\partial y_1} \sum_i \gamma_i x_2^i$$

$$p_2 = \frac{\partial D_2(y_1, y_2)}{\partial y_2} \sum_i \gamma_i x_2^i$$

Again, the right-hand sides are to be evaluated at the optimal $\{x_1^i, x_2^i : i = 1, \dots, n\}$. The partial derivatives can be calculated if the functions D_1 and D_2 are known explicitly, or estimated using on-line measurements. As we see in Section 9.4, we might also create the same rate of charges using sample path shadow prices. The ideas of Section 9.4.2 could be generalized to the model of this section. This approaches avoids the need for performance models.

9.3.3   A Blocking Model

This model shows how congestion prices can be used when congestion occurs because of blocking, i.e. when users are refused service. Blocking typically occurs because the call admission algorithm detects that there are not enough resources for new calls to be accepted.

See the discussion in Section 14.4. Suppose that there are n users and m types of calls. User i produces calls of type j at rate g_j^i, $j = 1, \ldots, m$. A call type is differentiated in terms of its duration and the amount of resources that need to be reserved at the time the call is set up.

The problem of the social planner is to choose the rates at which the users produce calls of the different types so as to maximize a social welfare function given by

$$W = \sum_{i=1}^{n} u^i (g_1^i, \ldots, g_m^i, B_1, \ldots, B_m)$$

where $B_j = B_j(g_1, \ldots, g_m)$ is the blocking probability for a call of type j and $g_j = \sum_{i=1}^{n} g_j^i$ is total arrival rate of calls of type j. Note that the blocking probability for calls of type j depends quite generally upon all the arrival rates of calls. Thus, we might model circumstances in which certain types of call have bandwidth reserved for them, or they are given priority. Blocking probabilities have interesting properties when calls are routed through a network and calls of the same type follow the same route. Increasing the rate of type i calls will definitely increase the blocking probability of the same call type, but may decrease the blocking probability of another type. This occurs because increasing blocking of calls sharing common links releases capacity in other parts of the network otherwise used by the blocked calls, and hence makes it easier for calls that would otherwise block to go through.

We would like to know if social welfare can be maximized by prices. Can the social planner post prices for each of the call types, so that when customers do local optimizations, they end up choosing arrival rates that solve the social welfare maximization problem? At the social welfare optimum

$$\frac{\partial W}{\partial g_j^i} = \frac{\partial u^i}{\partial g_j^i} + \sum_{k=1}^{n} \sum_{\ell=1}^{m} \frac{\partial u^k}{\partial B_\ell} \frac{\partial B_\ell}{\partial g_j^i} = 0, \quad 1 \le i \le n, \ 1 \le j \le m$$

Since B_ℓ depends upon g_j^i only through the sum $g_j = \sum_{i=1}^{n} g_j^i$, the above condition becomes

$$\frac{\partial u^i}{\partial g_j^i} + \sum_{k=1}^{n} \sum_{\ell=1}^{m} \frac{\partial u^k}{\partial B_\ell} \frac{\partial B_\ell}{\partial g_j} = 0, \quad 1 \le i \le n, \ 1 \le j \le m \tag{9.15}$$

Let $w_1, \ldots, w_m$ be the prices charged to the customers for the calls that are accepted (i.e. not blocked), which depend only upon the call type. Then customer i will choose arrival rates that solve his local optimization problem

$$\underset{g_1^i, \ldots, g_m^i}{\text{maximize}} \ u^i (g_1^i, \ldots, g_m^i, B_1, \ldots, B_m) - \sum_{k=1}^{m} w_k g_k^i (1 - B_k) \tag{9.16}$$

where the values of the blocking probabilities B_k are those which are observed for the current operating point of the link. They are considered as given (measured). This is an important assumption which leads to the above definition of the local optimization problem. If, instead, users have knowledge of the derivatives of the blocking probabilities with respect to their arrival rates then this leads to a different optimization problem, in which such prices might not exist. In any case, if the size of the system is large compared to individual users, then it is a reasonable approximation that a single user cannot have a significant effect on the blocking that takes place. We made a similar assumption in (9.8).

User i is faced with solving (9.16). Users choose arrival rates to satisfy

$$\frac{\partial u^i}{\partial g^i_j} - w_j(1 - B_j) = 0, \quad 1 \le j \le m, \ 1 \le i \le n \tag{9.17}$$

Observe now that, if we choose the prices w^*_j so that

$$w^*_j = -(1 - B_j)^{-1} \sum_{k=1}^{n} \sum_{\ell=1}^{m} \frac{\partial u^k}{\partial B_\ell} \frac{\partial B_\ell}{\partial g_j}, \quad 1 \le j \le m \tag{9.18}$$

we are guaranteed that the global conditions (9.15) and the local conditions (9.17) are equivalent, and hence the arrival rates that maximize social welfare are also an equilibrium for the system of prices (9.18). Note that w^*_j is a congestion price, in the sense that a customer pays the rest of the customers (including himself) for the marginal decrease of their utility because of the increase of blocking that is caused when he increases the rate of requests for calls of type j.

If we are to use (9.18) to compute the w^*_j, we need to know how the blocking probabilities vary with arrival rates. These values of $\partial B_\ell / \partial g_j$ might be obtained through more sophisticated modelling or by measurement. We also need explicit knowledge of the utility functions of the customers so that we can compute $\partial u^k / \partial B_\ell$. Perhaps the customers of the network fall into a small number of classes, for which the utilities are known. If this is so, then the network operator can compute prices once he knows the number of users in each class.

9.4 Congestion prices computed on sample paths

To compute the congestion price defined in (9.5), we need to take derivatives of the utility functions of the users. There are two practical problems. First, to take derivatives, one must know the explicit form of the utilities. Secondly, the performance of a network is measured in some average sense, such as average delay, average throughput, or percentage of blocked calls. Hence, to compute a price in an actual situation, one must estimate such quantities, estimate their derivatives, and then charge each user a price that is defined on a per unit of usage x_i. Such a process can be lengthy and inaccurate.

Let us look at a different approach. Instead of constructing a deterministic price that reflects the derivative of some average quantity, one may construct fluctuating prices that capture temporal congestion effects, and which result in the same average price. For instance, if the quantities x_i and y introduced in the previous sections are average flows of packets, it may be sensible to *charge each packet individually* for the exact amount of cost its existence imposes on the rest of the packets in the system while this packet travels through the network. Then one may expect that, although each packet is charged a different amount, in a reasonably large time window the source of the packets will see the same average congestion price as is given in (9.5). Such prices are computed on a sample path rather than on an average basis. It turns out that such price computation has important practical implementation advantages.

There is a possible problem with such pricing schemes on the 'supply' side. Who collects the charges? If it is the network provider then he has an incentive to increase congestion, perhaps by under-investing in capacity. To eliminate this possibility, there must be competition in supply. Competition induces social welfare maximization, which has been our objective throughout this chapter.

Sections 9.4.1–9.4.4 concern congestion charges that can be implemented as a per-packet charge. We begin with a model of a congested system in which the social welfare is a function of the rate of packet loss. In Section 9.4.2 social welfare is a function of packet delay. In both cases, the congestion charge is computed on a sample path basis.

9.4.1   A Loss Model

Suppose n users produce streams of packets as Poisson processes of rates $x_1, \ldots, x_n$. Time is slotted into unit length slots, so that in any given slot the number of packets that arrive from stream i is distributed as a Poisson random variable with mean x_i. Suppose that exactly C packets can be served in each slot. If more than C packets arrive then those in excess of C are lost. The network desires to maximize a social welfare function

$$\sum_i u_i(x_i) - c\left(\sum_i x_i\right)$$

where $c\left(\sum_i x_i\right)$ is the rate of cost due to lost packets. For simplicity, assume that each lost packet costs one unit. Then c is the expected number of packets that are lost per slot. The maximum is achieved when x_i is chosen such that $u_i'(x_i) = c'\left(\sum_i x_i\right)$.

We explain how the social welfare optimum can be obtained in a decentralized way. We begin by computing the value of c'. Let the random variable X denote the total number of packets that arrive during a slot. If the mean of X increases from $\sum_i x_i$ to $\sum_i x_i + \delta$, the cost of lost packets increases by about $c'\left(\sum_i x_i\right)\delta$. This is equal to the extra losses that would be seen if an additional independent stream of packets were to arrive as a Poisson process of rate δ. As δ is assumed to be extremely small, the probability of a loss from this stream within any given slot is just $\delta P(X \geq C)$, i.e. the probability that an arrival occurs in this small rate-δ stream times the probability that this arrival finds at least C packets from the other streams have also arrived in the same slot. Thus, $c'\left(\sum_i x_i\right) = P(X \geq C)$. Notice that the loss probability is $P(X > C)$, which is less.

Suppose the following charging scheme is adopted. A unit of charge (or mark) is sent to user i whenever a packet in stream i is lost, or a packet in stream i is not lost, but had it not been present then some other packet would not have been lost. In other words, whenever the number of packets in a slot exceeds C, each packet that arrives in that slot generates a charge mark to the user to whose stream it belongs. This is illustrated in Figure 9.1.

To compute the expected number of charge marks sent to user i per slot, think of the Poisson process of rate x_i as the superposition of N independent streams, each of rate $\delta = x_i/N$. Focus on one of these streams and observe that a packet from this stream produces a charge mark whenever the number of packets received from the other streams is at least C. Thus, the probability that this stream produces a charge mark is about

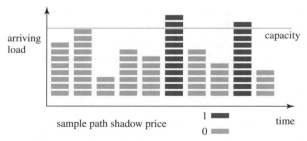

arriving load   capacity

sample path shadow price   1 ▬   time
                                             0 ▬

Figure 9.1   Sample path shadow price. A packet incurs a charge of 1 whenever it is lost, or if it had not been present another packet would not have been lost.

$\delta P(Y \geq C)$, where Y is a Poisson random variable with mean $\sum_j x_j - \delta$. Thus, the expected number of charge marks returned to user i per slot is $N\delta P(Y \geq C)$. However, N is arbitrary, so let $N \to \infty$. This gives $N\delta P(Y \geq C) \to x_i P(X \geq C)$, the expected number of charge marks sent to user i per slot. Note that this is greater than the loss rate of $x_i P(X > C)$.

User i seeks to maximize $u_i(x_i) - x_i P(X \geq C)$. As previously in this chapter, he treats $P(X \geq C)$ as a fixed probability, which he is individually too small to affect by his choice of x_i. Thus, to maximize his net benefit he chooses x_i to satisfy $u_i'(x_i) - P(X \geq C) = 0$. From the paragraph above, this is the same as $u_i'(x_i) - c'\left(\sum_i x_i\right) = 0$, namely the condition for social welfare maximization.

For simplicity, we have assumed that the congestion cost is the rate of the lost packets. If the cost of losing a packet is known, and the same for every packet and every user, then it makes sense to let the price per mark be the same to all users ('anonymous' or 'nondiscriminatory' in the economist's language), and equal to the cost of losing a packet. More generally, however, the cost of losing a packet (through loss or delay) may be different for different users, and known only to the users. What now are the properties of an equilibrium? On the basis of present research, the answer is unclear.

Thus, we have described a decentralized charging scheme that achieves the social welfare optimum. Note that it takes the general form of the congestion charges that we meet in this chapter: user i is charged for the losses that the presence of his packets inflict on other users. Because the charge is computed on-line, and sends a user a mark precisely when a small increase in capacity would reduce the rate of loss packets by one, we call this a *sample path shadow price*.

Such a marking scheme is simple to implement in packet networks. A router can charge each packet that contributes to an overflow the monetary value of a mark (the cost of a packet loss). It then notifies the source by sending back a mark; this is implemented by setting a special bit in the header of a packet that is travelling back to the destination. The user adjusts his rate of sending packets according to the rate of marks he receives, i.e. according to the rate of charge. Such schemes are analysed more in detail in Section 10.2.

9.4.2 A Congestion Model with Delay

As above, suppose n users produce streams of packets as Poisson processes of rates $x_1, \ldots, x_n$. These are queued and served by a single server. Suppose all packets have service time 1. We model the system as a $M/D/1$ queue. Suppose user i has a net benefit of the form

$$v_i(x_i) - \gamma x_i D\left(\sum_j x_j\right)$$

where D is the queueing delay of an average packet. Social welfare is

$$\sum_i \left[v_i(x_i) - \gamma x_i D\left(\sum_j x_j\right)\right] \tag{9.19}$$

and is maximized where

$$v_i'(x_i) - \gamma D\left(\sum_j x_j\right) - \gamma \left(\sum_j x_j\right) D'\left(\sum_j x_j\right) = 0 \tag{9.20}$$

We can calculate a sample path shadow price as follows. For each given packet that passes through the system we count the number of packets whose queueing delay would have been one unit less if the given packet had not been present. Note that this equals the number of packets that arrive after the given packet and are served during the same busy period. Although this number is not known until some time after the given packet has left the system, it is eventually known. User i receives this count for each of his packets, as a congestion signal that reflects the effect that his packet has on increasing the delay of other packets. Suppose the expected number of packets that arrive after any given packet and are served in the same busy period is Y. User i can be charged a congestion price $\gamma x_i Y$ and presented with the problem

$$\underset{x_i}{\text{maximize}} \left[v_i(x_i) - \gamma x_i D - \gamma x_i Y \right] \qquad (9.21)$$

Suppose user i is sufficiently small that neither D or Y changes appreciably with x_i. As in the previous section, we may consider a small increase in flow at rate δ and argue that the increase in rate of delay charge is $\delta Y = \delta D'(y)$, so giving $D'(y) = Y$, where $y = \sum_j x_j$. Thus when users are subject to this form of sample path congestion charging the solution to (9.21) occurs where (9.20) holds, and so the social welfare maximum in (9.19) can be achieved by users individually optimizing their net benefits in (9.21). Many of these ideas for constructing sample path congestion prices can be extended to the congestion charging of the differentiated services discussed in Section 3.3.7.

9.4.3 Bidding for Priority

Suppose that n packets are queued at a router. It is desired to schedule the packets to minimize a cost function of the form $\sum_i c_i^* D_i$, where D_i is the delay in the system experienced by packet i. This is done by the '$c\mu$-rule', which schedules packets in decreasing order of c_i^*/d_i, where d_i is the amount of time it takes to serve packet i. However, suppose the waiting costs c_i^* are unknown to the router. How can it ask the packets for these values and ensure that truthful answers are obtained?

Suppose each packet declares a c_i and packets are then served in decreasing order of c_i/d_i. Packet i is charged $d_i \sum c_j$, where the sum is taken over all j such that $c_j/d_j < c_i/d_i$. Thus, the charge is the extra delay charge that packet i imposes on those packets which come behind it.

The incentive issue is this. The true cost, c_i^*, is only known to packet i. It can declare any cost c_i. However, if it declares a cost different from c_i^*, its total cost (delay cost plus charge) is greater than if it declares $c_i = c_i^*$. That is, the scheme is incentive compatible.

To see this, suppose packet i declares c_i, with $c_i < c_i^*$. If it were to increase this by some small δ it might swap places with the packet immediately ahead of it, say packet j. It would save $c_i^* d_j$ in delay cost, but pay an extra charge of $c_j d_i$. So the swap is to packet i's advantage if $c_i^* d_j > c_j d_i$. Since we have assumed that packet j swaps positions with packet i, we must have $(c_i + \delta) d_j > c_j d_i$, and if $c_i^* > c_i + \delta$ this implies $c_i^* d_j > c_j d_i$. Thus c_i should not be declared any less than c_i^*. A similar argument shows that c_i should not be declared any more than c_j^*.

9.4.4 Smart Markets

The idea of a *smart market* is that of a bandwidth auction that occurs time slot by time slot. It is more of theoretical than practical interest, but provides another illustration of a way that congestion charges can be computed on a sample path basis.

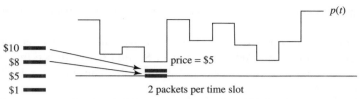

$10 ■
$8 ■   price = $5
$5 ■
$1 ■   2 packets per time slot

Figure 9.2   In a smart market bandwidth auction the packets bid for slots and pay the price of the highest bid of a packet not accepted. The price $p(t)$ fluctuates as a function of the competition in each slot.

Consider, as an example, a resource that can carry at most two packets in each time slot. This capacity is to be auctioned amongst several competing users. In a given time slot, all competing users state the amounts they would be willing to pay for the network to carry one of their packets. For example, four users might bid $1, $5, $8 and $10. In this case, the price would be set at $5 and the packets of the two users who bid $8 and $10 would be carried, (see Figure 9.2). The idea is that if one more unit of capacity had been available then additional revenue of $5 could have been obtained. Thus, $5 is the 'sample-path shadow price' of the resource. It is also a congestion charge, in the sense that accepting the bid of any given user decreases the total utility that could otherwise be obtained by all other users if that user's bid were not accepted, and this decrease equals the value of the highest unaccepted bid. It can be shown that this auction mechanism is incentive compatible, in that each user does best for himself by bidding the amount that equals his true utility for bandwidth (similarly as in the Vickrey auction, that we discuss in Section 14.1.2). This is a standard property of 'second-price' auctions, in which users do not pay their own bids, but pay a charge that is determined by the bids of the other users.

One can think of practical schemes in which the user limits his total bill by stating in advance the amount he is prepared to pay (perhaps by prepayment). The network decrements this amount as the user's traffic passes through successive routers. When the balance reaches 0, the user's packets become 'best-effort'. If the user still has a balance left at the end, he receives credits.

9.5   An incentive compatible model for congestion pricing

In concluding this chapter we give a model for congestion pricing that is even more general than in Section 9.1. It uses a form of Vickrey–Clarke–Groves mechanism to provides incentives to the users to choose socially optimal levels of demand and to reveal their true valuations of the service.

Suppose that $y = x^*$ maximizes the social welfare, $\sum_i u_i(y)$, for $y \in A$. Congestion is modelled by having u_i depend upon the complete vector y. Suppose user i demands x_i. Let x be the vector whose ith component is x_i and whose jth component is x_j^*, for each $j \neq i$. Let user i be charged

$$p_i(x_i) = \max_{y \in A} \sum_{j \neq i} u_j(y) - \sum_{j \neq i} u_j(x) \qquad (9.22)$$

We call view this a 'congestion charge' because it represents the greatest increase in benefit that other users might obtain if they were allowed freely to choose x, as against having to take the value of x_i that has been chosen by user i, and to take $x_j = x_j^*$, for $j \neq i$. In other words, $p_i(x_i)$ measures the reduction in the benefit obtained by other users because user

i is present and has chosen x_i, and the other components have been fixed at their social welfare optimal values.

A property of this method of charging is that it makes the social welfare optimal point, x^*, a Nash equilibrium, as defined at the start of Section 6.4.1. To see this, suppose that $x_j = x_j^*$ for $j \neq i$, and that user i seeks to maximize his consumer surplus by choice of x_i. He obtains

$$
\max_{x_i} \left[u_i(x) - p_i(x_i) \right] = \max_{x_i} \left[u_i(x) - \left\{ \max_{y \in A} \sum_{j \neq i} u_j(y) - \sum_{j \neq i} u_j(x) \right\} \right]
$$
$$
= \max_{x_i} \sum_j u_j(x) - \max_{y \in A} \sum_{j \neq i} u_j(y)
$$
$$
= \sum_j u_j(x^*) - \max_{y \in A} \sum_{j \neq i} u_j(y) \tag{9.23}
$$

which is achieved by taking $x_i = x_i^*$.

Let us write the charge $p_i(x_i)$ in one other way. Fix i. Let x^0 be the vector whose ith component is 0 and whose jth component is x_j^*, for all $j \neq i$. Then

$$
p_i(x_i) = \left[\max_{y \in A} \sum_{j \neq i} u_j(y) - \sum_{j \neq i} u_j(x^0) \right]
$$
$$
+ \left[\sum_{j \neq i} u_j(x^0) - \sum_{j \neq i} u_j(x) \right] \tag{9.24}
$$

The first term is fixed and is the opportunity cost to users j, $j \neq i$, of being forced to take $x_j = x_j^*$, even if user i creates no negative externality ($x_i = 0$). The second term is the extra congestion cost to the rest of the users when user i consumes x_i.

In the above we have assumed that the users' utility functions are all known. Suppose users are asked to state their utility functions. Then truthful revelation of utility functions is also a Nash equilibrium strategy for all users. For suppose user i may be untruthful, stating say $u_i^\dagger$, while all other users state their true utility functions, u_j, $j \neq i$. The network allocates bandwidth on the basis of the utility functions that are revealed. User i is allocated bandwidth x_i^*, where x^* maximizes the supposed social welfare function: $u_i^\dagger(x_i) + \sum_{j \neq i} u_j(x_j)$. User i is charged using (9.22). Under these conditions there is an incentive for user i to be truth-telling. This is because his surplus is (9.23), and this is maximized when x^* is chosen by the network to maximize $\sum_j u_j(x_j)$, which will happen if he states $u_i^\dagger = u_i$. Thus, truthful statement of utility functions is a Nash equilibrium in a game in which users may lie about their utility functions.

9.6 Further reading

The concepts of congestion pricing with applications to the Internet are nicely developed by MacKie-Mason and Varian (1994, 1995), who also proposed the idea of using smart markets in the networking context. Congestion pricing with priorities is addressed by Wilson (1990) and Mendelson and Whang (1990). Another interesting paper on concepts of congestion pricing is that of Gupta, Stahl and Whinston (1997). Congestion is a negative externality. The economics of externalities are studied in Chapter 24 of Varian (1992). Congestion has been always an important topic in transportation. An interesting web site that has references to congestion topics from this point of view is that of Metro Dynamics (2002).

In Section 9.2 the congestion model with delay cost of the $M/M/1$ queue is taken from Walrand and Varaiya (2000). The chapter on economics is particularly worth reading. The blocking model in Section 9.3.3 is from Courcoubetis and Reiman (1999). For further discussion of a delay-based sample path congestion charge, as in Section 9.4.2, see Key, Massoulie and Shapiro (2002). The model of bidding for priority in Section 9.4.3 is from private discussions with Pravin Varaiya.

FURTHER READING

10

Charging Flexible Contracts

Service contracts define guarantees that are to be met by both the service provider and the buyer of the contract. As we saw in Chapter 2, service contracts may have flexible guarantees in terms of performance that allow both the network operator and the user to vary important parameters, such as the rate at which the users may send data to the network. Both parties can benefit. If dynamic pricing is available, the user can dynamically vary the amount of resources he consumes by changing his willingness to pay. The network can control congestion by dynamically adjusting the share of resources that each user is allowed making use of the loose obligations in the contracts. Flexibility is particularly useful when the load on the network dynamically rises and falls as connections start and finish, or the available capacity fluctuates.

Flexibility can also benefit a user whose resource requirements vary with time. He need not predict and reserve at the start what he thinks his maximum resource requirements might be. The network also benefits because it need not reserve resources that might not be used, and so more service requests can be accommodated. Of course, by not reserving his maximum requirements, a user runs the risk that at a future time the resources he needs may be unavailable or too expensive to obtain. Since the time-varying availability of resources means that flexible contracts do not provide strict performance guarantees, users with applications that require strict guarantees may prefer contracts whose parameters are defined at the time the contract is set up and remain static throughout their lives. Such contracts can be priced using the ideas for guaranteed services described in Chapter 8.

Contracts with flexible guarantees are appropriate for elastic applications. An *elastic application* is one that can adapt itself to a varying availability of network resources. Examples are video and voice encoders and decoders that can change their data compression rate without significant loss of perceived quality. Of course, even elastic applications require some minimum resources and this may restrict the number of such applications that can be accommodated simultaneously. A network that shares its resources amongst elastic applications with flexible contracts must decide how to divide the network resources amongst them. It does this by setting the dynamic parameters of the contracts to make best use of the resources. If each user is allowed too large a peak rate then there may be unnecessary congestion. However, if each user is allowed only a small peak rate then resources might be wasted.

In general, any parameter of a contract can be flexible, but in practice it is the peak rate that is most often flexible. Let us focus on contracts like this. As there is only a single

Pricing Communication Networks C. Courcoubetis and R. Weber
© 2003 John Wiley & Sons, Ltd ISBN 0-470-85130-9 (HB)

traffic parameter that can dynamically change, such contracts are simple to define. Suppose that each user can always fully exploit the peak rate of his contract. So if he is allowed a peak rate x he will send at rate x. Since the peak rate is not fixed in the contract, but varies during the life of the connection, the user who buys such a contract must be an *elastic user*, that is, a user of an elastic application, who can adapt his sending rate. The notion of an elastic user can be related to the user's utility function. Suppose $u(x)$ denotes a user's utility for a flow of rate x. Let p be the price per unit flow, so the user's net benefit is $u(x) - px$. The user is said to be elastic if $u(x)$ is an increasing function of x and has no convex parts over the range $x > 0$. Under these conditions the maximizing value of x increases with p. By contrast, an inelastic user might require a definite rate, say $\bar{x}$. He buys that rate if $u(\bar{x}) > p\bar{x}$, but otherwise does not participate.

There are two main approaches for implementing flexible contracts for the peak rate. These influence the way such contracts are priced. The first is by creating a market for the peak rate, in which prices fluctuate to reflect demand and resource availability. In this *dynamic pricing approach*, users observe prices and react by adjusting the amount of resources they purchase, i.e. their peak rates. Such an approach requires dynamic price creation and propagation mechanisms, together with an accounting mechanism to monitor each user's response. It can be viewed as a generalization of flow control, in which flow adaptation decisions are based on price signals instead of traditional congestion signals.

The second approach is based on the differentiated services concept similar to that introduced in Section 3.3.7. The network operator simply creates several service classes, differentiated by the peak rates they are allowed (which in turn differentiates them by quality aspects such as packet loss probability, delay or blocking). He posts a price for each service class, and lets users self-select the classes to which they want to subscribe. Once a user chooses his contract, all parameters remain fixed for the duration of the contract and he is served in a 'best-effort' sense according to class of service he has chosen. The task for the network is to adjust the resources in each class and the corresponding prices so that each user eventually ends up using the type of service that is most appropriate for him. In this case, prices also fluctuate, but on much slower timescales, of hours and days, rather than on the timescale of a round-trip packet propagation time in the Internet, as they do under dynamic prices. This approach is very similar to that of versioning in price discrimination. It is simpler to implement than dynamic pricing, since prices remain fixed for longer periods of time and depend upon only the peak rate and the class of the contract (although one may also include a usage charge as an incentive to reduce waste). However, it has the disadvantage that users have no explicit quality of service guarantees and no means of dynamically changing the amount of resources they are allocated, unless they switch service classes by changing their contracts. In this regard, it is an example of flexible contracts with fixed price but unknown variable quality (since the number of customers that will subscribe and hence share the resources dedicated to each class is not known *a priori*). An important remark is that dynamic pricing mechanisms offer a broader possibility of services than do simple differentiated services. By varying the amount of peak rate he buys, the user can obtain a fixed charge per unit time and variable quality as prices fluctuate. Or, by always buying the same amount of peak rate, he can obtain fixed quality and variable charge per unit time. Finally, he may be able to obtain both a fixed price and fixed quality if he can buy some insurance against fluctuating prices from a third party who charges the average price plus some markup. Dynamic pricing signals can also be used in call admission control, as we shall see later.

The next few sections concern the problem of allocating rates of flows to elastic users. In Section 10.1 we discuss possible meanings of 'fair allocation'. Section 10.2 describes a decentralized method of using congestion signals to both efficiently and fairly allocating flows. It has similarities to the TCP algorithm that is used in the present Internet. Section 10.3 explains how this might be implemented in the Internet. Section 10.4 discusses a mathematical model of the present Internet in further detail. In Section 10.5 we extend the previous concepts to sharing effective flows, i.e. flows measured in terms of their effective bandwidth. Section 10.6 is about user agents (software that runs at the edges of the network and controls price reaction on behalf of the customers), and Section 10.7 is about pricing uncertainty. Section 10.8 discusses the differentiated services approach where the issue is the allocation of quality amongst users, where quality is measured in terms of congestion. We give an example to illustrate Paris Metro Pricing and show how it can increase social welfare. In Section 10.9 we discuss certain important issues that need to be considered when price mechanisms are to be used for managing network resources.

10.1 Notions of fairness

In this and the following two sections, we adopt the following model of a network of elastic users. We let L be a set of links and R be a set of routes, a route being a set of links. User r uses route r and is allowed a peak rate of x_r. We assume that he can use all this peak rate, and so his flow is x_r. The sum of the flows on the routes that use link j must not exceed the capacity of link j, denoted by C_j. Thus, we must allocate the flows $\{x_r\}$ subject to the set of constraints

$$\sum_{r:j\in r} x_r \le C_j, \quad \text{for all } j \in L \tag{10.1}$$

where $\{r : j \in r\}$ is the set of routes that use link j. Note that, in (10.1), we can give interpretations to x_r other than the peak rate; for instance, x_r may be the effective rate or the average rate. We return to this in a later section. Consider Figure 10.1. Here route 0 uses links $1, \ldots, n$, and route i uses just the single link i, $i = 1, \ldots, n$. Suppose that $C_1 = \cdots = C_n = 1$. If our goal is to maximize total throughput then we should take flows of $x_0 = 0$ and $x_1 = \cdots = x_n = 1$. However, this allocation of flows might be regarded as unfair, since route 0 is denied any flow. This suggests that throughput may not be the only consideration, but that fairness can also be important.

We have already discussed problems of fairly allocating cost in Chapter 7. One could proceed similarly for the problem of allocating flows. For example, just as we defined the characteristic function $c(\cdot)$ in Section 7.1.1, we could define the characteristic function, $v(S)$, as the maximal total flow that can be accommodated by the network if it is used only by the users in the set S, $S \subseteq N$, where $N = \{1, \ldots, n\}$. Notions such as Shapley value payoff, or nucleolus, that are concerned with ways to share $v(N)$, could be used

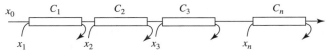

Figure 10.1 A problem of fairly sharing bandwidth. Here route 0 uses links $1, \ldots, n$, route i uses just the single link i, $i = 1, \ldots, n$, and the link capacities $C_1, \ldots, C_n$ are all 1. Throughput is maximized by taking flows of $x_0 = 0$ and $x_1 = \cdots = x_n = 1$. However, this allocation of flows might be regarded as unfair, since route 0 is denied any flow.

to define a fair allocation of flows. Of course, we prefer flow allocation methods that are easily understood, and which can be implemented by a simple algorithm or by pricing.

We say that an allocation satisfies the *max-min fairness* criterion if it is not possible to increase some flow without simultaneously decreasing another flow that is already smaller. This gives absolute priority to small flows. No increase in a larger flow can compensate for a decrease in a smaller flow. Equivalently, a flow allocation is max-min fair if for every route r there exists some link on that route, say l, such that l is filled to capacity and x_r is maximal amongst the flows that use link l. This condition clearly identifies a max-min fair flow, because if one were to try to increase the flow on route r it would necessarily require a reduction in the flow on some other route through link l, and that could only decrease the value of a smaller or equal flow. In Figure 10.1 the max-min fair allocation is $x_0 = x_1 = \cdots = x_n = 1/2$.

The following algorithm can be used to find the max-min fair allocation of flows. Start with all flows set at zero and then gradually increase them at the same rate, until some link (or set of links) becomes full. Fix the flows on the routes that use that link (or set of links). Note that these flows are equal. The flows on all other routes are allowed to increase equally, until some other link (or links) becomes full. The flows on the routes that use these links are now fixed. The algorithm continues in this manner until the values of the flows on all routes are fixed. At the point that the flow on a route becomes fixed, it is maximal amongst the flows that use that link. Thus, the condition of the previous paragraph is satisfied.

We can make other definitions of fairness. We say an allocation of flows exhibits *proportional fairness* when there is no change in the allocation of flows that can increase the sum of the proportional rate changes. That is, the allocation $\{x_r\}$ is proportionally fair, if for every other feasible allocation, $\{x_r'\}$,

$$\sum_{r \in R} \frac{x_r' - x_r}{x_r} \leq 0$$

Note that this is equivalent to $\{x_r\}$ maximizing $\sum_r \log x_r$, subject to $\{x_r\}$ feasible. In the example of Figure 10.1, the proportionally fair allocation is $x_0 = 1/(n+1)$ and $x_1 = \cdots = x_n = n/(n+1)$. Note that flow 0, which uses many links, is penalized.

Similarly, we say an allocation of flows exhibits *weighted proportional fairness* when there is no change in the allocation of flows that can increase a weighted sum of the proportional rate changes. That is, for every other feasible allocation, $\{x_r'\}$,

$$\sum_{r \in R} w_r \frac{x_r' - x_r}{x_r} \leq 0$$

Now $\{x_r\}$ must maximize $\sum_r w_r \log x_r$ subject to $\{x_r\}$ feasible.

Proportional fairness arises naturally in the solution of a number of interesting problems. Recall the solution of the Nash bargaining game described in Section 7.2.1. This concerns the choice of a point $u = (u_1, \ldots, u_k)$ within a bargaining set U. The choice of u is to be made subject to four reasonable assumptions about what constitutes a fair choice. If players have equal bargaining power then the Nash bargaining game is solved by maximizing $\sum_r \log u_r$ over $u \in U$. Taking $u_r = x_r$ and U equal to the set of feasible flows, we see that the Nash solution is the same as a proportionally fair allocation. If players have unequal bargaining powers then u should be chosen to maximize $\sum_r w_r \log u_r$, and this corresponds to a weighted proportionally fair allocation. The next example illustrates that proportional fairness arises naturally as a result of existing flow control procedures used in the Internet.

Example 10.1 (A fair window flow control) Suppose the flow x_r on route r is controlled by a window of size B_r bytes (as described in Section 3.3.7). In general, the use of window flow control leads to fluctuating rates. That is, x_r varies in time and the resulting traffic is bursty. However, as an approximation to the actual case, suppose that the network is equipped with additional mechanisms to smooth bursts so that there is a static regime in which x_r remains constant. Let T_r be the round-trip time on route r, excluding any queueing delay on the forward path or backward path due to packets waiting in buffers (i.e. the round-trip time if the system is lightly loaded). Let B_r and T_r be fixed. Also assume that inside the network packets are queued at the routers in order of their arrival, and buffers are large enough to imply no packet losses. In this model, sources do not react to congestion indication signals and send packets at constant continuous rates $x_r = T_r^{\text{total}}/B_r$, where T_r^{total} is the total round-trip delay on route r. It can be shown that in the corresponding fluid model there is a unique static regime and in this regime the resulting set of flows $\{x_r\}$ is the unique solution to the optimization problem

$$\text{maximize}_{x \geq 0} \sum_{r \in R} (B_r \log x_r - x_r T_r), \quad \text{subject to} \sum_{r:j \in r} x_r \leq C_j \quad \text{for all } j$$

This is similar to the problem of allocating flows in a weighted proportionally fair way, but with a penalty term of $-\sum_r x_r T_r$, that has the interpretation of the total rate of delay incurred by packets. If the round trip times are negligible, then the static flows will comprise a weighted proportionally fair allocation with weights equal to the window sizes.

Thus, we see that fixed size window control can achieve a fair sharing of the bandwidth according to this criterion, provided scheduling at each link is performed in an appropriate manner. In more general set-ups, it can also be shown that flow control mechanisms that use additive increase/multiplicative decrease tend to produce flow allocations that are proportionately fair. Such mechanisms control the window size so that in the absence of congestion the sending rate increases at a constant rate, while if capacity is exceeded on some link in the path and congestion occurs, then the sending rate decreases at a rate proportional to its size at the time congestion was sensed. Further details are given in Section 10.4.

10.2 The proportional fairness model

Let us continue with the model of the previous section. Recall that each route is associated with a user and requires some subset of the links. User r is an elastic user, whose utility for a flow of x_r is $u_r(x_r)$, which we assume to be increasing, strictly concave and continuously differentiable in x_r. Let us define SYSTEM as the global optimization problem of maximizing the social welfare, subject to the capacities of the links. It is

$$\text{SYSTEM}: \text{maximize}_{x \geq 0} \sum_{r \in R} u_r(x_r), \quad \text{subject to} \sum_{r:j \in r} x_r \leq C_j, \quad \text{for all } j \in L$$

We can rewrite this as

$$\text{SYSTEM}: \text{maximize}_{x \geq 0} \sum_r u_r(x_r), \quad \text{subject to } Ax \leq C \quad (10.2)$$

where $A_{jr} = 1$ or 0 as $j \in r$ or $j \notin r$, respectively.

Although it is easy enough to formulate SYSTEM, it is not practical for the network operator to solve it. The problem is that he does not know all the utility functions, u_r. Similarly, the users cannot solve SYSTEM because they do not know each other's utilities or the constraint set $Ax \leq C$.

Fortunately, if the network and users exchange a little information they can solve SYSTEM together. Suppose user r is willing to pay an amount w_r per unit time, and that if he does this he receives a flow $x_r = w_r/\lambda_r$. Here λ_r is the charge per unit flow on route r and is controlled by the network. Imagine that the network somehow chooses the λ_rs so that the resulting flows are feasible. Given that user r has been told the price λ_r his problem is

$$\text{USER}_r : \underset{w_r \geq 0}{\text{maximize}}\, u_r(w_r/\lambda_r) - w_r \qquad (10.3)$$

Note that he has all the information he needs to solve this problem.

What problem should the network solve? It cannot easily solve SYSTEM because it does not know the users' utility functions. However, suppose the users communicate their w_rs and the network then allocates flows so they are weighted proportionally fair. That is, it solves the problem

$$\text{NETWORK} : \underset{x \geq 0}{\text{maximize}} \sum_{r \in R} w_r \log x_r, \quad \text{subject to } Ax \leq C \qquad (10.4)$$

A key result is that the solution of SYSTEM occurs when the NETWORK and USER$_r$ problems are solved simultaneously, with their solutions in equilibrium. That is, there exist $\{\lambda_r\}$, $\{x_r\}$, $\{w_r\}$ such that $x_r = w_r/\lambda_r$ and these are simultaneously solutions to SYSTEM, NETWORK, and USER$_r$ for all r. This result is important because it means that if the users and network communicate their w_rs and λ_rs, then the solution of SYSTEM can be obtained using information that is available to each agent.

Let us prove the key result. It follows from solving SYSTEM by Lagrangian methods. Given the value of λ_r the solution of USER$_r$ is straightforward: w_r satisfies

$$u'_r(w_r/\lambda_r) - \lambda_r = 0 \qquad (10.5)$$

Consider the solution of NETWORK. Imagine that the $\{w_r\}$ are given and fixed. The fact that $\sum_r w_r \log x_r$ is concave in x and the constraints are linear, means that NETWORK can be solved by maximization of its Lagrangian, i.e. by solving

$$\underset{x,z \geq 0}{\text{maximize}}\, L, \quad \text{where } L = \sum_{r \in R} w_r \log x_r + \mu^\top (C - Ax - z) \qquad (10.6)$$

Since for this objective function $x_r = 0$ cannot be optimal, L is maximized where it is stationary with respect to x_r, and so

$$\frac{\partial L}{\partial x_r} = \frac{w_r}{x_r} - \sum_{j:j \in r} \mu_j = 0$$

Thus necessary and sufficient conditions for $\{x_r\}$ to be a solution to NETWORK are that we can find $\{\mu_j\}$ such that

$$x_r = \frac{w_r}{\sum_{j:j \in r} \mu_j}, \quad \mu \geq 0, \quad x \geq 0, \quad \text{and} \quad \mu^\top (C - Ax) = 0 \qquad (10.7)$$

These equations have a nice interpretation as defining a set of market clearing prices $\{\mu_j\}$ when user r has an initial amount of money w_r and purchases as much bandwidth as possible in route r. As necessary conditions these imply the existence of $\{x_r\}$ and $\{\mu_j\}$ satisfying (10.7). Similarly, consideration of the solution of SYSTEM by maximizing its Lagrangian gives that a sufficient condition for its solution is the existence of $\{x_r\}$, $\{\mu_j\}$ such that

$$u'_r(x_r) - \sum_{j:j\in r} \mu_j = 0, \quad \mu \geq 0, \quad x \geq 0, \quad \text{and} \quad \mu^\top(C - Ax) = 0 \qquad (10.8)$$

Now if USER$_r$ and NETWORK have solutions that coincide, we must have $\lambda_r = \sum_{j:j\in r} \mu_j$. This means that the sufficient condition (10.8) of SYSTEM can be satisfied using the necessary conditions for the NETWORK and USER$_r$ problems in (10.5) and (10.7).

We stress again that the advantage of approaching the solution of SYSTEM by way of the NETWORK and USER$_r$ problems is that the network and users only need information to which they have access. The network does not need to know the users' utility functions, and the users need not know the resource constraints of the links. The network communicates with the user r by announcing a new value of the rate x_r, or equivalently announcing a new price λ_r. Using this data, user r solves USER$_r$ and then tells the network his new selection of w_r.

In practice, however, it is not guaranteed that this iterative computation between the users and the network converges and that its solution is stable. This is because the computations are asynchronous and the communication of the values of $\{w_r\}$ and $\{x_r\}$ are subject to random delays which differ from user to user. It would be better if we could avoid having to solve the NETWORK problem in some central location, but could instead solve it in some decentralized way. Ideally, the network could communicate some information to the users that would give them the incentives to make their own adjustments to flows in a way to solve NETWORK themselves. This would be similar to the way that the TCP protocol operates in the Internet, in which software running on each user's computer adjusts the local sending rate and results in an allocation of flows inside the network by operating in a distributed fashion. The challenge is to understand how this can be done so that the network as a whole reacts intelligently to perturbations and the resulting allocation of flows solves the desired network optimization problem.

We begin by presenting two algorithms that solve NETWORK in a distributed fashion. We then discuss some convergence and stability results, which suggest that such a distributed solution to NETWORK, combined with users solving their own local USER$_r$ problem, does eventually converge to the solution of the SYSTEM problem.

10.2.1 A Primal Algorithm

One way to find $\{x_r\}$ and $\{\mu_j\}$ that satisfy the sufficient conditions (10.7) of NETWORK is by use of a so-called *primal algorithm*. In both this algorithm and the dual algorithm which follows, we assume the w_rs are held fixed by the users at constant values. Consider the system

$$\frac{d}{dt}x_r(t) = \kappa_r \left(w_r - x_r(t) \sum_{j:j\in r} \mu_j(t) \right) \qquad (10.9)$$

$$\mu_j(t) = p_j \left(\sum_{s:j\in s} x_s(t) \right) \qquad (10.10)$$

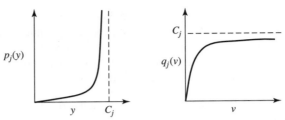

Figure 10.2 Functions $p_j(y)$ and $q_j(v)$ are inverse functions of one another. $p_j(y)$ is the rate of charge when the total flow through link j is y. $q_j(v)$ is the rate of flow in link j that produces a rate of charge in that link of v.

Here $\kappa_r > 0$ is a route-dependent parameter which affects the convergence rate of the algorithm, and $p_j(y)$ is a nonnegative, continuous, increasing function of y, not identically zero, which is close to 0 when link j is not full and rises rapidly to infinity as the flow in link j tends to C_j from below. See the left part of Figure 10.2.

There are some interesting interpretations to (10.9)–(10.10). We can think of resource j as marking a proportion $p_j(y)$ of packets with a feedback signal when the total flow passing through is y. These feedback signals are sent back to the users at the rate they are produced, indicating congestion. In actuality, there is a delay between the time a signal is produced and the time it is received at the edge of the network. This can affect the stability of the solution, an issue we return to later. User r responds to each congestion signal by reducing his sending rate by a constant amount.

Alternatively, we can think of $\mu_j(t) = p_j(t)$ as the price per unit flow charged by link j. These prices are communicated to the users. In fact, user r need only see the sum of the prices of the links along route r. In accordance with (10.9), user r decreases or increases x_r as his rate of charge is respectively more or less than w_r. Prices can also be communicated using the following mechanism. Assume that resource j marks passing packets in proportion to $p_j(y)$, but these marks have the meaning of a unit charge. Then the second term within the parenthesis on the right hand side of (10.9) is the total rate of charge observed by user r. Observe that using marks to convey price information is much easier to implement than actually informing users about price values. A mark can be conveyed by setting a bit in a packet header. Also, this interpretation is naturally related to the definition of sample path congestion prices discussed in Section 9.4. A packet is marked if it contributes to congestion. In this case, $p_j(y)$ refers to the probability of marking a packet.

One can show that the system defined by (10.9)–(10.10) converges to a stable equilibrium point. To do this, we consider the Lyapunov function

$$\mathcal{U}(x) = \sum_r w_r \log x_r - \sum_j \int_0^{\sum_{s:j\in s} x_s} \mu_j(y)\,dy$$

One can check that $\mathcal{U}$ is strictly concave in $\{x_r\}$, with a single interior maximum. Using (10.9) we find

$$\frac{d}{dt}\mathcal{U}(x(t)) = \sum_r \frac{\partial \mathcal{U}}{\partial x_r}\frac{d}{dt}x_r = \kappa_r \sum_r \frac{1}{x_r(t)}\left(w_r - x_r(t)\sum_{j:j\in r}\mu_j(t)\right)^2 \geq 0$$

Thus, x converges to the maximizer of $\mathcal{U}$. One can show that by sharpening the form of p_j this maximizer can be made an arbitrarily good approximation to the solution of

NETWORK. For example, one can take $p_j(y) = \max\{0, y - C_j + \epsilon\}/\epsilon^2$. Then maximizing $\mathcal{U}$ is an arbitrarily close approximation to the SYSTEM problem as $\epsilon \to 0$.

10.2.2 A Dual Algorithm

An alternative approach uses a *dual algorithm*, based on the Lagrangian dual problem to NETWORK. This problem is: $\text{minimize}_{\mu \geq 0} \max_{x,z \geq 0} L$, where L is defined in (10.6). It reduces to

$$\underset{\mu \geq 0}{\text{maximize}} \sum_r w_r \log \left(\sum_{j:j \in r} \mu_j \right) - \sum_j \mu_j C_j$$

and the objective function can be approximated by

$$V(\mu) = \sum_r w_r \log \left(\sum_{j:j \in r} \mu_j \right) - \sum_j \int_0^{\mu_j} q_j(v) \, dv$$

where we imagine that $q_j(0) = 0$ and $q_j(v)$ increases rapidly to C_j for $v > 0$, as shown at the right of Figure 10.2.

To find $\{x_r\}, \{\mu_j\}$, consider the system

$$\frac{d}{dt} \mu_j(t) = \kappa_j \left(\sum_{r:j \in r} x_r(t) - q_j(\mu_j(t)) \right) \tag{10.11}$$

$$x_r(t) = \frac{w_r}{\sum_{k:k \in r} \mu_k(t)} \tag{10.12}$$

Again, $\kappa_j > 0$, and $q_j(\mu_j)$ is interpreted as a rate of flow in link j corresponding to a rate of charge in that link of μ_j. This rate of charge adjusts itself proportionally to the excess demand (the term in the right-hand side of (10.11)). This is similar to the tatonnement process of Section 5.4.1. As with the primal algorithm, it can be shown that $V(\mu)$ is strictly concave in μ and that it is a Lyapunov function for the system (10.11)–(10.12). Thus, μ converges to the maximizer of V.

10.2.3 User Adaptation

So far, we have assumed that the w_rs are fixed on the timescales of the operation of the primal and dual algorithms. In this case NETWORK is solved by assigning each user r a rate x_r.

Now suppose that the w_rs are allowed to change. A simple iteration between users and network could take place as follows. The users could wait until the x_rs in NETWORK converge. Then, after computing the resulting prices based on their previous selections of w_rs and the newly defined x_rs, they could solve their local USER$_r$ problems and make new choices of w_rs. These could become new inputs to NETWORK, which could be resolved, and so on. A more interesting iteration is when both problems are solved at the same time. Observe that $x_r(t)$ is defined by the differential equation (10.9), which user r can affect only through his choice of $w_r(t)$. Suppose that he monitors $x_r(t)$ continuously and varies $w_r(t)$ so that

$$w_r(t) = x_r(t) u_r'(x_r(t)) \tag{10.13}$$

This choice is consistent with continuously maximizing his net benefit. To see this, suppose that at time t he observes the implied price $\lambda_r(t) = w_r(t)/x_r(t)$. For a δ that is imagined to be infinitesimally small, he chooses $w_r(t + \delta)$ for time $t + \delta$ so to maximize his net benefit, i.e. so $w_r(t + \delta) = \arg\max_{w_r}[u_r(w_r/\lambda_r(t)) - w_r]$. This happens if and only if $w_r(t + \delta)$ satisfies $u_r'(w_r(t + \delta)/\lambda_r(t)) = \lambda_r(t)$, which is (10.13) as $\delta \to 0$. Thus, we see that choosing $w_r(t)$ by (10.13) is consistent with his continually maximizing his net benefit at the presently implied price.

In summary, user r responds to the varying prices by solving USER$_r$ and using (10.13) in the place of w_r in (10.9). Using revised Lyapunov functions, one can again establish the stability of both the primal and dual algorithms in this case, i.e. when users solve their local optimization problems while they perform flow control. The appropriate Lyapunov function for the primal is

$$\mathcal{U}(x(t)) = \sum_r u_r(x_r) - \sum_j \int_0^{\sum_{s:j\in s} x_s} p_j(y)\,dy \qquad (10.14)$$

where again, this can be made an arbitrarily close approximation to the SYSTEM problem by appropriate choice of $p_j(\cdot)$

10.2.4 Stochastic Effects and Time Lags

Further things can be said when there are delays and stochastic effects. One can still prove convergence but the coefficient κ_r is important. When there are no delays or stochastic effects, the speed of convergence increases with both κ_r and the magnitude of the derivatives of $p_j(\cdot)$ at the point where $\mathcal{U}(x)$ is maximized. One can think of $p_j'(\cdot)$ as the sensitivity of the resource's load response, and of κ_r as the sensitivity of the response of end-systems to congestion marks. To model delays we would write $\mu(t - d_{jr})$ and $x_s(t - d_{js})$ on the right-hand sides of (10.9) and (10.10), where d_{jk} is a number expressing the time delay in communication between user k and link j. When there are stochastic effects, decreasing κ_r or increasing $p_j'(\cdot)$ leads to a smaller variance in the distribution of $x(t)$ at equilibrium. Increasing $p_j'(\cdot)$ and κ_r increases the speed of convergence but may make the system unstable. Thus, there is a trade-off between the speed of convergence and stability. In broad terms, the speed of convergence decreases as the magnitude of the d_{jr} increase, and the equilibrium point of the resulting system is asymptotically stable provided κ_r is sufficiently small (for a given choice of $p_j(\cdot)$s).

In our discussion so far, we have assumed that the choice of the function $p_j(\cdot)$ that is to be implemented at link j is simply an engineering decision, whose aim is to realize a robust and efficient flow control procedure that will lead to a good approximation to the optimal solution of SYSTEM. We now discuss how such a choice may be related to actual congestion costs.

10.2.5 Proportional Fairness with a Congestion Cost

In Chapter 9 we considered models in which social welfare is reduced by a congestion cost. A model with a congestion cost that is similar to (10.2) is

$$\text{SYSTEM}_c : \underset{x \geq 0}{\text{maximize}} \sum_r u_r(x_r) - \sum_j c_j \left(\sum_{s:j\in s} x_s \right) \qquad (10.15)$$

This problem is of interest in its own right, as it models a network, such as the Internet, in which social welfare is reduced because of congestion cost.

It can also be compared to SYSTEM, differing in that the constraint $\sum_{r:j\in r} x_r \le C_j$ in SYSTEM has been replaced by the congestion cost $c_j\left(\sum_{s:j\in s} x_s\right)$. In fact, SYSTEM$_c$ can be viewed as an approximation to SYSTEM if we imagine that $c_j(y)$ stays near 0 for $y < C_j$ and then rises rapidly to infinity as y approaches C_j from below. Then c_j acts as a sort of penalty function for violation of the constraint. Let us assume c_j is convex, strictly increasing and differentiable, and define $p_j = c'_j$. Then p_j is nonnegative, continuous and increasing, SYSTEM$_c$ is a good approximation to SYSTEM when p_j is close to 0 when link j is not full and then rises rapidly to infinity as the flow on link j tends to C_j from below (again see Figure 10.2).

SYSTEM$_c$ can be solved directly, using either the primal or dual algorithms. The same Lyapunov functions can be used to prove this. For instance, in the primal algorithm defined by (10.9)–(10.10) and (10.13), we let $p_j(y) = c'_j(y)$ and use (10.14).

10.3 An internet pricing proposal

We are now in a position to bring together all the above ideas in a proposal for price-sensitive flow control that can be used in IP-based networks such as the Internet. As in Section 9.4, the computation of $p_j(y) = c'_j(y)$ can be made using sample path shadow prices. We have seen how this can be done for both a loss system (Section 9.4.1) and a queue with delay cost (Section 9.4.2). Recall the basic ideas: charge marks are sent by routers to any source that produces a packet that contributes to a busy period in which there is an overflow. In an IP network this could be implemented by marking packets using the *Explicit Congestion Notification* (ECN) bits of the IP header. The users perform the flow control of (10.9), and the network computes (10.10) on a sample path basis. On slower timescales the users update their w_r using (10.13).

The problem of constructing the marking function p_j raises some interesting issue. In practice, p_j can only approximate the true sample path congestion marking function. This is because when a packet loss occurs, the identity of all the previously served packets that contributed to the packet loss is not available to the router (since it could not predict that such a future loss would occur when these packets were served). Hence all one can do is propose such marking functions and then argue that they so indeed have the right properties. Note that it is important both that $p_j(y)$ should send congestion marks at a greater rate than actual packet losses (since more than one packets contribute to a single packet loss), and also that these congestion marks should start appearing before actual congestion takes place. This allows sources to adapt early and avoid unnecessary and costly packet losses. Also, since bursty flows contribute more to congestion phenomena than do less bursty sources of the same mean rate, marks should be allocated in proportion to some burstiness measure such as the effective bandwidth of the flows. Predicting imminent congestion is not simple, and it cannot be done by simply observing packet losses. For instance, if many nearly deterministic flows (i.e. of small burstiness) are multiplexed, then packet losses will appear suddenly and will occur at very large rates when the load approaoches the link capacity. By contrast, more bursty flows generate packet losses that are visible at even low link utilizations, and so can provide adequate feedback signals earlier. However, if signals occur too early they may make the system overly conservative and result in underloading the links. A good choice of $p_j(y)$ should not be 'tricked' by the nature of the flows and should generate congestion signals at rates such that, in all circumstances, the system maintains stability and achieves a good utilization of its links.

We describe two mechanisms for implementing appropriate p_js. The first of these, called **RED**, is a proposal for preventing packet losses in the Internet. The basic idea is that the routers should monitor the average queue size of outgoing links, and when this size exceeds some threshold, they should randomly place ECN marks on outgoing packets as congestion indications, doing so with a probability that increases linearly in the average queue size. These ECN marks eventually reach the sender. Originally, RED was intended to be used in combination with TCP, the idea being that TCP should react to a congestion mark as if a packet loss had occurred. The second mechanism for implementing appropriate p_js, is the *virtual queue approach*, in which an algorithm runs an on-line simulation of a virtual queue of a proportionally smaller size, i.e. in which the buffer size and service rate are multiplied by some factor $\theta < 1$. It feeds the queue with the same traffic (or with a fraction θ of the traffic, randomly chosen, depending on the variant of the implementation). The algorithm waits until the virtual queue overflows, and then marks all subsequently arriving packets until it empties. The idea is that the virtual queue will overflow before the actual queue does, and so most packets that cause overflow in the actual queue will be marked in the virtual queue. Also, virtual queues produce larger rates of congestion signals. Using this approach, bursty flows receive more marks.

Both of the above algorithms have many parameters, and tuning them appropriately takes experiment, study and skill gained by experience. There are many subtleties. For instance, since in RED the burstiness of the marking process affects the burstiness of the traffic that results from the flow control, one may be tempted to reduce such burstiness by averaging the queue length process. The danger is that this may reduce stability margins by making the system slower to respond to congestion (because of increased extra delay in the feedback loop).

There are several nice consequences of the network flow control mechanism in (10.9)–(10.10). It essentially allows users to control the quality of service they obtain from the network (i.e. the value of their x_r). Since the possible values of such x_rs can be arbitrary, this means that a simple packet network that is equipped with this mechanism may be able to support an arbitrarily differentiated set of services by conveying information on congestion from the network to intelligent end-nodes, which themselves determine their demands on network. This is a radically different approach to that of differentiated services in Section 3.3.7, where the network must itself be engineered to provide service differentiation at a packet level using extra mechanisms at the routers. In the proportional fairness approach, the network treats all packets equally in respect of quality, while the incentives and the capabilities for service differentiation are moved to end-devices. It is analogous to the electricity distribution network, in which the same network is used to transport electricity for any type of use, instead of there being different networks for each of 110 V, 220 V, 360 V, 12 kV, etc.

One may even implement call admission control by measuring marking rates. For instance, suppose that we want to create a network service for real-time traffic such as voice calls. Edge-devices where such calls originate could first probe the network by sending a few packets along the path of the call and counting the number of congestion marks received. The call is rejected if this number is above some threshold, indicating congestion and hence a low bandwidth share. Low-priority non-adaptive traffic may also be treated similarly, by not allowing it into the network if the rate of congestion marks is above a certain level.

There are several design decision that must be taken to make the above approach applicable. For instance, in the client-server model of the Internet it is the receiver, rather than the sender, who obtains value from the flow, and hence it should be he who controls the sender's choice of w_r. Also, when several interconnected networks produce congestion marks, there should be mechanisms that allow the payments of the users to be shared fairly among these networks. There are more difficult issues when intermediate networks deploy different technologies for providing quality of service, which may create problems in propagating congestion marks generated by other networks. There are also interesting incentive issues. Networks may add congestion indications simply to collect more revenue, or simply because they like to stay in a congested mode. Users may not declare truthfully their best choice of w_r. The assumption of a competitive market solves the first issue, since networks having users as clients will choose to interconnect with transit networks that are less congested so they can offer lower prices to their customers. Individual users will tend to make truthful declarations if they are too small to affect the overall prices. There is another issue for traffic that originates from connections that are so short-lived that they cannot adapt to flow control decisions. In this case end-devices should use past history information to infer prices.

There are two interesting problems related to dynamic pricing. The first concerns the complexity of the decision to be made by end-devices in choosing the parameter w_r to optimize USER$_r$. In Section 10.6 we describe an approach in which we use software algorithms that can hide this complexity from the user. A second problem regards the inherent difficulty that such a system has in being able to offer a fixed quality service at a price determined beforehand. This may be thought as a serious drawback since users are usually risk averse and do not like unpredictability. A possible way to remedy this is by creating insurance contracts which remove such risks from users. Such insurance providers may be third parties who know the statistics of the price fluctuations at the various periods of the day and offer to pay the charge generated by congestion marks by charging users the expected value of the charge plus a markup. These ideas are discussed in Section 10.7.

10.4   A model of TCP

We have already mentioned that the quantity $\mu_j(t)$ that is defined in (10.10) can be viewed as a rate of congestion indication signals. These signals are generated at link j and passed back to user r, who then adjusts $x_r(t)$ according to (10.9). That adjustment involves a linear increase, proportional to w_r, and so is greater when he has a greater willingness to pay. It also involves a multiplicative decrease that depends on the rate at which congestion indication signals are received. These mechanisms make one think of Jacobson's TCP algorithm, which operates in the present Internet and which we described in Section 3.3.7. It also has several differences, which we now discuss.

Recall that a flow through the Internet receives congestion indication signals as dropped or marked packets. These occur at a rate roughly proportional to the size of the flow. The response of Jacobson's congestion avoidance algorithm to a congestion indication signal is to halve the size of the flow. Thus there are two multiplicative effects: both the number of congestion indication signals received and the response to each signal scale with the size of the flow. A further important feature of Jacobson's algorithm is that it is self-clocking: the sender uses an acknowledgment from the receiver to prompt a step forward and this produces an important dependence on the round trip time T of the connection. In more detail, TCP maintains a window of transmitted but not yet acknowledged packets; the rate x and the window size W satisfy the approximate relation $W = x\mathrm{T}$. Each positive acknowledgment increases the window size W by $1/W$; each congestion indication halves the window size. Crowcroft and Oechslin (1998) have proposed that users be allowed to set a parameter m, which would multiply by m the rate of additive increase and make $1 - 1/2m$ the multiplicative decrease factor in Jacobson's algorithm. The resulting algorithm, MulTCP, would behave in many respects as a collection of m single TCP connections; its smoother behaviour is more plausibly modelled by a system of differential equations. For MulTCP the expected change in the congestion window W per update step is approximately

$$\frac{m}{W}(1 - p) - \frac{W}{2m}p \tag{10.16}$$

where p is the probability of congestion indication at the update step. Since the time between update steps is about $1/x = \mathrm{T}/W$, the expected change in the rate x per unit time is thus approximately

$$\frac{1}{\mathrm{T}}\frac{\left(\frac{m}{W}(1 - p) - \frac{W}{2m}p\right)}{\mathrm{T}/W} = \frac{m}{\mathrm{T}^2}(1 - p) - \frac{x^2}{2m}p$$

Motivated by this calculation, we can model MulTCP by the system of differential equations

$$\frac{d}{dt}x_r(t) = \frac{m_r}{\mathrm{T}_r^2} - \left(\frac{m_r}{\mathrm{T}_r^2} + \frac{x_r(t)^2}{2m_r}\right)\sum_{j:j\in r}\mu_j(t), \quad r \in R, \tag{10.17}$$

where T_r is the round trip time for the connection of user r, and where $\mu_j(t) = p_j(t)$ is again given by equation (10.10) and viewed as the probability that a packet produces a congestion indication signal at link j. Note that if congestion indication is provided by dropping a packet, then the sum on the right-hand side of equation (10.17) approximates the probability of a packet drop along a route by the sum of the packet drop probabilities at each of the links along the route.

To ensure $x_r(t)$ remains nonnegative, let us interpret the left-hand side of (10.17) as zero if the right-hand side is negative and $x_r(t)$ is zero. It can be shown that

$$\mathcal{U}(x) = \sum_r \frac{\sqrt{2m_r}}{\mathrm{T}_r}\arctan\left(\frac{x_r\mathrm{T}_r}{\sqrt{2m_r}}\right) - \sum_j \int_0^{\sum_{s:j\in s}x_s} p_j(y)\,dy \tag{10.18}$$

is a Lyapunov function for the system of differential equations (10.17), which have stable point

$$x_r = \frac{m_r}{\mathrm{T}_r}\left(\frac{2(1 - p_r)}{p_r}\right)^{1/2}$$

where $p_r = \sum_{j:j\in r} p_j$. This is the unique value x maximizing $\mathcal{U}(x)$ to which all trajectories converge. We can view (10.18) as the social welfare of an economy in which the utility function of user r is

$$\frac{\sqrt{2m_r}}{\mathrm{T}_r}\arctan\left(\frac{x_r\mathrm{T}_r}{\sqrt{2m_r}}\right)$$

and as if the network's cost is the final term in (10.18). If in the expression (10.16) we were to approximate the factor $(1 - p)$ by 1 then the implicit utility function for user r

would be

$$-\frac{(2m_r)^2}{T_r^2 x_r}$$

and the stable point of the system would be

$$x_r = \frac{m_r}{T_r}\sqrt{\frac{2}{p_r}}$$

which shows the inverse dependence on the round trip time and square root of packet loss that is familiar from the literature on TCP.

The lesson from the above is that TCP-like flow control algorithms maximize social welfare for particular choices of user utility functions and cost functions, e.g. in the above, for a social welfare function of (10.18). The model in (10.9)–(10.10), (10.15) is more general since it allows for arbitrary utility functions and does not penalize users with long round-trip delays.

In this section, we have not assumed that p_j is the derivative of some cost function. It is merely the probability that a packet produces a congestion signal on link j. For example, suppose we model the probability of a packet loss on link j by the buffer overflow probability at a $M/M/1$ queue that has a finite buffer of size B_j, is served at rate 1, and at which the arrival rate is x, $x < 1$. Then we have $p_j = x^{B_j}$. If every lost packet costs 1 unit, then the congestion cost on link j is $c_j(x) = xp_j(x)$ and $c_j'(x) = (B_j + 1)x^{B_j}$. Now equation (10.9) says that for a problem of maximizing the sum of users' utilities minus a sum of congestion costs on the links, x_r should decrease at a rate that depends upon x_r times the derivative of the cost on route r with respect to x_r. If lost packets are also used as congestion signals, so that μ_j is the probability that a packet is dropped at link j then (10.17) makes a quite different prescription; it says that the rate of decrease is to depend upon x_r times the congestion cost on route r.

This disparity in prescriptions can have some very interesting consequences. The model of Section 10.2 can be generalized to incorporate the possibility that users may choose between alternative routes. It turns out that the solution to the resulting SYSTEM problem has Pareto efficient flows, in the sense that it is impossible to increase the utility of any one user (i.e., any term within the first sum of (10.15)), or decrease the congestion cost (the second sum in (10.15)), without decreasing the utility of some user or increasing the congestion cost. This Pareto efficiency property does not hold for the TCP-like algorithm analysed here. It is possible to find examples in which the equilibrium point that results when TCP is combined with routing decisions is not Pareto efficient. Moreover, there are examples in which, when one runs a TCP-like algorithm, the addition of an extra link can actually decrease the social welfare! This is similar to what happens in examples of the so-called *Braess's paradox*. The crux of the matter is that in performing routing and flow control, decisions should be based on charges reflecting the derivative of the cost at each resource rather than the actual cost. If congestion signals are generated in proportion to the actual cost, rather than its derivative, then Pareto efficiency may be violated and Braess's paradoxes may occur. See the references at the end of the chapter for more details.

10.5 Allocating flows by effective bandwidth

In Sections 10.1, 10.2 and 10.2.5, flows are treated as simple one-dimensional parameters denoting peak or average rates. A simple extension can be made to a network in which flows have more complex statistical properties and are subject to a fixed quality of service requirement, such as a maximum packet loss probability.

Given the parameters of a flow and a quality of service constraint, the flow of user r has an effective bandwidth. Users can be allocated flows of specified effective bandwidths, subject to constraints of the form $\sum_{j:j \in r} x_r \leq C_j^*$, where C_j^* is the effective capacity of link j. Allowing user r a flow of effective bandwidth x_r is interpreted as allowing him a flow with certain traffic contract parameters. In this manner, we can capture the effects of many dynamic contract parameters upon the burstiness of the traffic. For instance, two connections can have the same peak rate, but if other parameters of burstiness differ, they will not be treated the same.

To illustrate these ideas, consider traffic contracts that specify the peak rate by h, and burstiness by leaky bucket parameters ρ, β. Suppose only the peak rate can be varied. A simple formula for an effective bandwidth of such a connection is (4.20), which for multiplexing parameters s, t is

$$\alpha(m, \mathbf{h}) = \frac{1}{st} \log \left[1 + \frac{tm}{H(t)} \left(e^{sH(t)} - 1 \right) \right]$$

where

$$H(t) := \min\{\rho t + \beta, ht\}$$

This formula can be used to specify a connection's effective bandwidth. If m is unknown, it can be replaced by its upper bound ρ. Prices are defined exactly as before, but it is the effective bandwidth rate that is priced. Users are allocated effective flows. If user r is allocated flow x_r this actually means he is allowed a peak rate h_r for which the effective bandwidth is x_r. More refined allocation schemes could take account of the mean rates of the connections, these being either measured directly or revealed by a tariff choice.

There are many ways to implement such a mechanism. One is to auction the effective link capacity and have the users adjust their traffic contracts to reflect the effective bandwidth they obtain. Another is for the network to post prices and for the users to choose their peak rates. A link may raise the price if the effective capacity is exhausted. Alternatively, the users may post their willingnesses to pay, and the network may choose the peak rates. On slower timescales, a user can change his willingness to pay after consulting his utility for effective bandwidth. Indeed, users may value the possibility of sending bursty traffic. There may be applications of high burstiness, but low throughput, that need such a mechanism to express their preferences for traffic contract parameters. Lastly, note that flow control at the effective bandwidth level can be applied on a slower timescale than the timescale of the burstiness in the traffic sources. Hence it may be well suited to networks that multiplex large amounts of traffic, but whose links have such large round trip delays that burst level feedback is impossible.

10.6   User agents

In Section 10.2.5 we examined a network that provides elastic services and communicates price information in the form of a rate of charge. This charge serves as a control that induces users to adjust their input rates (and demands) so that the available network bandwidth is shared in an economically optimal fashion. Given a current price per unit of flow, users have the problem of optimizing the net benefit they obtain from using the network; they do this by choosing the rates at which they send data. A user r who sees a price p_r solves the problem

$$\underset{x_r}{\text{maximize}} \, [u_r(x_r) - p_r x_r]$$

He assumes that his traffic represents a small percentage of the overall traffic, and so varying his rate does not affect the total congestion price along route r. In essence, the value of the congestion price is the value of the 'network state' to which the user must adapt his behaviour.

Because users may enter or depart the system, or modify their utility for bandwidth during their connection, the demand for bandwidth varies with time. Moreover, the capacity that is available for providing elastic services can be changed by the network's capacity allocation policy. Thus, for a given ongoing elastic connection, it is generally not optimal in terms of 'utility-for-money' to send at a constant rate. The elastic user will wish to vary x_r: either because increased demand or decreased supply makes the present rate no longer 'worth the money', or because decreased demand or increased supply allows a greater rate to be obtained at little extra cost. These ideas apply to the transport of video on demand over elastic traffic connections, with variable-rate encoding of movies (e.g. MPEG). There are video servers that can adapt the quality of the video to the instantaneously available bandwidth x, either by selectively discarding frames, or by varying a quality factor; see Bolot and Turletti (1998). When the transport of video is charged, this introduces an additional component to the feedback loop of video adaptation. As we see towards the end of this section, these ideas also apply to Web browsing over ABR, and to other applications in which the user's perceived quality of service is mainly related to the mean bandwidth offered to his connection.

In choosing his rate, a user need not know explicitly his utility function $u_r(\cdot)$. He simply observes how his application's performance, and so his net benefit, varies as he perturbs the data rate up and down. He selects the rate that is best for him in the present network state.

In a large system with many users, a user may find it necessary to make rather frequent variations in his sending rate. However, it may be hard for him to spot that the network state has changed, and difficult (and perhaps annoying) to manually and frequently re-set this rate. Such monitoring and re-setting is therefore a suitable task for an *Intelligent Agent* (IA), that is, a piece of software residing at the user side. Such an agent can take on the job of constantly maintaining a good level of the user's net benefit. Of course, the user should be able to bypass the agent if he is not satisfied by its selections; alternatively, the agent's selection may be presented to the user as a recommendation.

An intelligent agent algorithm

The job of the Intelligent Agent (IA) is to tune the user's sending rate, so that as the network state changes the user's net benefit is constantly maximized. The information set of the IA generally comprises both prior information about the user's preferences and dynamic information on the network state. Ideally, this information would be complete and 'noiseless' knowledge of the curve

$$x^*(p) = \arg\left[\max_x u(x) - px\right]$$

and complete dynamic information of the network, i.e. the present value of p. In this case, the IA would simply select the data rate that is optimal under the present network state, i.e. $x = x^*(p)$. However, this is unrealistic, because the explicit form of $u(\cdot)$ is usually not known.

It is more realistic to suppose that the Intelligent Agent has only partial knowledge of user preferences; in particular it has a record of a set of points $\mathcal{R}$ that have been previously selected by the user. Since the user actually makes trial-and-error adjustments, $\mathcal{R}$ contains both outliers and near-optimal points. The IA must filter out the noise from the set $\mathcal{R}$ and fit to it a curve $\mathcal{R}_{\text{fit}}$. If only optimal points were recorded then these would belong to a single curve. Each point of the curve $\mathcal{R}_{\text{fit}}$ corresponds to a value of x for a price p, which we denote as $x_{\text{fit}}^*(p)$ to emphasize that it is not the same as $x^*(p)$.

There is an interesting curve fitting approach that can be used when $u(\cdot)$ is concave, since its derivative is then a nonincreasing function of x, and it follows that $x^*(p)$, which is the inverse function of $u'(x^*(p))$, must be a nonincreasing function of p. This means that $x_{\text{fit}}^*(\cdot)$ should be restricted to the set of nonincreasing functions and can thus be solved by means of a special algorithm.

Let P be a set of prices for which the user's bandwidth selections have been recorded, and let $x(p)$ denote the bandwidth recorded for price p. A function $g(\cdot)$ is called an *antitonic regression* of $\{p, x(p)\}$, with weights $h(p)$, if $g(p)$ minimizes

$$\sum_{p \in P} [x(p) - f(p)]^2 h(p)$$

over functions f that are nonincreasing in p. Such a function $g(\cdot)$ is a natural candidate for $x_{\text{fit}}^*(\cdot)$. The weight $h(p)$ can be chosen to reflect the 'age' of the information. As more points become available, the older ones can be given lesser weights than more recent ones.

It turns out that $g(\cdot)$ is a *step* function and it can be found using the so-called *Pool-Adjacent-Violators algorithm*. After applying this algorithm, the antitonic regression function g partitions P into subsets on which it is constant, i.e. into level sets of $g(\cdot)$, called solution blocks. On each of these solution blocks, the value of $g(\cdot)$ is the weighted average of the value of $x(p)$ over the set of prices within the block, weighted with the $h(p)$. Note that $g(\cdot)$ is defined only for isolated points of the set P. However, $g(p)$ can be extended to a piece-wise constant function, by associating the value corresponding to a solution block to all the values of p between the extreme points of the block. Prices not belonging to any of the blocks are treated later.

To find the solution blocks, the Pool-Adjacent-Violators algorithm proceeds as follows: Assume that $p_0 < p_1 < \cdots < p_k$. If

$$x(p_0) \geq x(p_1) \geq \cdots \geq x(p_k)$$

then this partition is also the final partition and $g(p) = x(p)$ for all $p \in P$. In this case, each p_i is a solution block. Otherwise, the algorithm selects a pair of *violators*; that is, it selects a j such that $x(p_j) < x(p_{j+1})$. A new block $\{p_j, p_{j+1}\}$ is then formed by replacing the ordinates of the points $(p_j, x(p_j))$ and $(p_{j+1}, x(p_{j+1}))$ with their weighted average value

$$\frac{x(p_j)h(p_j) + x(p_{j+1})h(p_{j+1})}{h(p_j) + h(p_{j+1})}$$

and associating with them the weight $h(p_j) + h(p_{j+1})$.

The algorithm continues by finding another pair of violators (if any), taking into account all the solution blocks already formed. It can be shown that the order in which violators are considered does not affect the final solution. If no other violators can be found, then

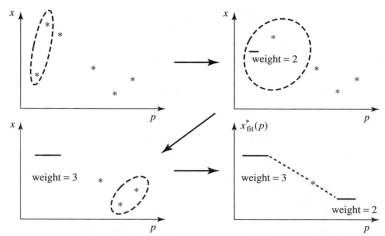

Figure 10.3 The Pool-Adjacent-Violators algorithm is used to fit an antitonic regression to six points. This is shown as $x^*_{\text{fit}}(p)$ in the final picture, with the interpolation between the three blocks is shown by the dotted line.

the present set of blocks (with their associated values) yields the desired function. Since $g(p)$ is only defined for prices within solution blocks, the agent should make a linear interpolation to provide selections for other prices. Note that the algorithm is simple to implement and has a small computational overhead, because it only makes comparisons and computes weighted averages. Figure 10.3 illustrates the operation of the algorithm on six points, terminating after three steps.

One can think of several ways to improve the algorithm. Since early selections made by the user may not be that successful, it is desirable that they do not affect significantly the output of the antitonic regression algorithm. To this end, we could assign to each point a weight that decreases exponentially with the 'age' of this point. At the time a point is recorded it is given a weight of 1. When a new measurement is taken, the weight of each old point is multiplied by $e^{-\alpha \Delta t}$, where Δt is the time that has elapsed since the last measurement. This means that the impact of the original selections will be diminish as new ones are made. The value of α should be small enough, so that each point has a considerable weight for some time. Other weights could also be used, but exponentially decaying ones have the advantage that they are easily updated.

Antitonic regression can be performed for a pre-specified large number of points. Measurements can be taken prior to activating the IA, until the mean square error,

$$\text{MSE} = \frac{\sum_{p \in P} [x(p) - g(p)]^2 h(p)}{\sum_{p \in P} h(p)}$$

is considered to be small enough.

After the IA is activated, more measurements can be taken by letting the user make the decision from time to time. Upon receiving a new measurement the antitonic regression can be updated with starting from scratch. If the p for the new point does not fall into any of the intervals spanned by the solution blocks and it does not violate monotonicity, then it simply constitutes a new solution block. If monotonicity is violated, it suffices to run the Pool-Adjacent-Violators algorithm and group the new point with a previously derived solution block, and then continue until there are no more violations. If the p falls within an interval spanned by an existing solution block, then this block should be decomposed into its constituent points and the Pool-Adjacent-Violators algorithm run, starting with these points, the new one, and the remaining solution blocks. These rules can be shown to follow from the fact that the order in which violators are considered does not affect the final outcome. Thus, one can pretend that the new point (the one that triggered the updating of the blocks) was available from the beginning, but was not yet involved in the pooling.

Note that when one adds a point to a set of measurements this can completely alter the solution blocks. Suppose we start with $x(p) = 5, 4, 3, 2, 1$ for prices $p = 1, 2, 3, 4, 5$; since $x(p)$ is already decreasing, we have $x^*_{\text{fit}} = x$. If we add the measurement $x(6) = 21$, which violates monotonicity, then the antitonic regression curve becomes a single block with $x^*_{\text{fit}} = 6$ (namely the average of $5, 4, 3, 2, 1, 21$). Thus, introduction of a new point may cause multiple blocks to be pooled into one. However, no more than one block must be decomposed. Moreover, as this example indicates, major modifications to the solution blocks are required only when outlier points are introduced. One could use a heuristic algorithm specifying that solution blocks be modified by means of only local changes, and only when a 'reasonable' new measurement is taken. Even if this does not yield an antitonic regression curve, the associated MSE can be satisfactory. The heuristic can be applied a few times, and then the Intelligent Agent can run the Pool-Adjacent-Violators algorithm from the beginning, with all points available.

10.7 Pricing uncertainty

In this chapter, we have been considering the problem of pricing flexible contracts. The price of bandwidth fluctuates with the level of usage and a customer can balance the bandwidth he obtains against the amount he is willing to pay. This is acceptable to a user whose application can tolerate some unpredictable variation in the bandwidth. However, the user still has the problem that the charge is unpredictable and that at a future time the resources that he needs for his application may not be available.

In a market for a commodity such as pork bellies or olive oil these problems of risk management can be tackled using the financial instruments of forward contracts and options. Unfortunately, bandwidth differs from olive oil in that it cannot be stored. Like electricity or airline seats, network capacity at a given time is either used or wasted. Nonetheless, it is possible to envisage a market for instantaneous bandwidth. We saw one way to do this in Section 9.4.4, by the proposal for a smart market in which the bandwidth price is set by auction. A mechanism for spot-pricing the flow rate sold under flexible contracts has been described in Sections 10.2 and 10.2.5.

We will only pose some problems. Consider a user who at time t knows that he will need X Mb of bandwidth over some future interval $[u, v)$ (perhaps for a videoconference call). He wants to guarantee that this bandwidth will be available. He also wants to protect himself against uncertainty in what he will be charged. Let us write the price of this 'future contract' as $p_t^X(u, v)$. In general, this is not linear in X, though we certainly expect to have $p_t^X(u, v) > p_t^Y(u, v)$ for $X > Y$.

Now let p_t^X denote the 'spot price' of X Mb at time t. By this we mean that $p_t^X dt$ is the cost at time t of X Mb of bandwidth over the small interval $[t, t + dt)$. The price of the future contract, $p_t^X(u, v)$, clearly depends upon expectations about the stochastic process $\{p_\tau^X : u \leq \tau < v\}$.

If it were possible for future contracts to truly be bought and sold, then it would be possible to create an options market. For example, a user might purchase at time 0 the option to buy at time t, a contract for X Mb of bandwidth over a future interval $[u, v)$. However, there is a problem with this. Traditional option pricing relies on the fact that a seller of a call option can hedge his position by purchasing the underlying security and then selling it at a later date. As we have said, it is not possible to buy bandwidth at one time, store it, and then sell it later.

Example 10.2 (A model for a forward price) Some problems in pricing bandwidth futures have been described by Upton (2002). These combine nicely with our model of dynamic prices and elastic users. In particular, he looks at the problem in which there is a single link of bandwidth C. At time 0 a large user wishes to reserve a constant bandwidth of size c, $c < C$, for use over a future interval $[u, v)$. The total bandwidth consumed by all other users obeys

$$\dot{x}_t = -x_t + D(p_t) \qquad (10.19)$$

where p_t is the price at time t and $D(\cdot)$ is a decreasing function. The right hand side of (10.19) is the excess demand at price p_t, i.e. the difference at time t between the demand justified by price p_t and the actual demand, x_t. These users are small elastic users who can adapt their sending rate in response to price. Note that for a constant price p the equilibrium of (10.19) is $x = D(p)$, so $D(\cdot)$ can be regarded as the aggregate demand function of the small users. It is assumed that the system is in equilibrium at time 0, so $p_0 = D^{-1}(C)$.

One way that the large user could ensure that there is free bandwidth of c at time u is by increasing his willingness to pay. As he increases his willingness to pay, the network increases the price seen by all users, the small elastic users reduce their sending rates and x_t decreases. It is reasonable that the large user should pay to reserve bandwidth c for $[u, v)$ the same amount it would cost him to drive up the price so that at time u the small user consume bandwidth of no more than $C - c$. Assuming a zero interest rate, this means that his problem is to control $p_t(\cdot)$, through choice of his willingness to pay, to solve the problem

$$\operatorname*{minimize}_{p_t(\cdot)} \int_0^u (C - x_t) p_t \, dt \qquad (10.20)$$

subject to

$$x_0 = C, \quad x_u \leq C - c \quad \text{and} \quad \dot{x}_t = -x_t + D(p_t) \qquad (10.21)$$

Suppose that $D(\cdot)$ is a concave function, p_t is constrained by $p_t \leq D^{-1}(0)$, and (10.21) has a feasible solution. Then it is an exercise in optimal control theory to show that (10.20) is minimized by setting $p_t = D^{-1}(C)$ for $t \in [0, \tau)$ and $p_t = D^{-1}(0)$ for $t \in [\tau, u)$, where τ is chosen so that (10.21) holds with $x_u = C - c$. That is, the large user waits until the last possible moment before u that he can begin to buy up bandwidth and yet obtain an amount c by time u.

What would be a reasonable amount to charge the large user for reservation of capacity c over $[u, v)$? We have found the minimum cost at which he can ensure free capacity of c at time u by solving (10.20)–(10.21). To this, must be added the cost of holding this capacity over $[u, v)$. Assuming the link is profit maximizing, and is able to extract all the benefit of the smaller users, it will be able to charge at least the utility lost by the small users who are displaced. Recall from (5.2) that the inverse demand function $D^{-1}(q)$ is the derivative of the utility function, so the total utility obtained by small users when the price is such that demand is q is $\int_0^q D^{-1}(q')\,dq'$. This suggests that a reasonable reservation price is

$$p_0^c(u, v) = D^{-1}(0)\left[C\log\left(\frac{C}{C-c}\right) - c\right] + (v - u)\int_{C-c}^{C} D^{-1}(q)\,dq$$

The first term on the right-hand side is the minimized value of (10.20). The second term is simply the utility lost to the population of smaller users over $[u, v)$ because the bandwidth available for them has been reduced to $C - c$. The large user has nothing to pay after time v because we assume that after that time the price is set so small that x_t increases back to C near instantaneously; thus the large user is assumed to have no further effect on the system after time v.

Note that we are discussing the market for instantaneous transfers of data between two points over a timescale of milliseconds to a few hours. A quite different market exists for multi-year contracts for the infrastructure of fibre optic cables and local access links.

10.8 The differentiated services approach

In this approach the network creates several versions of the service differentiated by quality and price. Users self-select taking account of the price-quality difference of the various service classes. In many cases, the network provider offers no strict quality guarantees (see, for example, the technology in Section 3.3.7. It just ensures, by allocating resources appropriately, that higher prices do indeed correspond to some better average measure of quality. Such resource allocation cannot be static since the network cannot predict in advance the number of customers who will subscribe in each class, and may be performed by network management at reasonably slow time scales, after measurements indicate that performance has deteriorated below some acceptable level. However, it is reasonable to keep prices constant so that users have the time to experience the price-quality differences and decide which class of service to join. At slower time scales, prices may vary so that demand for the service classes that tend to be overloaded is reduced and more users subscribe to cheaper substitute services so that the desired quality levels are maintained. The fact that prices adjust slower to demand than they do when we use dynamic prices, and that there is a small finite number of service classes (instead of the infinite one in our proportional fairness flow control model), suggests that in practice it may be more difficult to achieve social optimality. Instead, the designer of such a scheme aims to obtain a clear improvement in economic performance compared to when no service differentiation is used.

To illustrate the above concepts we use the simple example introduced in Section 9.3.2. It can be easily generalized to an arbitrary number of service classes. The SYSTEM problem is (9.13). An interpretation of (9.14) is that each user i is faced with the decision to make a different contract x_t^i for each class $t = 1, 2$ (where x_t^i is the rate allowed by the contract). His utility is equal to the total rate sent to the network minus the delay cost. If service classes are implemented as priority classes, then the optimal prices (per unit load of the corresponding contract type) are congestion prices that satisfy

$$p_1 = \frac{\partial D_1(y_1)}{\partial y_1}\sum_i \gamma_i x_1^i + \frac{\partial D_2(y_1, y_2)}{\partial y_1}\sum_i \gamma_i x_2^i$$

$$p_2 = \frac{\partial D_2(y_1, y_2)}{\partial y_2}\sum_i \gamma_i x_2^i$$

where the right-hand sides are to be evaluated at the optimal $\{x_1^i, x_2^i : i = 1, \ldots, n\}$. At the optimum a user does not benefit by using both types of services and hence $x_1^i = 0$ or $x_2^i = 0$. This simple example suggests that even in more general circumstances there are optimal prices under which social welfare is maximized. An important issue is how such prices can be computed. In our simple example above, the selections available to the users are continuous (as they are not forced to choose a single class and hence to create discontinuities by switching among classes while prices vary). Consequently, a tatonnement-like procedure may converge. The system posts prices and the users make their decisions. Then, based on the new load of the system, the prices are recomputed, and so on. In practice, allowing users to tune their performance-cost tradeoffs by switching between service classes at arbitrary times during their contract (in addition to varying their sending rates) may create instabilities, even when prices are fixed. The advantage of such flexibility is that it may increase the value of the service to users.

An interesting feature of the differentiated services approach is that even though the optimal prices may be hard to compute, the mere existence of versioning and service differentiation can increase the social welfare of the system compared to the same system without service differentiation. We illustrate this in the next section. It further argues for the use of a service differentiation approach, due to the simplicity of its implementation compared to dynamic price-based flow control. As we see, differentiation can take place through users' choices alone, with no need for the network to implement any kind of complex priority mechanism in the routers.

10.8.1 Paris Metro Pricing

A novel method proposed for Internet pricing is so-called Paris Metro Pricing (PMP). The name derives from a time when the Paris Metro had two types of metro car. The cars were identical, except that it was more expensive to buy a ticket for one type of car than the other. Naturally, the more expensive type of car was less crowded than the cheaper type of car and so attracted those customers who were willing to pay for the comfort of riding in a less crowded car. Similarly, we might partition one physical network into two or more logically separate networks, such that the only difference in the networks is the price charged per packet. Each user chooses for himself the network he will use. Levels of congestion in the networks depend upon the numbers of users they attract. A network that charges more than another network will attract less users and be less congested. Customers with differing sensitivities to congestion now have the option to choose amongst different congestion levels; effectively, we have created a market in which customers can trade congestion amongst themselves.

Example 10.3 (A model of Paris Metro Pricing) Let us make a simple model of PMP and illustrate how it can increase social welfare. Suppose there are just three users. We have available a resource of size $2C$ which we can use to build either (a) a single network with capacity $2C$, or (b) two networks, each with capacity C. Suppose user i has a utility of the form

$$u_i = v - \theta_i \frac{n_j}{C_j}$$

where j is the index of the network to which user i is assigned, n_j is the number of users who use network j, and C_j is that network's capacity. In this utility, $\theta_i n_j / C_j$ is a congestion cost, in which θ_i models the fact that users have different sensitivities to congestion.

Suppose $\theta = [0, 0.5, 1]$. In case (a) the social welfare is

$$v + \left(v - (0.5)\frac{3}{2C}\right) + \left(v - \frac{3}{2C}\right) = 3v - 9/4C$$

Now consider case (b), and suppose users 1 and 2 are assigned to network 1, while user 3 is assigned to network 2. The social welfare is now

$$v + \left(v - (0.5)\frac{2}{C}\right) + \left(v - \frac{1}{C}\right) = 3v - 2C$$

This is greater than before, by $C/4$. Thus, social welfare is increased by constructing two networks and arranging that network 1 serves the two users who are least sensitive to congestion, while network 2 is reserved for the user who is most sensitive to congestion.

We can arrange for the optimal allocation of users in case (b) to occur simply by charging an additional price, p, to any user who chooses network 2. To see this, we argue as follows. First, note that since user 1 suffers no congestion cost, he will choose network 1 irrespectively of which networks are chosen by the other users.

Suppose that the users were to partition themselves between networks 1 and 2 as $\{1, 3\}$, $\{2\}$, respectively. User 2 increases his net benefit by moving to network 1 if

$$v - (0.5)\frac{3}{C} > v - (0.5)\frac{1}{C} - p$$

That is, if $p > 1/C$. So suppose $p > 1/C$, so that $\{1, 3\}$, $\{2\}$ cannot be an equilibrium of the system.

Similarly, suppose that the users were to partition themselves between networks 1 and 2 as $\{1, 2\}$, $\{3\}$, respectively. User 2 has no incentive to move to network 2 since he will pay more. User 3 has no incentive to move to network 1, provided

$$v - \frac{1}{C} - p > v - \frac{3}{C}$$

That is, if $p < 2/C$. Thus if $1/C < p < 2/C$ and users maximize their individual net benefits, then they will partition themselves between networks 1 and 2 as $\{1, 2\}$, $\{3\}$, respectively. No other partition is an equilibrium. As we have seen, a greater value of social welfare is obtained than if we had built a single network.

The next example is similar to the above and shows that it is also possible to increase social welfare by defining and selling priorities of service.

Example 10.4 (Differentiating service using priority) Suppose the same three users as in the previous example now share a single network of capacity C, but may be allocated to priority classes, labelled 1 and 2. A user i in priority class j has utility

$$u_i = v - \theta_i \frac{\sum_{k \leq j} n_k}{C}$$

where n_k is the number of users in priority class k. That is, a user's congestion cost depends only upon the number of users who are of the same or greater priority. It is easy to check that social welfare is maximized by placing user 3 in the high priority class 1 and users 1 and 2 in the low priority class 2. This can also be arranged by charging a price, p, for priority class 1. As with the model of PMP above, an analysis of cases shows that if $1/C < p < 2/C$ then the only equilibrium is one in which user 3 purchases priority class 1, but users 1 and 2 content themselves with priority class 2.

10.9 Towards a market-managed network

We have presented several ideas about how to control resource allocation in a network through demand. Such a network, in which users can express their preferences and acquire the amount of resources for which they are willing to pay, is a *market-managed network*. Economic signals play a crucial role in the operation of such a network. These signals allow for a real-time market in which users and resource providers interact. We discuss certain important issues that such a concept involves:

- *Enhance user flexibility*. The end-user is the only one who knows the value of the service and the type of customization he prefers. Traditional approaches to providing QoS place the network at the centre of making decisions about how to customize and differentiate services. However, end-devices may benefit greatly from the ability to 'construct' their own services. This is in line with the 'end-to-end principle', in which the internal network nodes are kept simple and complexity is moved to the edges where information about user utility resides. If users can obtain greater surpluses then the network can probably obtain more revenue. It is reasonable to expect that ISPs will be forced to offer more flexibility to their customers if they are to survive in the highly competitive Internet services market.

- *Handle rapid price fluctuations*. Dynamic pricing during congestion results in the most efficient market, but customers do not like dealing with rapidly fluctuating prices. The network should provide its customers with software that 'absorbs' the rapid price fluctuations and implements their buying policies. It may also offer insurance services by selling communications services at constant prices to customers. Building such a layer of services above the rapidly fluctuating dynamic prices is important for user acceptance.

- *Use existing network mechanisms as much as possible*. It is difficult to incorporate new protocols in existing commercial networks such as the Internet. If possible, the mechanisms for market-management should only use already existing mechanisms. The example of using the ECN bit in the IP header for conveying congestion information is an example of such a mechanism.

- *Separate commercial decisions from technology*. Network engineering suggests that any technology, function or protocol that implements commercial decisions should be abstracted from the network infrastructure or protocols, or any other more generic part of the system that is involved in offering the communications services. Preferably such functions and protocols should be placed at a policy layer (conceptually between the user and the top of the application layer) so that it is easy for providers to differentiate themselves by constructing their own policy. This principle is primarily required because of the long timescales necessary for infrastructure changes and their standardization. Ideally, the whole charging system should be as separate as possible from the transmission system. Pricing should merely be applied to events already occurring in the network, rather than introducing elements into network protocols for charging purposes.

- *Use split-edge pricing. Split-edge pricing* is a way to avoid dealing with your customers' customers or your suppliers' suppliers. In this type of pricing, the price that a provider offers his customer for service on one side of his network takes account only of onward charges from onward networks in providing the service. However, onward charges are not passed on unaltered directly to the customer. Instead, the provider consolidates his pricing schedules, and so avoids things becoming increasingly complex as network paths becomes longer. This requires that for each invocation of a service every contract-related relationship (pricing, responsibility for service failure, charge advice, etc.) must be bipartisan, that is, between a single customer and a single provider. No provider should be able to pass the blame to a more remote provider for any of the charges it passes on, or for some technical requirement that another provider imposes (e.g. international roaming, carrier selection).

- *Provide customized network services to information service providers.* In order to sustain and expand the value of the Internet to its business customers, it is necessary that the network infrastructure enables new Internet services and business models to be implemented and provided rapidly. That means that the network infrastructure must provide the means for information service providers and network service providers to easily establish collaborations. Information service providers must have the capability to express their requirements for quality of service to network service providers, since they need to bundle their information services with the appropriate network service. It is crucial that an information provider can choose and, in many cases, customize the appropriate network service to be bundled with particular content.

10.10 Further reading

The definition of elastic user in terms of utility function is due to Shenker (1993). See also Shenker (1995). For further details of Example 10.1, see Massoulié and Roberts (1999). They show how to select the source behaviour so as to achieve an equilibrium distribution concentrated around rate allocations that are considered fair. Details for the stability of both the primal and dual algorithms of Section 10.2 are in Kelly, Maulloo and Tan (1998). Kelly (2001) gives an example in which the flows determined by TCP have the nature of Braess's paradox. See also Braess (1968). The impact of random effects and of time lags on stability together with arrivals and departures of users are further treated in Johari and Tan (2001). Other interesting formulations of the social welfare optimization problem with elastic users are in Low and Lapsley (1999) and Paschalidis and Tsitsiklis (2000b). Gibbens and Kelly (1999) describes ways in which the transmission control protocol of the Internet can evolve to support heterogeneous applications. They introduce the idea of computing congestion costs over sample paths where congestion expresses packet losses. The material on the mathematical model of the Internet is taken from Kelly (2001) and Kelly (2000).

The Pool-Adjacent-Violators algorithm used in Section 10.6 is described by Barlow, Bartholomew, Bremner and Brunk (1972). An up-to-date collection of papers related to proportional fairness and the Internet is in the web page Kelly (2002a). A similar collection of papers regarding Internet modelling and stability is in the web page Kelly (2002b).

The TCP algorithm in its present version is due to Jacobson (1998). Odlyzko (1997) is the proposer of Paris Metro Pricing described in Section 10.8. An interesting paper on the evolution of pricing in telecommunication networks is Odlyzko (2001). A proposal for a billing architecture for pricing the Internet is in Edell, Mckeown and Varaiya (1995).

For an example of *a screen-based global commodities exchange for the trading of international wholesale telecoms minutes and bandwidth*, see Band-X (2002).

Approaches for charging differentiated services can be found in Paschalidis and Tsitsiklis (2000a) and Altmann, Daanen, Oliver and Sánchez-Beato Suárez (2002).

Many new ideas on how to design a market-managed Internet have been developed by the IST project M3i (see Market Managed Internet (2002)). The discussion in Section 10.9 summarizes some of the key principles addressed in the above project.

Split-edge pricing has been suggested by Shenker, Clark, Estrin and Herzog (1996), and Briscoe (1999).

Part D

Special Topics

11

Multicasting

A unicasting service is one that requires the network to provide point-to-point transport between just one information source and one receiver. A *multicasting service* extends this idea by requiring the network to provide transport between one or more information sources and a group of receivers. Multicasting services can be used for teleconferencing, software distribution and the transmission of audio and video. A key characteristic of a multicasting service is that it its cost must be optimized for the particular group of receivers to which it provides service. This poses important resource management and control problems, which add new complexity to pricing issues.

A special case of multicasting is *broadcasting*. Broadcasting is simple, in that the same information is continually made available to all potential receivers, and so there is no need to optimize network resources to the subset of receivers that is presently listening. The transmission rates and network resource allocation are fixed, and the transmission cost is independent of the customer group. If broadcasting technology is in place, then we can multicast information by broadcasting it, but only granting the subscribers of the multicast the permissions to access or decode it.

Multicasting over a data network such as the Internet requires far more complex resource management than does broadcasting. This is because there are different mechanisms available at the network level, and the identities of the end receivers can influence routing decisions about which links of the network should carry the multicast traffic. Also, whereas satellite broadcasting typically uses constant bit rate channels, applications that use data network multicasting services may produce bursty data flows and have more flexible quality of service requirements. In this chapter, we investigate the issues of resource allocation and pricing that arise when multicasting services are to be provided over a data network like the Internet. We see that the final resource allocation may depend upon decisions taken by a large number of participants. This contrast with unicast, where one of the two connected parties makes all the decisions about the properties of the connection and is responsible for paying the bill. Hence, if one is to achieve globally efficiency by giving appropriate incentives to the various decision-makers, there are many delicate gaming aspects that can make pricing very complex. Of course, we can always view a single unicast connection as the simplest case of a multicast service, in which the sender and the receiver make independent decisions and so must agree on common features of the connection, such as the bit rate and how to split the network charge.

Pricing Communication Networks   C. Courcoubetis and R. Weber
© 2003 John Wiley & Sons, Ltd   ISBN 0-470-85130-9 (HB)

In Section 11.1 we set out some requirements for multicasting. In Section 11.2 we describe some basic technologies for it. Section 11.3 considers mechanisms for providing quality of service and Section 11.4 addresses flow control. Starting from a model for allocating bandwidth to elastic multicast traffic, Section 11.5 considers issues of cost sharing and the formation of the multicast tree. Section 11.6 is about settlement.

11.1 The requirements of multicasting

Multicasting is potentially a very promising network service for IP technology networks. Great efficiency can be achieved by arranging that only one copy of the data transverses any common paths on its way towards multiple destinations. For example, in satellite broadcasting there is a single common path; all receivers share the same set of broadcast channels, all of which are transmitted over the same link.

Multicasting services provide positive network externalities. Since a customer shares common cost with other customers he can access services that he would otherwise find too expensive. However, there is a negative externality, since a customer may not be able to choose the precise type and quality of the service that he desires. His choices are restricted because other customers in his multicast group value service differently or have different technological capabilities. These issues make the pricing of multicasting services interesting, but complex. As for unicast services, pricing plays an important role in controlling the way network resources are shared. A pricing policy must fairly reflect the externality effects and provide the right incentives for customers to join or leave a multicast session when it is economically justified from the viewpoint of the multicast group as a whole.

Before looking at the economic aspects of a multicast service model, we consider the technology aspects. Clearly, multicast services can provide savings in network resources. Savings occur because network routers and switches can, at no cost, copy incoming packet flows and direct resulting identical flows to more than one output link. The network gains by taking information that is destined for multiple receivers and forwarding it over paths for which receivers have common parts. An inefficient network could always use traditional unicast technology to support a multicast service. However, this would lead to unsustainable prices since a competitor who uses multicasting would have lower transmission costs and so could offer lower prices.

The resource savings of multicast come at the cost increased complexity. Some difficult tasks are the scheduling of the multicast packets at the routers, the routing of the packets inside the network, addressing, congestion control, and quality of service issues, such as the reliability and variability of transmission. These are the subject of undergoing research. Furthermore, many decisions depend upon the assumptions that are made about the semantics of the multicast service, and these are often not precisely defined. The optimal solution of some fundamental multicast problems, such as constructing the least cost multicast tree, are very difficult and cannot be solved under practical assumptions.

It is important to distinguish between multicasting's network implementation and multicasting viewed as a service. For instance, multicasting's network implementation through IP is based on standard IP unicast concepts, which allow IP packets originating from a set of sources to reach a set of destinations. The semantics of such a lower-level network service are similar to IP: packets are transported unreliably, with no guarantee on synchronization or delay. By contrast, a multicasting service at the application layer may have requirements for reliability, an upper bound on packet delay, and a minimum rate guarantee. It may also require there to be mechanisms for group management (controlling who joins or leaves the multicast group), negotiating various economic and service specific terms, and charging customers. The network provider may use a lower-level service, such as IP multicast, and then add some additional protocols to meet these requirements, or he can use other mechanisms. For instance, he may substitute unicast services if these can better provide the desired service quality, although the economies of scale produced by the specialized multicasting technology should make this rarely advantageous. In practice, multicasting services as seen by the end customer, are mostly defined in a bottom-up fashion. They are not shaped by the demand for some killer application, but by the capability of the multiplexing-specific technology that have been implemented at the various network devices.

The problem with such a 'technology-centred' approach, compared to one that is 'application-centred', is that present multicasting technology has limited capability for supporting real-world, revenue-generating services. For example, the simple IP multicast service model, based on existing IP multicast technology, is suitable for simple low quality teleconferencing applications. However, it seems overcomplicated for software distribution or TV-like applications, where only one source exists and group management and payment capabilities are of great importance. Indeed, the presently deployed 'open' IP multicast service model was not defined with a commercial service in mind, and poses severe technical restrictions to most reasonable charging mechanisms. The absence of group management from the model and the increased routing complexity, are perhaps the most important reasons why there has been slow deployment of multicast services in the Internet thus far. The fact that there are as yet no well-defined multicast services at the application layer leaves research in these areas truly open.

As far as pricing is concerned, some very important questions must be answered. Cost-based pricing for multicast services is strongly tied up with game-theoretic notions of bargaining and arbitration. Since transmission cost is partly common, how should it be shared amongst the members of a multicast group? Should a customer who obtains greater value from a service pay a greater fraction of the common cost? Clearly so, since customers who obtain less value will leave if they obtain negative net benefit when they are asked to pay equal share of the cost. What pricing mechanisms will make users reveal their true utilities? A multicast service may be offered in an uncertain and dynamically changing environment. The number and identity of the receivers may not be known in advance, but fluctuate during the multicast session. How can one reduce the risk of customers paying exceedingly high prices when there is small participation, or reduce the risk to the provider if prices are fixed at the start? If the service can be offered at various quality levels, but only one will be actually selected, perhaps because of a constraint imposed by the receiver with the slowest access link, how should one charge such a receiver? How could one give an incentive for that receiver to leave if that would increase the value of the service to all other receivers? These questions illustrate the diversity of the issues that must be addressed, and the complexity of designing a sound pricing policy.

In the following sections we extend the models that we have used for unicast flows to discuss the state-of-the-art in different research areas that are relevant to pricing multicast services.

11.2 Multicasting mechanisms at the network layer

The range of applications that multicasting can support depends strongly on the capabilities of the network technology. Important high-level features include mechanisms for knowing the exact number of receivers, controlling access and transmission, providing security, and obtaining the quality of service required for the transport of the packet flows. In practice, the network provides some basic mechanisms with simple features, and other mechanisms must be built on top to fit each particular application.

The basic multicasting technology proposed for the Internet is *IP multicast*. In keeping with the Internet philosophies of openness and simplicity, its service model defines the notion of a 'multicast group' as a group of computers connected to the Internet, which at the network layer is identified by one specific IP address. Any host on the Internet (member or not of the multicast group) has an equal right to send packets to, or receive packets from, members of the group. A packet received by one member of the group (i.e. with an IP address of the multicast group) will be forwarded by to all other members of the group in a best-effort fashion, similar to a standard IP packet. The packet will follow a tree of routers (i.e. a 'multicast tree' that connects the sending computer to all other members of the group), and will be duplicated at each branch of the tree.

There is no control over who joins or leaves the group, who transmits to the group, and there is no knowledge at the network layer of the identity and number of the subscribed members. A receiver makes its subscription request to its edge-routers (i.e. the router in its LAN that is a node in the multicast tree of routers) using the Internet Group Membership Protocol (IGMP). The wide area multicast tree is constructed by a network layer multicast routing algorithm/protocol such as PIM, DVMRP, CBT or MOSFP.

Two approaches have been adopted for constructing this multicast tree, each with many variations. The basic difference between the two approaches is in whether the routing tree that is used to distribute the traffic is the same or different for each sender in the group. If it is to be the same for all senders, the so-called 'group-shared approach', then one might like it to be the tree of routers that connects all the members of the multicast group at least cost. However, finding this tree is related to the *Steiner tree problem*, which is known to be NP-complete. Instead of trying to find an optimal tree, one could use the 'centre-based approach', which constructs the shared tree by first identifying a centre router (the 'core' for the multicast group. Subsequently, each edge-router in a LAN with a host that is a member of the multicast group sends a 'join' message to the core router. The paths followed by the join messages define the multicast tree. If each sender is to have its own specific routing tree to all destinations, the so-called 'source-based approach', then each such tree should ideally approximate the optimal Steiner tree. We already mentioned that solving such problems is hard. To provide a practical solution, some protocols in the Internet use existing underlying unicast mechanisms to set up trees from each source to the destinations, and then merge these where they overlap. An example of both approaches is shown in Figure 11.1.

In practice, network operators have been slow to deploy IP multicast because they are reluctant to risk the stability and efficiency of their network to such an uncontrolled service without the opportunity to extract corresponding revenues. Note that there is no flow control on information transmitted over the multicast tree: a single source could end-up flooding a large part of the network.

Some important features that are missing from the present multicast service model, but which are necessary for a simple and efficient pricing structure, are receiver counting, authentication and access control. There are addressing issues, which arise because multicast addresses are not controlled by a central Internet authority and so a newly created multicast group may choose an address already used by another group. There are inter-domain routing difficulties, due to different ISPs using different routing algorithms for constructing their parts of the multicast tree. These issues are presently motivating a search for new multicast routing approaches based on different service models.

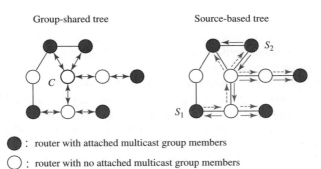

Figure 11.1 Group-shared and source-based multicast trees. In the group-shared tree there is a single common tree for the entire multicast group, and it could be constructed by defining router C as the core. In the source-based approach, separate trees are constructed for senders S_1 and S_2 (shown by dotted and solid line, respectively), and every other potential sender of the group.

11.3 Quality of service issues

Quality of service is a generic notion, but it is most commonly associated with the ability of a network to provide deterministic or statistical delay and bandwidth guarantees, especially for multimedia real-time applications. Present proposals for improving existing IP network technology are mostly unicast in nature, and do not address the requirements of such multicast applications

Although multicasting could benefit if quality of service mechanisms were available, their slow deployment in the present best-effort Internet discourages the use of multicasting for high-value commercial services such as TV and video broadcasting. As a result, these services continue to be delivered over specialized networks of satellite or fibre, which offer guaranteed quality; they are then combined with broadband access to the customers (by, for example, using cable). However, new encoding mechanisms, which require lower bit rates, improved network technology such as Differentiated Services (see Section 3.3.7), and intelligent end-devices, are beginning to make the Internet attractive for multimedia transmission when applications have less strict quality requirements, and can adapt to varying network conditions.

We now turn consider a number of questions. What are the intrinsic differences between multicast and unicast applications? Do applications that rely on multicasting services have similar quality of service requirements as typical unicast applications? By what mechanisms can the network to provide the performance that multicast applications require?

11.3.1 Multicast Application Requirements

It is useful to distinguish between multicasting applications for which either reliability or timing is the more important. Take, for example, the distribution of a database. Here, reliability is paramount. No data should be lost or altered. Timing may be relatively unimportant, as there may be no hard constraint on when all receivers should have received their copy of the database. In contrast, if one is distributing a real-time video of a sports event, then timing is key; loss of a small fraction of the information may not noticeably degrade the quality of the video, but the information must travel regularly, incurring small delays and jitter. Thus the network must ensure a regular and even flow on all links of the multicast tree. This is not required if the content is not real-time, for then any positive rate allocation along the links of the tree (not necessarily uniform) can suffice.

One can imagine even more 'exotic' requirements. For instance, it could be essential for a real-time, cooperative work application that data be delivered to all members of the multicast group simultaneously. In this case, the delay jitter of the information across receivers in the group may be important.

In practice, things are complicated further by the fact that members of a multicast group may differ. For example, suppose a video transmission can be encoded by the sender as a 1 Mbps (low quality) or as a 2 Mbps (high quality) stream. There are several types of receivers: some are linked to the network with access bandwidths of less than 2 Mbps, and so while all can decode the low quality encoding of video only some can decode the high quality. One solution is to use a single multicast tree with enough resources to satisfy the minimum common requirements, i.e. to distribute low quality video. Another solution is to use two independent multicast trees, for the low and high quality, each reaching the relevant members of the group. This solution maximizes the value of the service to the customers, but costs more. Cost of the second solution can be reduced using a 'layered' encoding technology, in which the added quality in the high quality video is transmitted as an extra 1 Mbps stream on top of the low quality stream. Accessing both streams allows a decoder to offer the high quality playout, whereas accessing only the low quality stream remains a valid possibility. Now a single tree can be used. All receivers receive the low quality stream. The high quality stream passes only through nodes that eventually reach receivers who want high-quality video.

If the above idea of layered trees is used, then maintaining the appropriate trees is likely to be very complicated, since a fluctuating congestion level will make the higher bit rate more or less expensive as receivers join or leave. If dynamic pricing is used then the cost of supporting the various quality layers over the links of the tree may change during the multicasting session. The receivers should be notified of the ongoing cost of the service and be allowed to choose the number of layers (and hence the quality) that they wish to receive. This may be a complex task for a receiver. Even more difficult is the problem of sharing the cost amongst the receivers. This could be addressed in the more general context in which prices are designed to offer incentives for resource usage that achieve maximum economic efficiency for the group as a whole (see Section 11.5).

11.3.2   Network Mechanisms

Other mechanisms can be used to complement the simple service provided by IP multicast. Most of these are extensions of mechanisms for unicast connections. We have made a distinction between requirements for reliability (when distributing content that is not real-time) and for timing (when transmitting real-time content). The latter also usually has some requirement for a minimum average information rate over all links of the multicast tree.

In unicast, reliability is achieved at the transport layer using the TCP protocol, or at the application layer with various mechanisms (if UDP is used instead of TCP). In multicast, there are two problems that make reliable transmission at the transport layer very difficult. The first problem is that of 'feedback implosion', which occurs when one packet loss causes many members of the same multicast group to generate loss signals. A solution is to aggregate/suppress loss signals on their way up-tree towards the sender. The second problem is that of 'efficient retransmission', which has to do with suppressing the re-transmission of lost packets to receivers that have already received them. This problem can be addressed either by local lost packet recovery (through designated receivers, routers, or repair servers) or by making the routers remember the links from which loss signals arrived, so that retransmission can be efficient.

There are some other interesting technologies that can be used for reliability. For instance, the Digital Fountain technology eliminates the need for specific retransmissions. The file is first encoded using a special encoding scheme, and then divided into n packets which are continuously transmitted in a circular fashion. A receiver can reconstruct the complete file if he receives correctly any m out of the n packets, for some $m < n$.

Multimedia applications have different requirements. When information is transmitted in layers, each layer having its own bit rate, then the network must reserve the appropriate bandwidth over the links of the multicast tree to transmit the information. If the network is best-effort, as in the present Internet, there is no mechanism for guaranteeing such average rates. The problem can be solved if the network implements some bandwidth reservation protocol, such as RSVP (see Section 3.3.7), or has a dynamic pricing mechanism, so that any amount of bandwidth can be obtained by paying the appropriate price (see Chapter 10).

However, in a best-effort network, even if such a desired average rate can be achieved over the links of the multicast tree, one has to compensate for the fluctuations that cause packets to queue at the routers and so reach the receivers as an irregular stream of packets. A simple way to eliminate this delay jitter is to use buffering at the receiver end. The sender time-stamps each packet with the time it is transmitted. The application at the receiver end looks at the time stamps and estimates the average rate and its variance. Arriving packets are stored in a buffer, from which the application picks packets at regular intervals and feeds them to the display device. This is known as *streaming*. Streaming allows playout to start before all the data is transferred from the source to the receiver. Obviously, the average rate at which packets enter the buffer must equal the rate at which they are picked out. By knowing the statistics of the rate at which the buffer fills, the receiver can compute how large the buffer must be initially, so that once playout starts the probability that the application should ever request a packet and find the buffer empty is very small.

An alternative method to streaming would be for a receiver to wait until it has received from the sender the complete data for the video and then play it from a file in its memory. The drawback is the delay in starting to view: it may take much longer to transfer the complete file than to transfer the initial small part of the file that is required by streaming. There already exist commercial streaming products, such as QuickTime and RealOne Player. The information needed for streaming is standardized through the Real Time Protocol (RTP), and feedback statistics about the connection's losses, jitter, and so on, are sent back to the sender using the Real Time Control Protocol (RTCP). Of course, streaming could become obsolete if there were abundant bandwidth, both inside the network and in the access part.

11.4 Flow control mechanisms

Flow control is used to control the rates at which information flows in the network. In practice, it is implemented with two components. The first component is part of the network technology: it generates flow control signals and communicates them to the entities that are responsible for generating or receiving the traffic. These entities are usually the applications that contract the network services. The second component resides in these applications. It receives the flow control signals and reacts by appropriately adjusting the rate of the information flow. Note that flow control signals could be prices, giving the price per unit flow at the given point in time. In this case, a rational application will adjust its use of the network so as to maximize its net benefit, that is, the value of the service minus the charge as a function of the information rate.

Flow control signals could be sent to senders or receivers. It is a matter of implementation details as to which parties receive them. It can be helpful to think of all parties as constituting a single application. For instance, in a multicast session, the application could be taken as all senders and receivers. These would internally decide how to react. In unicast the sender simply reacts by adjusting his sending rate to the unique receiver. In multicast, many actions are possible. One could decide temporarily to drop some receivers from the group, hence resulting in a smaller multicast tree. Or, in the case of layered multicast, one could constrain some receivers to subscribe to a smaller number of information layers, thereby reducing the total data rate in certain parts of the multicast tree. This can be accomplished by assigning a different multicast group to each layer, and dynamically forcing receivers to subscribe and unsubscribe to the corresponding groups.

Let us investigate the multicasting flow control in more detail. For simplicity, assume that there is a single sender in the multicast group, and that each link of the multicast tree generates flow control signals that reflect congestion of the link. We can distinguish applications in respect of their ability to adapt to flow control signals. Suppose that data is transmitted into a single layer, in which receiver membership is fixed, and so the only available control is the sending rate. In this case, the sender must explicitly compute one common rate for all receivers, perhaps by adjusting the sending rate to the congestion signals generated by the most congested path in the multicast tree. A way to implement this is as follows. Congestion signals are implicitly modelled by packet losses. Receivers produce negative acknowledgments (NACKs) upon packet loss, which are sent to upstream routers. A router filters such information and forwards up-tree NACKs at the maximum rate these are received from any of its downstream links. Clearly, the sender sees a rate of NACKs corresponding to the path to the worst receiver, and adjusts his rate accordingly. For instance, he might use the TCP-like rate adaptation scheme PGMcc. The sender computes the worst receiver according to loss rate and round trip time information that is added in the NACK packets, nominates that receiver as the 'representative receiver', and runs a TCP-like window based congestion control algorithm with it. This is the only one that sends positive ACKs upon a packet reception. Note that a receiver with a slow access link will restrict the rate of transmission to the whole group.

Now suppose a multi-rate layered scheme is used, in which each receiver can choose to receive a subset of the layers. The sender need make no adaptation, and all the control can be exercised at the receiver end. Receivers are faced with flow control information and it is up to them to react by increasing or decreasing the number of layers they receive. Note that when a receiver subscribes to a layer, the information in this layer must reach the receiver. This increases the total flow of the multicast tree over a path connecting some internal node of the tree to the receiver. This node is the root of the largest subtree to which the given receiver is the only subscriber to the particular layer. The advantages are that there is no need for feedback to the sender nor a compromise in quality due to a 'slow' group member. However, it is not obvious how congestion information generated in some internal link of the tree should be propagated down through the tree to the receivers. For instance, if congestion signals reflect congestion cost, then this cost should be shared by the receivers. But how should this cost be shared? In more general terms, what incentives should be given to the receivers through the flow control signals to subscribe or unsubscribe to the various layers, and what is the underlying optimization problem? We return to these questions in Section 11.5.

11.5   The economic perspective

11.5.1   A Model for Allocating Multicast Bandwidth

Let us extend the proportional fairness model of Section 10.2 to multicasting. Suppose a multicast group consists of a single source and a fixed set of receivers and multicast tree. There is a single rate that is to be sent to all receivers and which can be adjusted by the sender. There is contention for bandwidth at each link of the multicast tree, since there are other connections that share the resources of the network with the multicast group. As before, we seek to use prices to regulate flows and optimize the overall economic efficiency of the network. We discuss the similarities with the case of unicast flows and indicate the new issues that arise.

The multicasting problem differs from that of unicasting because many entities are involved, each of whom obtains a different value from the service and contributes to a common cost. Suppose the members of the multicast group agree to pick a representative, who is given full information about the utility functions of all the group members, and is delegated to make choices that maximize the net benefit of the group as a whole. Faced with the prices that the network communicates through the flow control signals, the representative will make a rational decision and choose the optimal rate for the sender to transmit. Given more authority, he could decide that some receivers ought temporarily to leave the group if they add more to the common cost than the extra utility they bring. Furthermore, he could decide, according to some pre-agreed fairness criteria, what contribution each receiver should make towards the total cost of the service.

In practice, some of the members of the group, knowing how the cost will be shared, may have an incentives not to cooperate. They may feel they are in a stronger position to obtain a larger share of the overall net benefit. Interestingly, it is the hidden information about utilities of the members of the group that makes the problem hard. In our unicast model we did not face such issues, since we assumed there was a single entity, the sender, who has full information and full control. We can make these clear with a model.

Consider a model of a network in which there are sets of links and routes. Each route is associated with either a unicast user or a multicast group user and requires some subset of the links. The route associated with a multicast group is a tree of links rather than a path. A unicast user r has a utility for a flow of rate x_r along route r of $u_r(x_r)$. A multicast group r consists of a set of users, $r_1, \ldots, r_{n_r}$, these being the sender and the receivers of the group. Each member r_ℓ has its utility $u_{r_\ell}(x_r)$ for a multicast flow rate x_r, and we denote by $u_r = \sum_l u_{r_\ell}(x_r)$ the total utility of the group r. The unicast and multicast users have elastic utilities; that is, their utility is assumed to be increasing, strictly concave and continuously differentiable. Our aim is to maximize the social welfare subject to constraints on the capacities of the links. Similarly as in Chapter 10, we have the SYSTEM problem

$$\text{SYSTEM} : \underset{x \geq 0}{\text{maximize}} \sum_r u_r(x_r), \text{ subject to } Ax \leq C$$

where $A_{jr} = 1$ or 0 as $j \in r$ or $j \notin r$. We will use the terminology of user r to refer to a unicast user or to a multicast group associated with route r.

Suppose user r may choose an amount to pay per unit time, w_r. Then he receives a flow $x_r = w_r / \lambda_r$, where λ_r is a price per unit flow on route r. The network chooses the price

λ_r in such a way as to make the flow feasible. The user's problem is

$$\text{USER}_r : \text{maximize } u_r(w_r/\lambda_r) - w_r , \text{ subject to } w_r \geq 0$$

If r is a multicast group, then the above is consistent with a delegate of the group choosing the rate of payment w_r so as to sustain a rate $x_r = w_r/\lambda_r$ that is optimal for the group.

Suppose the network solves the problem

$$\text{NETWORK} : \underset{x \geq 0}{\text{maximize}} \sum_{r=1}^{n_r} w_r \log x_r , \text{ subject to } Ax \leq C$$

This is equivalent to allocating feasible flows in a weighted proportionally fair way. As in Section 10.2, SYSTEM is solved when the USER$_r$ problems and NETWORK problems are solved simultaneously, with their solutions in equilibrium. That is, there exist $\{\lambda_r\}$, $\{x_r\}$, $\{w_r\}$ such that $x_r = w_r/\lambda_r$ and these are simultaneously solutions to SYSTEM, NETWORK and USER$_r$, for all r.

At this point, let us return to the multicast group r. We assumed that a delegate chooses the optimal rate x_r for the group by solving the problem

$$\underset{x_r}{\text{maximize}} \sum_{\ell=1}^{n_r} u_{r_\ell}(x_r) - \lambda_r x_r \tag{11.1}$$

where λ_r is obtained by summing the prices for a unit of bandwidth over all links of the multicast tree. Here the delegate seeks to maximize the net benefit of the group viewed as a coalition of agents, and this requires a complete knowledge of the utilities of the participants. This formulation raises the following important problems:

- How should the total net benefit be shared among the participants? We could require participant r_ℓ to pay the delegate θ_{r_ℓ}, so that $e_{r_\ell} = u_{r_\ell} - \theta_{r_\ell}$ becomes his net benefit. This internal payment mechanism must be agreed by all the group. A payment scheme is only stable if no participant has an incentive to leave the coalition because he feels unfairly charged.
- Under which conditions will the internal payment mechanism cover the total cost, i.e. $\sum_\ell \theta_{r_\ell} = \lambda_r x_r$?
- If a participant knows how his payment θ_{r_ℓ} will be determined, does he have the incentive to declare his true utility u_{r_i} to the delegate, or will stating some other utility allow him to profit by making a smaller payment? Note that to obtain the maximum profit for the coalition as a whole the delegate must know the actual utilities. Otherwise, some participants may benefit at the expense of others. Clearly there is a game here. A payment rule is incentive compatible if participants do best by declaring their true utilities, and so provide the information that is in the best interests of the group as a whole.
- Is there payment scheme that is both an incentive compatible and covers the total cost? Interestingly, there is not. One must sacrifice net benefit to cover cost.

In the next section we investigate the above questions in terms of a simple example.

11.5.2 The Problem of Sharing Common Cost

The left of Figure 11.2 shows a multicast group consisting of two identical receivers with a price $p = 1$ per unit bandwidth on the shared link, and a price of 0 on each of the

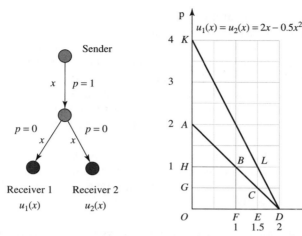

Figure 11.2 An example of cost and benefit sharing in multicast. AD is the demand curve of each receiver and KD is the demand function of the two receiver group. For price $p = 1$ the two receivers benefit by forming a coalition and sharing the common net benefit. However, declaring their actual utility function is not an equilibrium strategy; if receiver 2 tells the truth about u_2, then receiver 1 can benefit by declaring a smaller utility function. He pays a smaller fraction of the common cost and so obtains greater net benefit.

two links feeding the individual receivers. The receivers have identical utility functions, $u_1(x) = u_2(x) = 2x - 0.5x^2$, with demand curves corresponding to the line AD in this figure. The line KD is the demand curve of the joint utility $u_1(x) + u_2(x)$.

The receivers can form a single multicast group, or form two separate multicast groups (each consisting of a single receiver). In the first case, they will obtain a joint net benefit of

$$\max_{x} [u_1(x) + u_2(x) - px] = 2.25 \tag{11.2}$$

where the maximum occurs at $x = 1.5$. Note that this is the area of the triangle HLK, and twice the area of triangle GAC. Each receiver obtains net benefit equal of 1.125, i.e. the area of $OACE$ (one receiver's utility) minus the area of $OGCE$ (half the total payment).

Since the receivers have identical utility function it is reasonable that the net benefit should be shared equally. If their utilities were not equal then the total net benefit might be shared following some bargaining that takes account of the fact that the receivers make different contributions to the coalition. One possibility is that each receiver should receive the net benefit he could obtain on his own, plus half of the extra net benefit that is obtained through coalition. For instance, if q_1 and q_2 are the net benefits of the receivers acting individually, and $q_{12} > q_1 + q_2$ is the net benefit they can obtain by coalition, then receiver 1's net benefit should be $e_1 = q_1 + 0.5(q_{12} - q_1 - q_2)$. This is equivalent to his making a payment of $\theta_1 = u_1 - e_1$, where u_1 is the utility he obtains in the coalition. In this case, the sum of the payments exactly covers the cost, since $\theta_1 + \theta_2 = u_1 + u_2 - q_{12}$, which equal the total cost.

If each receiver were to act individually, each would obtain a net benefit equal to the area of HAB, i.e. 0.5. Hence they would rather form a single multicast group, since this will increase each receiver's net benefit by 0.625 (the area $GHBC$).

Interestingly, a receiver can do better by not declaring his true utility. Suppose that, knowing receiver 2 will declare his true utility, receiver 1 falsely states his utility as ϵx, where ϵ is close to 0. The optimal flow for this coalition is $x = 1 + \epsilon$. Receiver 1 obtains net benefit of almost 1.5 since θ_1 will be only about 0.5ϵ. This suggests that it may not be possible to achieve the maximum in (11.1) when a receiver only knows his utility function and the internal payment mechanism of the coalition is designed to fully cover the cost. Indeed, the payment rule above gives the incentive for a 'free ride'. This example illustrates a general result from game theory that applies in the context of multicasting because of the particular structure of its cost functions. It states that any incentive compatible internal payment rule (that provides incentives for truth telling) will not in general recover the full cost incurred by the coalition. Hence, one must adopt a payment rule that either maximizes the economic efficiency of the system, but fails to recover the full cost of the multicast group through charging its participants, or makes participants pay for the cost, at the expense of net benefit of the overall group. With the payment rule above, participants have the incentive to 'shade' the declarations of their utility functions, with the result that a smaller choice of x is made than is optimal.

Can we think of a simple payment scheme that gives participants the incentive to truthfully declare their utility functions? This should happen if the payment a participant receives does not depend on his utility declaration. Consider again our two receiver problem. Suppose the flow has a single possible value x_0, which can be obtained at cost $c = px_0$. Receiver i knows his utility u_i for flow of x_0, and must make a declaration u_i^d to the manager of the group. The manger then adopts the following charging rule. First, he checks if $u_1^d + u_2^d > c$. If the answer is no, then the multicast group is not built, and both receivers obtain zero net benefit. Otherwise, the flow is set at x_0 and receiver i is made to pay $\theta_i = [c - u_j^d]^+$, $i, j \in \{1, 2\}$, $i \neq j$.

It is easy to see that this scheme is incentive compatible, since if receiver 1 falsely declares his utility function this can only decrease the net benefit he receives. Suppose $u_1, u_2 < c$, so no receiver alone can afford the service. If both tell the truth and $u_1 + u_2 > c$, the service will be provided and each receiver i will incur a payment $\theta_i = c - u_j$, leaving each with a positive net benefit of $u_1 + u_2 - c$. Suppose receiver 2 tells the truth but receiver 1 does not. If $u_1^d < u_1$ and the actual utilities satisfy $u_1 + u_2 > c$, then there is the risk of $u_1^d + u_2 < c$ in which case the manager will decide against the provision of the service, which will reduce the net benefit of receiver 1 from $u_1 + u_2 - c$ to zero. If $u_1^d > u_1$ and $u_1 + u_2 < c$, there is the risk of $u_1^d + u_2 > c$ in which case the service will be provided and receiver 1 will have to pay $\theta_1 = c - u_2 > u_1$, and so incur a negative net benefit. In all other cases, the net benefit of receiver 1 remains the same. Note that the sum of the receivers' payments is $\theta_1 + \theta_2 = 2c - u_1 - u_2 < c$.

The above example can be generalized to circumstances in which the flow x takes one of a large set values, in which case the manager will first compute the optimal value of the flow, and then define the corresponding payments.

The incentive compatible payment mechanism that we have described above is an example of the well-known Vickrey–Clarke–Groves (VCG) demand revealing mechanism. This mechanism defines

$$e(S) = \max_{T \subseteq S} \left[\sum_{i \in T} u_i^d - c(T) \right] \tag{11.3}$$

as the maximum net benefit that can be obtained by using the declared utilities to choose, from within an initial set S, the subset of participants, say T, for which the net benefit is greatest. As before, $c(T)$ is the cost of the optimal multicast tree for a multicast group T.

The payment required from a participant i, who belongs to S, is now fixed at

$$\theta_i = e(S \setminus \{i\}) - \left[e(S) - u_i^d \right]$$

where $S \setminus \{i\}$ is the set S, minus participant i. The payment θ_i is the loss in net benefit to other users that is due to i participating.

In this section we have investigated the various incentive issues that arise because of imperfect information in the simplest case of shared cost In the following sections we investigate the extra complications that occur because of the structure of the multicast tree, and discuss several alternatives for sharing the common cost.

11.5.3 Formation of the Optimal Tree

Consider the two trees in Figure 11.3. All links have zero cost unless specifically marked. We take the same utilities as in the example above, i.e. $u_1(x) = u_2(x) = 2x - 0.5x^2$.

If a single coalition is formed, the net benefit in both (i) and (ii) is given by (11.2) and equals 2.25. However, the structure of the cost in the two trees are quite different. In tree (i) the cost is common, but in tree (ii) the flow to receiver 2 is solely responsible for cost. Were it not for receiver 2, receiver 1 could obtain maximum net benefit of 2 units by taking $x = 2$ (i.e. the maximizer of $u_1(x) = 2x - .5x^2$). If receiver 2 were on his own he would obtain a net benefit of 0.5. Since the sum of these is greater than 2.25 the coalition of the two receivers obtains a smaller net benefit than the sum of the benefits the receivers obtain by acting individually. So there is no incentive for receiver 1 to join a coalition with receiver 2 since he can gain more on his own. This suggests that net benefit is actually maximized by forming two multicast tree: one for receiver 1 with rate $x = 2$, and one for receiver 2 with rate $x = 1$. This could be implemented by encoded information in two layers, each of rate 1. Both receivers obtain the basic rate, and only receiver 1 subscribes to the higher quality and receives the additional rate required.

This example demonstrates that the complete net benefit optimization problem is difficult. The solution must take account of optimum coalition formation, and then the optimum selection of rates within each coalition.

11.5.4 Cost Sharing and Multicast Trees

Let us investigate further the issue of cost recovery. For simplicity assume that the flow level through the tree is $x = 1$. We can share the cost of the multicast tree, say $c(T)$, amongst the participating receivers by a cost sharing function that makes $c_i(T)$ the contribution that

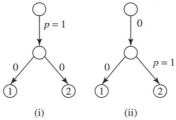

Figure 11.3 How cost structure effects formation of a multicast tree. Flows have costs rate on the links of 0 or 1, as marked. The users utilities are $u_1(x) = u_2(x) = 2x - 0.5x^2$. In (i) receiver 1 benefits by joining a coalition with receiver 2. In (ii) he does not.

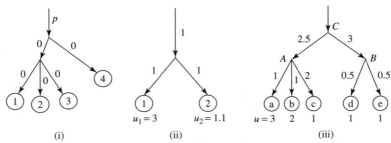

Figure 11.4 The formation of the optimal multicast group. In (i) user d might pay $p/2$ or $p/4$, depending on the rule used to apportion cost. In (ii), if the common cost is shared equally then user 2 will drop out, leaving a socially inefficient solution. In (iii) a VCG allocation has a paying 2.5 and b paying 1.5, which does not cover the cost of 4.5.

receiver $i \in T$ must make to the cost. By definition, $\sum_{i \in T} c_i(T) = c(T)$. There are many possible cost sharing functions. To take two examples: one could propagate the cost of each link down the tree by splitting its cost either (a) equally amongst the immediate downstream branches of the tree, or (b) in proportion to the number of receivers in the subtree descending from the link. The results of applying these two functions can be dramatically different. Consider, for instance, tree (i) of Figure 11.4. Here (a) results in receiver d paying for half the cost, which may force it to leave the coalition and not receive service. Under (b) each receiver will pay $1/4$ of the cost. Note that (b) corresponds to each receiver paying its Shapley value proportion of the cost and so certain natural fairness criteria will be satisfied (see Section 7.1.3). However, it may be more expensive to implement (b) because one has to obtain information about the number of down-stream receivers.

There is an implicit iteration in the above procedure: once all receivers have been notified of the costs they are asked to pay, they must each decide if they want the service. A receiver will drop out if his net benefit, of $u_i - c_i(T)$, is negative. Subsequently, costs are recomputed for the new multicast tree that is obtained by pruning those parts of the previous tree that do not have subscribers. Note that since the common cost is shared by fewer receivers at each round the share of the cost paid by any given receiver cannot decrease. More and more receivers drop out until a stable point is reached. However, convergence may involve a prohibitively long number of steps, since at each step only a single receiver may drop out.

Another disadvantage of such a fixed cost-sharing procedure is that it can end with a multicast group that is smaller than is optimal. Consider tree (ii) in Figure 11.4. If common cost is shared equally to the down-stream receivers, receiver b will leave since his cost share is $1.5 > u_2 = 1.1$. Consequently, the stable coalition has only receiver a, with net benefit of 1. However, receiver b could pay for the cost of its directly attached link, leaving a surplus of $1.1 - 1 = 0.1$. It could share this surplus with receiver a by contributing, say 0.05, to the cost of the common link. This is better for both since the net benefit of each receiver increase by 0.05. If we had $u_1 = 1.95$, then even receiver a would drop out after receiver 2 had dropped out. But this will not happen if they agree that receiver 2 should contribute 0.075 to the cost of the common link, leaving each with positive surplus of 0.025 each. This example suggests that to obtain an efficient coalition, cost sharing must depend not only on the structure of the multicast tree, but also on the utilities of the participants. In most cases, such information is not easy to obtain. This is reflected in social welfare loss.

A last issue concerns the cost of implementing an incentive compatible payment scheme such as VCG in a general multicast tree. We illustrate one possible implementation with an example. Consider tree (iii) in Figure 11.4. Let message $[i, u]$ denotes a message from receiver i with value u. Messages are propagated up the tree. In the first step, each of the receivers a–e sends its utility up-tree in a message. Each message passes through a link that carries data to one receiver alone, so the value in the message is decreased by the value of the cost of the link; if this value become negative the message is dropped and the corresponding receiver is deleted from the multicast group. Thus messages $[a, 2]$, $[b, 1]$ reach node A, messages $[d;0.5]$, $[e;0.5]$ reach node B, and receiver c is dropped.

Messages $[d;0.5]$ and $[e;0.5]$ have now reached node B and must share the cost of 3 in the link from B to A. Since the cost of 3 is greater than the sum of the two values 0.5 and 0.5, this subtree is pruned. Similarly, messages $[a;2]$ and $[b;1]$ have reached node A and must now share the cost of 2.5 in the link from C to A. Since the sum of their two values, 2 and 1, exceeds 2.5 they can share the cost of the common up-tree link. They do this by having their message values reduced by amounts of the VCG payment rule for sharing the cost of the link, given in (11.5.2). Here a pays 1.5 and b pays 0.5. New messages of $[a;0.5]$ and $[b;0.5]$ are propagated up-tree to node C.

So constructed, the final tree maximizes net benefit, and the final payments can be found by subtracting the values in the messages that reach the root node from their initial values. Thus the final tree has only receivers a and b, which must pay 2.5 and 1.5 units, respectively. Note that the total cost is 4.5 units, which exceeds the sum of these payments.

In summary, sharing the cost of service in a multicast tree can be a complex task and require many steps of computation. If the cost is to be covered, then the overall net benefit of the group may be reduced. An incentive compatible payment scheme is easy to implement, but may cover only part of the cost. These issues become even more problematic when the set of receivers changes dynamically during the multicast session.

11.6 Settlement

The *settlement* problem for multicasting is that of deciding who actually pays and then apportioning charges in line with how members of the multicast group value the service. We have already discussed several issues of how cost might be shared between various links of the tree, and how to give the participants incentives to declare their true values for the service.

Settlement is also concerned with how the charges collected by the edges of the tree, i.e. the revenue generated by the customers, is distributed to the various links of the tree, remembering that each link potentially belongs to a different network. Note that there may be incentives for links to report higher costs, or misreport revenue collected from downstream links or group members. There exist two main approaches, paralleling the two different philosophies that exist in the Internet today.

'Split-edge pricing', follows an open, distributed, optimistic, edge-centric approach. Both sender and the receivers initially pay a share of the cost of the transmission to their access networks, and claims over redistributing the value of the transmission within each participant's group are settled later. Network providers agree prices at which they will offer transport service to neighbouring networks as a way to collect locally the cost of their part of the multicast tree. All entities exchange payments at their interfaces, so that in the end, each entity is left with the amount that correct settlement would dictate. Both sender and receiver need to pay, since otherwise a downstream provider could collect more money by lying about the number of receivers. One could design the edge prices to implement many potential settlement models. For instance, if the data that is transmitted is advertising, the sender might pay a large amount to its access network, which would then keep a portion of this amount and distribute the rest according to the internal edge prices of downstream networks. At the leaves of the tree no charge would be made to the receivers, and all intermediate networks would have collected a share of the payment.

A second approach assumes there is a service level manager, who is responsible for logging new receivers, permitting access, paying the cost and charging customers. One of his tasks may be to manage the risk due to uncertainty about the total number of receivers who will subscribe to the multicast. Since the group must start with only few receivers, the initial cost per receiver might so great as to discourage receivers from joining the group. To attract customers, the manager must bear some risk by offering prices that reflect the anticipated size of the group once it reaches its steady state, which may itself depend on the prices he sets initially.

11.7 Further reading

IP multicast was introduced by Deering and Cheriton (1990) and a discussion of the first audiocast experiment by the IETF is given by Casner and Deering (1992). The connection between multicast tree complexity and Steiner tree problems is addressed by Cormen, Leiserson and Rivest (1995) and Andrews, Khanna and Kumaran (1999). Shapiro, Towsley and Kurose (2002) explain how proportional fairness can affect the distribution of the network resources between multicast and unicast flows. Several protocols for streaming are described by Schulzrinne, Casner, Frederick and Jacobson (1996). A nice introduction to the multicast IP protocols is provided by Kurose and Ross (2001). Information on the Digital Fountain technology can be found at the company's web site, and for RSVP, see Braden, Zhang, Berson, Herzog and Jamin (1997). The PGMcc flow control scheme is due to Rizzo (2000) and is based on the reliable multicast protocol PGM of Farinacci, Lin, Speakman and Tweedly (1998) for feedback suppression/aggregation. Layered multicasting issues and flow control are discussed in McCanne, Jacobson and Vetterli (1996) and Rizzo, Vicisano and Crowcroft (1998). Some game-theoretic aspects of sharing the multicast tree cost can be found in the pioneering papers of Herzog, Shenker and Estrin (1995), Feigenbaum, Papadimitriou and Shenker (2000) and Moulin and Shenker (2001).

Economists define a *private good* as one that can be consumed by just a single economic agent, and a *public good* as one that can be consumed by many agents. In our multicast model, the bandwidth consumed in the common part of the multicast tree is a public good. More precisely, since some agents may be excluded from consuming it, it is a type of a *club good*. Chapter 23 of Varian (1992) discusses issues of public goods and the VCG demand revealing mechanisms we have presented here. Settlement issues are discussed by Briscoe (1999) and Henderson and Bhatti (2000).

12

Interconnection

Two parties that are connected to different networks are able to communicate with one another if there is *interconnection* of the two networks. The positive network externalities that result when networks are connected are an economic force driving interconnection between network providers. Modern access technology strengthens internetwork connectivity by providing universal access. A customer using a wireless access system can access the network from any physical location.

Interconnection is important if networks are to offer truly global services. Network service providers often negotiate special interconnection agreements and tariffs so they can provide the best service to their end-customers. These play a vital role in ensuring the smooth operation of today's worldwide Internet. They also provide substantial income for network operators who have invested in building large backbone networks. By buying interconnectivity, a small network can appear larger to its customers, without investing in costly and rapidly changing infrastructure. This helps it to offer services that can compete with those offered by a network that has invested in greater geographical coverage.

There are important incentive issues involved in offering interconnection services. Unless prevented by the contract, the network that provides the service may be tempted to discriminate in favour of traffic that originates from its own end-customers and against traffic that originates from end-customers of its 'customer' network. If that customer network is a direct competitor in the market for retail services, then the interconnection service provider has an incentive to offer a poorer quality of transport to the interconnection traffic than to his own internal traffic. A carefully chosen interconnection charge can correct this inequality by making transport quality a measured part of the contract.

In this chapter we introduce some important concepts in interconnection services. Section 12.1 reviews types of interconnection agreement and pricing. In Section 12.2 we briefly consider the effect of competition on service differentiation and in Section 12.3 we consider the factors that motivate networks to interconnect or not. Section 12.4 is about the asymmetric information problem that can arise when one network buys interconnection service from another. Section 12.5 describes how an incentive contract can be used to solve this problem.

12.1 The market structure

12.1.1 Peering Agreements

Once interconnection is in place, a network service provider can use the infrastructures of a number of other networks to provide services to any of his customers. However,

Pricing Communication Networks   C. Courcoubetis and R. Weber
© 2003 John Wiley & Sons, Ltd   ISBN 0-470-85130-9 (HB)

it is only reasonable that he should transfer part of the charge that he makes to his customer to those other providers whose networks are used to provide the customer's service. Traditional telephone networks use the so-called *accounting rate system* to share charges. Interconnection charges are computed on a per call basis, and the network in which the call originates pays a predefined charge to the network that terminates the call (and possibly to intermediate networks). This is implemented by each carrier computing his 'traffic balance' with the other carriers over a certain time period, and then paying in proportion to it.

In today's data networks, things are different. First, since customers are connected to the Internet and data flows in all directions, there is no notion of charging on a per call basis. Second, interconnection is achieved by there being a number of *Network Access Points* (NAPs), at which many different networks interconnect with each other. As a function of the interconnection agreements between network providers, and routing decisions in the interior of the network (which may depend on how network congestion and topology changes), data can flow unpredictably through intermediate networks. In present Internet practice there are two ways that traffic is exchanged between data network providers. The first is *peering*, in which traffic is exchanged without payment, and the second involves interconnection charges for *transit traffic*.

Peering agreements have some distinct characteristics. Peering partners exchange traffic on the bilateral basis that traffic originates from a customer of one partner and terminates at a customer of the other partner. This allows customers of the two networks to exchange information. Note that peering agreements are only bilateral; so a peering partner does not agree to act as an intermediary to accept traffic from a partner and transmit it to a third network. Peering traffic is exchanged on a settlement-free basis, also known as 'sender-keeps-all'. The only costs involved in peering arise from the equipment and transmission capacity that partners must buy to connect to some common traffic exchange point. Peering agreements do not specify that a network should provide any minimum performance to the traffic originating from his peer; such traffic is usually handled as 'best-effort'.

Network providers consider several factors when negotiating peering agreements. These include the customer base of their prospective peer and the capacity and span of the peer's network. Clearly, some providers have greater bargaining power than others. It may be of no advantage for a provider with a large customer base to peer on an equal terms with a provider with a small customer base.

The second type of interconnection agreement is a *transit agreement*. It has important differences with peering. Now one partner pays another partner for interconnection and so becomes his customer. The partner selling transit services will route traffic from the transit customer to its own peering partners as well as to other customers. He provides a clearly defined transport service for the transit traffic of the first network, and so can charge for it in a way that reflects the service contract and the actual usage. This charge, if rightly set, is billed to the customer of the network in which the traffic originated, and becomes one of the components of his total charge. Observe that transit is not the same service as peering. Refusing peering in favour of transit is not a means of charging for a service that was otherwise provided for free. When regional ISPs pay for transit they benefit from the infrastructure investments of national or global backbones without themselves having to make the same investments. Transit gives an ISP customer access to the entire Internet, not just the customers of its peering partners. To fulfil his obligations, a transit provider must either maintain peering arrangements with other backbones or pay for transit to another backbone provider who maintains peering relationships.

The Internet connectivity market is hierarchical, with three main levels of participants: end-users, ISPs and Internet Backbone Providers (IBPs), as shown in Figure 3.12. At the bottom of the hierarchy are the end-users, individuals and business customers. They access the Internet via ISPs. At the top of the hierarchy are the IBPs, who own the high speed and high capacity networks which provide global access and interconnectivity. They primarily sell wholesale Internet connectivity services to ISPs. ISPs then resell connectivity services, or add value and sell new services to their customers. However, IBPs may also become involved in ISP business activities by selling retail Internet connectivity services to end-users.

Two markets can be identified in the Internet connectivity value chain: the wholesale market, and the retail market for global access and connectivity to end-users. There are two main types of contracts and they can be distinguished by their pricing: contracts for primary Internet access between end-users and ISPs, and contracts for interconnection between ISPs and IBPs. In the early days, when the Internet exclusively served public sector purposes of research and education, interconnection was a public good and its provision was organized outside competitive markets. Today interconnection is primarily commercial, yet its basic architectures remain the same. Network externalities generate powerful incentives for interconnection. They also provide incentives for potential opportunistic exploitation.

12.1.2   Interconnection Mechanisms and Incentives

To better understand the difference between transit and peering, let us examine the mechanisms which networks use to exchange traffic. One such mechanism is the routing protocol BGP (*Border Gateway Protocol*). It is primarily used at the borders of carriers' networks, i.e. at the gateways (the edge routers) between a carrier's network and other networks that are attached to it. Consider two networks, A and B, which are connected with a link running BGP. There are two principal types of information carried between the two edge routers of the networks: announcements and withdrawals. Announcements (also called 'advertisements') are packets containing lists of new destinations (network addresses) which are reachable via this link. Withdrawals are packets containing lists of destinations which can no longer be reached via this link. In making announcements, one network is soliciting from another network packets whose destination can be reached through the first network. So if network A advertises network C to network B, it signals its willingness to receive traffic from network B that is destined for network C. By advertising destination networks, and hence by being willing to receive and forward traffic to these networks, network A is offering a service to network B. This service may be offered for a price. Let us see now the difference between transit and peering.

In the case that A offers a transit service to B, network A advertises to B all the destinations he can reach (probably the whole Internet). Such information can be used freely by B and advertised to any other neighbouring network, say network D, that he can also reach all the above destinations (through A). In that respect, this routing information is transitive, since all destinations reachable through A are also reachable through B from any network connected to B (if B allows it). For this service A charges B a fee; in most cases, this charge is based on measuring the volume of data crossing the link AB, having as origin or destination the networks advertised by B. Similarly, B advertises to A all its destinations, so that A can advertise these further and the rest of the world can know how to reach them.

In the case of peering between A and B, a similar and bidirectional process takes place, the difference being that routing information is not transitive. This means that B will use the information about the destinations reachable through A only for the benefit of its own customers (and its customers from transit agreements), and will not further advertise these

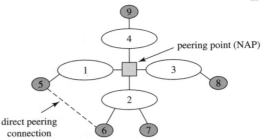

Figure 12.1 An example of interconnection agreements. Networks 1, 2, 3, 4 offer transit services
to networks 5, 6, 7, 8, 9, and peer at a peering point. A peering agreement must be established
between each pair of peering networks. As a provider of a transit agreement a network advertises
all networks it can reach (through other peers or customers). In a peering agreement a network
advertises only those networks he can reach which are also his customers. If the traffic between
networks 5 and 6 is substantial, it may be cost-effective for them to peer and incur the cost of the
direct connection 5–6. By doing so, the traffic between 5 and 6 will follow the direct route instead
of going through (and hence be charged by) the transit providers 1 and 2.

destinations to other neighbouring networks. Observe that B has no incentive to advertise
A's destinations to other than his own customers. To do so would result in B carrying for
free traffic that is destined for A's customers, but which does not benefit B's customers.

Consider the example in Figure 12.1. Four large networks, 1–4, offer transit services to
smaller networks, 5–8. For instance, network 2 offers transit service to networks 6 and 7. To
do so, networks 1–4 peer at a peering point. Each network must make a peering agreement
with each of the other networks. Let us consider networks 1 and 2 and the information
exchanged between their edge routers. Network 1 advertises network 5 to network 2, and
network 2 advertises networks 6 and 7 to network 1. Similarly, when networks 2 and 3
peer, network 2 advertises networks 6 and 7 to network 3, and network 3 advertises network
8 to network 2. Observe that network 2 does not advertise to network 3 the information
obtained from network 1 concerning the reachability of destination 5. To do so would be to
invite to carry traffic at no charge from 8 to 5. Thus, under peering, full-meshes of peering
relationships must be created between peers even if these are peering at the same peering
point. However, as network 2 charges its customers 6 and 7 for transit service, it does
advertise to 6 and 7 all the destinations that are advertised through its peers.

When should one prefer peering to transit? First, note that peering cannot completely
substitute for transit. This is because transit services usually offer global Internet access
(advertise all possible destinations), whereas peering is useful for only a fixed set of
destinations. Nonetheless, peering relationships can reduce the amount of traffic that is
served through transit agreements and so reduce transit charges. Of course there is a cost
to establishing a peering relationship. A network provider must decide whether a peering
agreement is worthwhile by estimating its effects on the traffic serviced under the transit
agreements. Consider again Figure 12.1. Networks 5 and 6 observe that they have substantial
traffic between one another. If networks 5 and 6 peer then they can save transit charges
that they would otherwise have to pay to networks 1 and 2. But to peer, 5 and 6 must pay
for a new connection, either directly connecting them (private peering), or through some
peering point. They should decide to peer if the cost of doing so is less than savings they
can make because their local traffic no longer needs to flow through the transit providers
1 and 2. After establishing a new peering relationship, the transit contract can continue to
act as a backup, to can carry traffic between peers whenever the direct peering connection
is out of order. This means that the peering connection can tolerate a certain amount of
downtime and so can be safely implemented with inexpensive facilities and technology.

12.1.3 Interconnection Pricing

Much study has been given to interconnection pricing for telecommunications networks and many intricate issues have been identified. Many of these issues remain equally relevant for communication networks which offer a broader set of services. Let us look as some details.

Two common regimes for interconnection pricing are *calling-party's-network pays* (CPNP) and *bill-and-keep*. CPNP regimes presently account for the majority of interconnection agreements for traditional voice traffic. In this regime, the calling party's local exchange carrier (LEC) or interexchange carrier (IXC) pays the called party's local network for the cost of processing and allocating resources to terminate the call. More specifically, the calling party's LEC collects a charge from the calling party and then pays a transport and termination charge to the called party's network. In the case of long-distance, the calling party's IXC collects a charge from the calling party and pays an originating access charge to the calling party's LEC and a terminating access charge to the called party's LEC.

In 'bill-and-keep' regimes the calling party's carrier does not pay any termination charge to the called party's carrier. Instead, the termination cost is recovered from the end-customer. This means that an end-customer has the incentive to choose among competing carriers. The resulting competition means that bill-and-keep regimes achieve better economic efficiency than do CPNP regimes (where the terminating carrier is a *de facto* monopolist). 'Bill-and-keep' regimes are also fairer. As both the calling and called parties benefit from a call, it is fair that they should share its cost. Of course there are subtleties as to how the cost should be shared. Let us investigate these issues further.

As we say, a fundamental problem with CPNP is that the LEC who terminates calls is a *de facto* monopolist. Because end-customers usually receive their access service from just one provider, any interconnecting LEC or IXC who carries the call prior to it reaching the terminating network has no choice over the terminating carrier. Thus, each terminating carrier, no matter its size, has monopoly power over termination to its own customers. A long distance IXC does not enjoy such a monopoly because there are usually a number of competing IXCs that can carry the call between the LECs of the calling and called parties. The problem with CPNP is that a terminating carrier can safely raise termination prices without losing customers, since the called party does not share any part of the termination charge. Furthermore, there are usually geographic rate-averaging requirements, which mean that call prices cannot depend in a very detailed way upon precisely how they are terminated. For example, calls from Greece to all UK destinations should be priced the same, even though they may be terminated by different local carriers in the UK. If one of such terminating carrier raises his price, then this will raise the average termination price, but his identity will be hidden. Since it is only the average termination charge that affects the bill of the calling party, that party will have little or no incentive to complain to the called party and encourage him to switch to a local carrier with a smaller termination charge. Thus, CPNP denies customers the ability to choose the most cost-efficient network provider. Customers must pay part of the charges of all the networks involved in the interconnection service, even though some of these may be artificially inflated and uncompetitive. By comparison, 'bill-and-keep' provides better incentives, because a LEC with high access charges will lose customers when they see those charges in their bills.

Another issue with CPNP is that interconnection charges tend to be traffic sensitive, typically being calculated on a per-minute or per-call basis. These inter-network charges are typically reflected in the traffic-sensitive retail prices charged to customers. Customers facing such tariffs are given traffic-sensitive incentives to reduce usage, even when the network may actually have large amounts of unused capacity and so there is no need for congestion pricing. This may significantly reduce the economic efficiency of the overall system.

To prevent LECs behaving as monopolists, it is common for access charges to be regulated. Typical regulatory frameworks are rather complex and treat different classes of interconnecting parties and types of services in different ways, even when there may be little difference in the costs that they generate. For instance, the regulatory regime can depend on whether the interconnecting party is another local carrier, an interexchange carrier, or a subscriber. This complexity in the regulatory framework creates regulatory arbitrage opportunities that motivate entrepreneurs to invent new ways to provide services. The availability of new services can be highly beneficial unless these are motivated solely by artificial differences in regulatory rules. For instance, Internet telephony is not subject to LEC access charges (either originating or terminating) for that part of the call that is placed over the IP protocol. In this respect (besides being more cost-efficient), IP telephony is more competitive than traditional long-distance telephony (where the long-distance carrier must pay access charges). Pressure from Internet-based technologies should cause interconnection regimes based on CPNP to collapse. Looked at another way, so long as interconnection regimes based on CPNP continue to exist, they act as a spur to the introduction of new disruptive technologies such as IP telephony.

'Bill-and-keep' may be unfair to a large network that interconnects with smaller networks. A smaller network, with smaller operating costs, may be able to offer lower prices to its customers. Yet because of interconnection with the large network its customers can reach the same population of customers as those of the large network. A way to remedy this could be to split the cost of the interconnection facilities so that customer prices are the same for both networks. This is the idea of *facility-based interconnection cost sharing*, which contrasts with the usage-based prices that CPNP computes on a per call basis. Another interesting idea is to make the calling party's network pay all the cost of the call up to the point that it reaches the called party's network, which then does not receive any payment for terminating the call. That final part of the cost is paid by the called party. In this scenario, the originating network also pays for the long-distance part of the call, and so has the incentive to choose a lost-cost IXC, since the long-distance charge will be seen in the bills of its customers.

The reader may wonder why charging for interconnection has evolved differently in the Internet than in traditional telephony. A principal reason is the difference in the market structure. The market for local Internet service that is offered by ISPs is highly competitive, whereas the market for backbone connectivity is less competitive, reversing to some extent the trends of the telephony market. No ISP can survive by charging high access prices. So peering, which is a type of 'bill-and-keep', is widely used. In the market for backbone connectivity, competition encourages IBPs to adopt a similar peering strategy for terminating each others' traffic, except that their customers are now ISPs. The limited competition in the IBP market justifies nonnegligible prices in the transit contracts paid by ISPs to IBPs. Note that ISPs have the incentive to operate as efficiently as possible, since they pass on the cost of their local network and its transit agreements directly to their customers, who can easily switch ISP if they feel they are not receiving the best value for money.

12.2 Competition and service differentiation

We can use standard models of oligopoly to analyse competition in networks that offer guaranteed services. In these networks, capacity determines the quantity of services that can be sold. However, when networks offer elastic services, then we must be more careful in modelling competition, as congestion must now be taken into account.

The desire of competing networks to discriminate between consumers who differently value various aspects of the offered services motivates the production of services with different qualities of service. These differing services can be realized by dividing a network into subnetworks with different congestion levels and profit can be increased thereby. However, when more services are offered they will be partly substitutable, and the resulting increase in competition can reduce the profits of all the competing network operators. It is therefore interesting to ask to what extent a competitive market induces service differentiation by making it advantageous for competing networks to offer many types of service. The answer is very sensitive to assumptions. Consider the market for access services. If a customer can subscribe to multiple services, and so benefit from multiple levels of quality, then it is probable that competing networks will wish to provide services at multiple quality levels, i.e. levels of congestion. However, if a customer can subscribe to just one quality level, then competition effects can outweigh service differentiation effects, and each competing network will wish to offer just one class of service, at a price that depends upon its congestion level. Of course this assumes competition. If network operators collude, then they can maximize profits by each producing at multiple quality levels. In any case, if an access network wishes to distinguish itself by a certain quality level, it must guarantee that quality by buying appropriate interconnection agreements. Thus, the intermediate networks' quality of service can be a constraining factor on the competitiveness of an access network.

12.3 Incentives for peering

Whether or not peering between two networks is beneficial depends on how their customers value those things that differentiate the networks, such as size and location. Network size is very important to users who wish to access a large customer base and buy or sell services through the network. Similarly, location is important to customers that find it easier to access one network than another. A network provider can make his network look more attractive by providing good performance to the traffic of his own customers, and worse performance to traffic that originates from outside.

Simple economic models of competition suggest that, as a function of customer preferences, either all or no competing networks may want to peer, or smaller networks may want to peer while larger ones do not. The case in which no network wants to peer occurs when most customers are more interested in network size than location. Here, the market is modelled by a game whose equilibrium solution is asymmetric, in the sense that competing networks grow to different sizes. However, if customers are more interested in location, then networks may wish to peer, since by increasing their customer bases, they add value to what they provide and can charge more for it. If both size and location are important then peering can benefit smaller networks, but not larger ones. This is because, in a competitive scenario, smaller networks can introduce access charges. Peering eliminates the advantage of network size and can encourage customers of larger networks to move to smaller and cheaper ones.

In practice, it is typical for a network provider to specify conditions for peering that depend on the other network's size and geographic span. He might also specify a 'peering charge' that compensates him for his loss of income when he peers with another network. Of course it is very difficult to determine this charge. In practice, it is often made a function of the access speed of the connection between the two networks at the NAP where peering takes place.

12.4 Incentive contract issues

Interconnection agreements may not always provide sufficient incentives for partners to collaboratively realize the full potential of positive network externalities. In the present best-effort Internet, interconnection agreements tend to be rather simple, specifying a maximum rate and perhaps a volume charge. However, newer Internet applications increasingly require specific network performance guarantees, and so new types of interconnection contract are needed that can account for both quality and volume. These contracts must give the peering network the appropriate incentives to allocate the effort required for the contracted quality. This contrasts with the present practice of flat contracts that do not include incentives for effort.

It is difficult to devise interconnection contracts because of *asymmetric information* about variables. We can discuss this using the terminology of the *principal-agent model*, in which a *principal* (the contractor who sets the terms of the contract) wishes to induce some action from an *agent* (the contractee who executes the contract). There are variables, such as peak rate, average throughput and number of bytes, that can be observed and verified by both principal and agent. However, there are other variables that cannot be observed by the principal. For example, a principal who buys from an agent a contract for interconnection may not be able to tell what minimum bandwidth the agent dedicates to his traffic, or the priority class to which his traffic is assigned. These are variables of the 'effort' provided by the agent. It is technology that dictates what is observable and what not. Sometimes, the effort of the agent may be observable, but the context in which this effort is exercised may not be known at the time the contract and the incentives are defined.

Information asymmetry can provide significant advantage to the contractee, who naturally tends to expend the least effort he can to fulfil his contractual obligations. The contractor takes a risk known as *moral hazard*. There is an *adverse selection problem* when, at the time the contract is agreed, the agent knows some important information that the principal does not. For instance, if the principal is the provider of the interconnection service and the agent is the network generating the traffic, the agent knows how he values 'heavy' or 'light' use of the contract. If he intends to make heavy use of the interconnection service, it is to his advantage not to reveal this to the principal. He would rather be charged the cost of a contract that is targeted at the average customer.

In practice, there are many important ways that information asymmetry can occur and can influence the performance that is obtained from an interconnection contract:

- Perhaps an ISP signs an interconnection agreement, but subsequently does not maintain or upgrade his network capacity. The result is that the interconnection traffic receives poor service. As peering agreements are presently based on best-effort services, one party cannot easily tell whether or not the other party is properly managing his network.
- An ISP carrying a high load of local traffic might actively discriminate against packets that enter his network from an interconnected partner. The damaging effect of the discrimination may be camouflaged as natural congestion, and it can be hard for his partner to detect the true cause.
- A client party cannot easily predict the traffic load that a network offering interconnection service carries on its backbone. It is hard for that party to know the other party's available spare capacity, his resource allocation and routing policies, or whether he effectively uses statistical multiplexing and overbooking. Resource allocation and traffic multiplexing can strongly affect network performance.

In negotiating peering or transit agreements, all the above are critical. However, information about these issues is not readily available, and ISPs have little incentive to reveal it. Present market practices only partly address the problem. Large ISPs exert their bargaining power to extract information from smaller partners. However, the requirements and terms of their agreements are private and undisclosed to third parties.

12.5 Modelling moral hazard

To model asymmetric information problems in the market for Internet connectivity, three fundamental parameters must be defined: effort, outcome, and the cost of providing effort. The effort of a network service provider is defined in terms of how he treats his client's traffic; e.g. how an IBP treats the traffic of client ISPs. Quantitatively, it can be described in terms of the resources that he allocates and the scheduling policies he applies to serving the client's traffic. When multiplexing traffic from different sources and applications, the network manager can assign different priorities to different flows of packets according to subjective criteria, such as the type of application being served (e.g. email vs. videoconferencing), the identity of the sender or recipient, and the revenue generated by the traffic transferred.

The dangers inherent in being unable to verify the level of effort can be reduced by using pricing mechanisms that provide the IBP with suitable incentives to exert the effort required to ensure the required performance. In effect, such mechanisms make the IBP responsible for the effort he provides by making his profit depend upon the outcome, after accounting for uncertain conditions. Performance indicators, such as average delay or packet loss, could be used to measure the observable outcome in an interconnection agreement.

Effort has a cost. This cost could be defined as the opportunity cost of not serving (or reducing the quality of service for) other client ISPs of the same network. An alternative but equivalent definition of this cost is based on the negative externality (congestion) imposed on the network and its other users. It is quite difficult to estimate this cost, as it depends on parameters that an IBP may not reveal. Often, a key component in the cost of serving the interconnection traffic is the load of 'local traffic' in the network, i.e., the traffic that originates from the network's other customers and which it is already contracted to carry. Information about this load may be available to the network provider before he must decide how to treat transit traffic from an ISP with whom he peers. The cost of allocating effort to the traffic of the new contract is negligible when the local traffic load is small, but increases quickly as the local load becomes greater and exceeds a certain threshold. This threshold may depend upon the total available capacity, the multiplexing algorithms used, and the burstiness of the traffic. In principle, the greater the amount of effective bandwidth that is allocated to the specific contract, the less bandwidth is available for the rest of the traffic, resulting in some opportunity or congestion cost.

For an incentive contract to be successful, one must be able to quantify reasonably well the expected cost to the contractee of the required effort, and the value of the resulting quality to the contractor. These issues are illustrated in the following example. There is information asymmetry at the time the contract is established. A rational service provider will provide the minimum possible effort, unless he is given appropriate incentives.

In our simple model, we assume that some network conditions are unobservable (implying an unobservable cost to the agent), but that the provider's effort is observable. The latter assumption is reasonable since interconnection contracts are typically of long duration, and so a customer ISP should be able to rather accurately estimate the parameters that he needs to infer the effort allocated by the contracted ISP. Only if contracts were of short durations, say a connection's life, might such estimation be inaccurate and effort unobservable. For simplicity, we focus on the modelling issues and the resulting optimal incentive schemes, omitting the complete analysis.

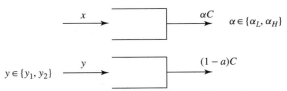

Figure 12.2 A model for an agent's effort. He operates a link serving two queues: one for the transit traffic and the other for his internal traffic. The effort given to the transit traffic is measured by the fraction of capacity α dedicated to serving the first queue. The rate of internal traffic at the time the contract is instantiated is random, taking values y_1, y_2 with probabilities p_1, p_2, respectively, with $y_1 < y_2$.

Example 12.1 (A principal-agent problem) Consider a transit agreement between two network service providers, using the formulation of the principal-agent model. Suppose a principal, P, contracts with an agent, A, for transport of a packet flow through A's network. We model A's network by two queues; one is dedicated to A's internal traffic and the other is dedicated to P's transit traffic (see Figure 12.2).

The service capacity of the network is C, of which αC is allocated to the P's transit traffic. For simplicity, we restrict the choice of α to two values, α_L, α_H, where $\alpha_L < \alpha_H$. Thus, α is the effort that is provided by A in the context of his contract with P. We suppose that A has no control over the rate of his internal traffic at the time he begins serving P's traffic. He can control the fraction of his capacity that he will allocate to it, and he knows the distribution of the future rate of his internal traffic at the time he agrees the contract with the principal. These are reasonable assumptions for many practical situations. The contract defines a service to be provided at some later point in time, and statistical information is available on the future state of the network. Let us denote the rate of the internal traffic by y, and suppose that it is known that it will take one of the two values y_1 and y_2, with probabilities p_1 and $p_2 = 1 - p_1$, respectively, where $y_1 < y_2$.

The cost of allocating capacity to P's flow is the extra delay experienced by packets of A's internal flow. Assuming, for simplicity, that this is a $M/M/1$ queue, we can calculate the cost using the fact that if a flow of rate y is served at rate C then the average packet delay is $1/(C - y)$. Taking γ as the monetary value of the cost of one time unit's delay, this implies a rate of delay cost of $\gamma y/(C - y)$ per unit time. Thus, the cost of allocating a fraction α of the available effort to the contract with P is

$$c(y, \alpha) = \gamma y \left[\frac{1}{(1 - \alpha)C - y} - \frac{1}{C - y} \right]$$

Let $c(ij)$ denote $c(y_i, \alpha_j)$, $i \in \{1, 2\}$, $j \in \{L, H\}$. It can be proved that

$$c(2H) - c(2L) > c(1H) - c(1L) > 0$$

In other words, a change from low to high effort is more costly to A when the system has a greater internal load. Of course, such a change benefits P, since it reduces the average delay of his packets. Denote by r_L and r_H respectively the monetary value of the service received by P when the effort levels are low and high.

Our task is to design an incentive contract in which P pays A an amount $w(\alpha)$. This payment is determined after the completion of the service and depends on the level of effort α allocated by A, which we suppose P can estimate both accurately and incontestably. Perhaps P measures the average delay of his traffic and then uses the delay formula for the $M/M/1$ queue to compute the effort that was provided by A.

Let the contract specify that P pays A amounts w_L or w_H as A provides low or high effort respectively. Once these are known, A needs to decide whether or not to accept the contract. His decision is based on knowledge of the distribution of the rate of the internal traffic at the point that service will be instantiated. At that point, he observes the rate of internal traffic and decides what level of effort to provide to P's traffic. This decision is rational, and is based on the information available. He maximizes his net benefit by simply computing the net benefit that will result from each of his two possible actions. This is easy to find for any given w_L and w_H. First, observe that if the value of the state is i, the rational action for A is $j = \arg\max_\ell \{w_\ell - c(i\ell)\}$, and the payoff is $w_j - c(ij)$. Thus, the sign of $w_L - w_H - [c(iL) - c(iH)]$ determines the most profitable action for the agent. The participation condition (i.e. the condition under which A will agree to accept P's traffic) can be written as

$$p_1 \max\{w_L - c(1L), w_H - c(1H)\} + p_2 \max\{w_L - c(2L), w_H - c(2H)\} \geq 0$$

Depending upon the parties' risk preferences, different incentive schemes can result. For example, P might be risk-averse, while A is risk-neutral. This could happen if A, who is perhaps a backbone provider, has many customers and so can spread his risk. His expected utility is then the utility of his expected value. The ideal contract for P is one that induces A to choose the efficient action, so maximizing total surplus from the interconnection agreement; and then extracts this entire surplus from A. (Note that A has to be willing to sign the contract – the participation condition must be satisfied – so that this is the best that P can achieve.) Simple convexity arguments suggest that a franchise contract is best for P. He keeps a constant amount F for himself, regardless of the outcome, and offers the surplus from the interconnection relationship minus the franchise payment F back to A. F is set so that A receives zero expected net benefit (or some tiny amount).

Suppose our risk-averse principal has a utility function of the form $U(r - w)$, where U is assumed concave, and the random variables r and w are respectively the value obtained by the principal and the value of his payment to A. These are well-defined for each pair w_L, w_H. The principal's problem is to maximize $E[U(r - w)]$ over w_L, w_H, subject to A's participation, and we know that this is achieved using a franchise payment F to P. For instance, if both actions L, H are enabled by the optimal incentive scheme, w_L, w_H must satisfy $r_L - w_L = r_H - w_H = F$ for some F which should be equal to the difference between the average value generated for the principal and the average cost to the agent as a result of the incentive scheme w_L, w_H. Observe that there are finitely many candidate Fs, since the number of different incentives provided by any choice of w_L, w_H is finite (in our case four). This suggests that we first compute all possible values for F and then choose w_L, w_H to realize the largest. This optimal F will depend on the values of the parameters $r_L, r_H, c(1L), c(1H), c(2L), c(2H)$. There are four cases to consider:

1. Always select high effort. Then $F_H = r_H - [p_1 c(1H) + p_2 c(2H)]$.

2. Always select low effort. Then $F_L = r_L - [p_1 c(1L) + p_2 c(2L)]$.

3. In state 1 select high effort, and in state 2 select low effort. Then $F_{HL} = p_1 r_H + p_2 r_L - [p_1 c(1H) + p_2 c(2L)]$.

4. In state 1 select low effort, and in state 2 select high effort. Then $F_{LH} = p_1 r_L + p_2 r_H - [p_1 c(1L) + p_2 c(2H)]$.

Let us restrict attention to the interesting case, $r_L < r_H$ and determine the optimal value of F as a function of r_L and r_H. In the region marked F_H in Figure 12.3, where $r_H - r_L \geq$

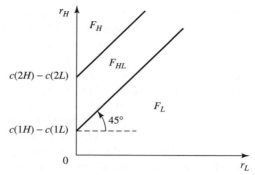

Figure 12.3 Optimal franchise contracts. There are three regions in which the principal's optimal franchise contract is different. Here r_L and r_H are the monetary value of the service received by P when the effort levels are, respectively, low and high, $r_L < r_H$.

$c(2H) - c(2L)$, F_H is the best franchise contract and $w_L = 0$, $w_H = p_1 c(1H) + p_2 c(2H)$. In the region marked F_L, where $r_H - r_L \leq c(1H) - c(1L)$, F_L is optimal and $w_H = 0$, $w_L = p_1 c(1L) + p_2 c(2L)$. In the region marked F_{HL}, where $c(1H) - c(1L) \leq r_H - r_L \leq c(2H) - c(2L)$, F_{HL} is optimal, and $w_L = -p_1(r_H - r_L) + p_1 c(1L) + p_2 c(2L)$, $w_H = p_2(r_H - r_L) + p_1 c(1L) + p_2 c(2L)$.

The intuition is that, given r_L, when r_H is sufficiently large we would like to provide incentives so that high effort is always used. As r_H decreases, it becomes economically sensible to use high effort only when the cost of providing it is not too great, which is when the system is in state 1. If r_H decreases even further and becomes close to r_L, then the greater cost of high effort does not justify its choice, regardless of the state of the system. It is only when $c(1H) - c(1L) \leq r_H - r_L \leq c(2H) - c(2L)$ that one needs to design a nontrivial incentive contract, i.e. one in which the provider's effort depends on network conditions.

12.6 Further reading

The interconnection issues addressed in the first part of this chapter are covered by Huston (1998), Huston (1999a), Huston (1999b) and Metz (2001). The web site of EP.NET (2002) provides information regarding Internet NAPs. Atkinson and Barnekov (2000) address facilities-based interconnection pricing issues. Mason (1998) discusses the international accounting rate system and the reasons this may be affected by Internet telephony. An interesting discussion of ISP interconnection agreements and whether regulation should be government-led or industry-led is given by Cukier (1998).

The ideas about competition and service differentiation in interconnected networks at the end of Section 12.2 are pursued by Gibbens, Mason and Steinberg (2000), Cremer, Rey and Tirole (2000) and Lafont, Marcus, Rey and Tirole (2001). The information asymmetry issues in Section 12.4 and Example 12.1 were introduced by Constantiou and Courcoubetis (2001). The book of Macho-Stadler and Perez-Castillo (1997) is also a good source on asymmetric information models for incentives and contracts.

13

Regulation

The regulator's job is to supervise a market so that it operates efficiently. He acts as a high level controller who, taking continual feedback from the market, imposes rules and incentives that affect it over the long term. In the telecoms market the regulator can influence the rate of innovation, the degree of competition, the adoption of standards, and the release to the market of important national resources, such as the frequency spectrum.

The efficiency of an economy can be judged by a number of criteria. One criterion is *allocative efficiency*. This has to do with *what* goods are produced. The idea is that producers should produce goods that people want and are willing and able to buy. Another criterion is *productive efficiency*. This has to do with *how* goods are produced. The opportunity cost of producing any given amounts of products should be minimized. Resources should be used optimally. New technologies and products should be developed as most beneficial. Finally, *distributive efficiency* is concerned with *who* things are produced for: goods should be distributed amongst consumers so that they go to people who value them most.

In general, competitive markets tend to produce both allocative and productive efficiency. However, in cases of monopoly and oligopoly firms with market power can reduce efficiency. We say there is *market failure*. In this case, regulation can provide incentives to the firms with market power to increase efficiency. The incentives can either be direct, by imposing constraints on the prices they set, or they can be indirect: for example, by increasing the competitiveness of the market. There is no single simple remedy to market failure.

Sometimes competition actually reduces allocative efficiency. In the case of a natural monopoly, social welfare is maximized if a single firm has the exclusive right to serve a certain market. This is because there are large economies of scope and scale, and because the rapid creation of industry standards leads to efficient manufacturing and also to marketing of complementary products and services. We see this in traditional telephony, and other public utilities, such as electric power, rail transportation and banking. The job of the regulator is to ensure that the monopolist operates efficiently and does not exploit his customers.

Information plays a strategic role in the regulatory context, because regulated firms can obtain greater profits by not disclosing full information about their costs or internal operations. A principal difficulty for the regulator is that he does not have full information about the cost structure and the production capabilities of the firm, nor does he know the actions and effort of the firm. This is another example of the problem of *asymmetric information*, already met in Section 12.4 in the context of interconnection contracts. We illustrate this in Section 13.1, with some theoretical models, and then explain ways in

Pricing Communication Networks   C. Courcoubetis and R. Weber
© 2003 John Wiley & Sons, Ltd   ISBN 0-470-85130-9 (HB)

which the regulator can achieve his goals despite his lacking full information. The firm's information about the future behaviour of the regulator may also be imperfect; this leads to intriguing gaming issues, especially when decisions must be made about large, hard to recover investments. In Section 13.2 we describe some practical methods of regulation. Section 13.3 considers when a regulator ought to encourage competition and how he can do this. In Section 13.4 we discuss the history of regulation in the US telecommunications market and describe some trends arising from new technologies.

13.1 Information issues in regulation

13.1.1 A Principal-Agent Problem

In this section we present a simple model for the problem of a regulator who is trying to control the operation of a monopolist firm. Unless he is provided with the right incentives, the monopolist will simply maximize his profits. As we have seen in Section 5.5.1, the social welfare will be reduced because the monopolist will tend to produce at a level that is less than optimal. The regulator's problem is to construct an incentive scheme that induces the firm to produce at the socially optimal level.

We can use the *principal-agent model* with two players to illustrate various problems in constructing incentives and the importance of the information that the regulator has of the firm. Recall, as in Section 12.4, that the *principal* wants to induce the *agent* to take some action. In our context, the principal is the regulator and the agent is the regulated firm. The firm produces output x, which is useful to the society, and receives all of its income as an *incentive payment*, $w(x)$, that is paid by the regulator. In practice, firms do not receive payments direct from the regulator, but they receive them indirectly, either through reduced taxation, or through the revenue they obtain by selling at the prices the regulator has allowed. To produce the output, the firm can choose among various actions $a \in A$, and these affect its cost and production capabilities.

There are two types of information asymmetry that can occur. The first is known as *hidden action asymmetry* and occurs when the regulated firm is first offered the incentive contract and is then free to choose his action a. The level of output x takes one of the values $x_1, \ldots, x_n$, with probabilities $p_1^a, \ldots, p_n^a$, respectively, where $\sum_i p_i^a = 1$ for each $a \in A$. The firm's cost is $c(x, a)$. Think, for example, of a research foundation that makes a contract with a researcher to study a problem. Once the contract is signed the researcher chooses the level of effort a that he will expend on the problem. 'Nature' chooses the difficulty of the problem, which together with the researcher's effort determines the success of the research. Note that the researcher does not know the difficulty of the problem at the time he chooses his level of effort. He only knows the marginal distribution of the various final outcomes as a function of his effort, for instance, the probability that he can solve the problem given that he expends little effort. The research foundation cannot with certainty deduce the action a, but only observe the output level. This is in contrast to the full information case, in which the regulator can observe a and make the incentive payment depend upon it. One way that full information can be available is if each output level is associated with a unique action, so that the regulator can deduce the action once he sees the output level.

Another possibility is that the regulator does not know the firm's cost function at the time he offers the incentive contract. We call this *hidden information asymmetry*. Now a denotes the type of the firm, and $c(x, a)$ is its cost for producing output x. At the time the contract is made, the firm knows its own $c(\cdot, a)$, but as we will see, it can gain by not disclosing it to the regulator. It turns out that information asymmetry is always to the advantage of the firm, who can use it to extract a more favourable contract from the regulator. By trying to 'squeeze' more of the profits of the firm from the contract, the regulator can only have negative effects on social efficiency.

Let us investigate the problems that the regulator must solve in each case. In the case of hidden action asymmetry the principal knows the cost function $c(a)$ (where for simplicity we suppose this cost depends only upon the action taken), but he cannot directly observe a. The principal's problem is to design a payment scheme $w(x)$ that induces the socially best action from the agent. Let $u(x)$ be the utility to the society of a production level x. The problem can be solved in two steps. First, compute the socially optimal action by finding the value of a that solves the problem

$$\underset{a}{\text{maximize}} \left[\sum_{i=1}^{n} p_i^a u(x_i) - c(a) \right]$$

Now find a payment scheme that gives the agent the incentive to take action a rather than any other action. Since there may be many such payment schemes, we might choose the one that minimizes the payment to the agent. This is the same as minimizing his profit. Let $v(w)$ be the agent's utility function for the payment he receives. In most practical cases, v is concave. The principal's problem is

$$\underset{w(\cdot)}{\text{minimize}} \sum_{i=1}^{n} p_i^a w(x_i) \tag{13.1}$$

subject to

$$\sum_{i=1}^{n} p_i^a v(w(x_i)) - c(a) \geq 0 \tag{13.2}$$

$$\sum_{i=1}^{n} p_i^a v(w(x_i)) - c(a)$$

$$\geq \sum_{i=1}^{n} p_i^b v(w(x_i)) - c(b), \quad \text{for all } b \in A \setminus \{a\} \tag{13.3}$$

Condition (13.2) is a *participation constraint*: if it is violated, then the agent has no incentive to participate. Condition (13.3) is the *incentive compatibility constraint*: it makes a the agent's most desirable action. The solution of (13.1) provides $w_a(\cdot)$, the best control. As a function of the observable output only, it induces the agent to take action a. Observe that at the optimum (13.2) holds with equality; otherwise one could reduce w by a constant amount and still satisfy (13.3). Hence, we must have that $\sum_i p_i^a v(w(x_i)) = c(a)$.

In the full information case, in which the principal observes a, a simple *punishment policy* solves the problem. Constraint (13.3) is ensured by taking $w = -\infty$ if any action other than a is taken. When a is taken, the optimal payment is $w(x_i) = w^*$ for all i, where $v(w^*) = c(a)$. Such a payment provides *complete insurance* to the agent, since he is recovers the cost of a, no matter what the outcome x_i.

Unfortunately, such a simple policy will not work if the action cannot be observed. If we use a complete insurance policy the agent will pick the policy with the least cost (as he has a guaranteed revenue). To guarantee (13.3), the payment must depend on the outcome, that is, $w(x_i) \neq w(x_j)$, $i \neq j$. Additionally, we must guarantee (13.2), with $\sum_i p_i^a v(w(x_i)) = c(a)$. Since $v(w)$ is concave in w, the payment vector will need to have a greater expected value than in the full information case, that is, $\sum_i p_i^a w(x_i) \geq w^*$. Thus, under information asymmetry, the principal must make a greater average payment.

Notice that other incentive schemes also work. Let us assume a simple hidden information model with $u(x) = v(x) = x$. The incentive payment

$$w(x) = x - F \tag{13.4}$$

for some constant F, has a nice interpretation. Firstly, we can interpret it as a 'franchise' contract: the agent keeps the result x and pays back to the principal a fixed amount F, the 'franchise fee'. Secondly, the participation and the incentive compatibility conditions imply that the agent solves the problem

$$\underset{a}{\text{maximize}} \, [x(a) - c(a) - F], \quad \text{subject to } x(a) - c(a) - F \geq 0$$

Hence, the agent will choose the socially optimal action a^* if F is small enough to motivate participation, i.e. if

$$F \leq x(a^*) - c(a^*)$$

If the principal knows $x(\cdot)$ and $c(\cdot)$ then he can set F equal to the maximum allowable value, say $F^* = x(a^*) - c(a^*)$, and so push the profit of the agent to zero. However, if these functions are not known, then the principal cannot take chances, and must choose F less than F^*. This illustrates how hidden information means greater profits for the regulated firm. We say there are *informational rents*. If the goal is to maintain the output that maximizes social welfare (allocative efficiency), then hidden information increases producer surplus (and decreases distributive efficiency). A possible way to solve the problem of defining a reasonable F is through auctioning (*monopoly franchising*), in which the agents bid for the least value of F that they can sustain, hence indirectly revealing information to the principal.

In another simple illustrative example of hidden information, which shows more clearly the trade-off between lower profits for the regulated firm and economic efficiency, the firm is one amongst a number of possible types, differing in the cost function, $c_i(x)$. Again, the agent's type is unknown when the contract is signed. The regulator knows only the probability distribution of the various agent types. In practice, such a model makes sense since it is hard for the regulator to construct the actual cost function of the regulated firm; moreover, it is to the advantage of the agent to hide his cost function from the regulator unless he is very inefficient. The possibility that the firm might have a high operating costs forces the regulator to offer him a high compensation. Similar examples from other contexts concern contracts between a firm and workers with different efficiencies, and between an auto insurer and drivers of different propensities to accidents. Again, an efficient worker benefits from the existence of inefficient workers, since these force the firm to offer him a greater incentive.

The principal wants to construct a payment scheme that maximizes economic efficiency under uncertainty about the agent's type. Suppose the principle posts a payment scheme that is a function of the output x. Given his cost $c_i(x)$, an agent of type i selects the optimal level of output. Although this is straightforward when there is a single type of agent, there is a complication when there are multiple types, since an agent of type i could find it profitable to impersonate an agent of type j, and produce the corresponding output level. To avoid this, the optimal payment scheme must allocate greater average profits to the agents that it would do if it could make the payments depend on agent type. This again illustrates the power of information in the regulatory framework. Any attempt to reduce the profits will result in different output levels, and so reduce economic efficiency.

More precisely, suppose an agent can choose any positive level of output x. Suppose it is desired to maximize social welfare. In the complete information case, it is optimal to offer a type i agent a payment of $c_i(x_i^*)$ to produce x_i^*, where x_i^* maximizes $x - c_i(x)$, i.e. $c_i'(x_i^*) = 1$. Suppose there only two types of agent, of equal probability, and that type 1 is the more efficient, in the sense that its marginal cost function $c_1'(x)$ lies below $c_2'(x)$ for all x, as shown in Figure 13.1. We say that the cost functions have the *single crossing property*, since even if $c_2(0) < c_1(0)$, the functions can cross at most once. Then a candidate payment scheme is given by two pairs: $(c_1(x_1^*), x_1^*) = (A + B, x_1^*)$ and $(c_2(x_2^*), x_2^*) = (A + D, x_2^*)$, as shown in (a) of Figure 13.1. Note that this is the optimal incentive payment scheme in the full information case, in which the principal knows the agent's type when he offers a contract. In this case, he offers an agent of type i only one possible contract: *make x_i^* for a payment of $c_i(x_i^*)+\epsilon$*, where ϵ is a small positive amount that gives the agent a small profit.

This payment could be optimal in the hidden information case if the incentive compatibility conditions were to hold, i.e. if an agent of type i were to choose output level x_i^* after rationally choosing the contract that maximizes his net benefit. Unfortunately, this does not happen. An agent of type 1 is better off to produce x_2^* and so receive a net benefit of D, instead of zero. The only way to prevent him from doing this is to add D to the payment for producing x_1^*, and hence provide incentives for socially optimal output. One can check that this works, and that each type will now produce at the socially optimal level. Note that the inefficient agent obtains zero profit, while the efficient agent is rewarded by obtaining a profit of D.

The principal cannot reduce the agents' profits without reducing economic efficiency. However, he can reduce the payment made to an agent of type 1, and so increase his own surplus, if he reduces the payment for output level x_1^* from the initial value $A + B + D$ to some value $A' + B' + D'$, and reduces x_2^* to x_2^{**}, as shown in (b) of Figure 13.1. By reducing distributive inefficiency (through reducing the incentive payment) he also reduces social efficiency, since a type 2 agent does not now produce at the socially optimal level. Note that this example is very similar to that given and for second degree price discrimination in Figure 6.4.

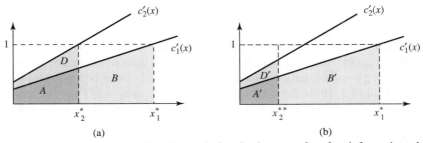

Figure 13.1 A principal-agent problem in regulation. In the case of perfect information, shown in (a), it is optimal to offer $A + B$ to agent 1 to produce x_1^*, and $A + D$ to agent 2 produce x_2^*. Each just covers his cost. However, if offers cannot be tailored to agents (because their types are unknown) then agent 1 will choose to produce x_2^* and obtain net profit of D. Now (b) shows how the principal increases his surplus. He reduces the target output level of the high cost agent to x_2^{**}, below the socially optimal level x_2^* so as to decrease D to D'.

13.1.2   An Adverse Selection Problem

We have seen above how the principal may experience an *adverse selection problem* because he lacks the information to discriminate amongst types of agents and make them distinct offers. Adverse selection occurs when some type of agent finds it profitable to choose the offer that was intended for another type of agent. We have seen that when the goal of the principal is to make agents choose actions that maximize social welfare, the effect of adverse selection is to force the principal to make a larger payment than he would if he had full information.

A consequence of adverse selection is that there may be no prices that a regulator can prescribe to a firm such that the firm can recover its cost. More generally, adverse selection can destroy a market, as we see in the following example.

Example 13.1 (A market for used cars)   Consider a market for used cars, in which the principal (the buyer of a car) can check the quality of the car only after he has purchased it from the agent (the seller). Suppose that cars have qualities uniformly distributed on $[0, 1]$, and that a seller of a car of quality x is willing to sell only if the offered price s exceeds x, which is perhaps an amount he owes on a loan and must repay. A buyer of a car of quality x values it at $u(x) = 3x/2$.

Since the buyer cannot observe the quality of a car before making an offer, he must make the same offer for every car. His problem is to maximize his net benefit. He does this by choosing his offer s to maximize $E_x[u(x) - s \mid x \leq s]$, where the expectation over the random variable x is conditioned by the participation constraint $x \leq s$. For a given s, the expected quality of a purchased car is $s/2$, and so the buyer must choose s to maximize $3s/4 - s$, giving $s = 0$. Thus no cars are sold.

Note that, if the quality of a car is in the interval $[2s/3, s]$, then both buyer and seller can benefit from a transaction. If the quality of a car is in the interval $[0, 2s/3]$ then a transaction profits the seller, but not the buyer. The average quality of a car is $s/2$, which is less than the lowest acceptable level of $2s/3$ for which the buyer would wish to participate. This adverse selection phenomenon causes market breakdown. Although there are social welfare gains to be made by matching some pairs of buyers and sellers, the lack of information makes such interaction impossible. Of course, if the distribution of the quality were such that the average quality were greater than $2s/3$, then the market would not break down, and there would be a positive value of s that it would be optimal for a buyer to bid.

The problem is that the buyer is unable to distinguish between high and low quality cars. If he were able to obtain information about the quality of a car he could adjust his bid appropriately. Hence, it benefits both the seller and the buyer if the quality can be signalled. The seller could allow the buyer to take the car for a test drive, or to have the car checked by a mechanic. As a simple illustration, suppose the buyer can check whether the quality of a car is more or less than $1/2$. It is easy to see that such a simple signal of 'high' or 'low' quality is enough to create a stable market in which both sellers and buyers profit. For instance, offering $s = 3/4$, but only for cars with $x > 1/2$ is a policy that gives the buyer an average profit of $3/16$. In fact, the optimal choice of s is $s = 1/2 + \epsilon$ for an arbitrarily small ϵ. Now the buyer has nearly full information as he knows the actual quality of any car he purchases must lie in the interval $[1/2, 1/2 + \epsilon]$.

Similar to the above example, let us consider a model of an ISP who sells Internet connectivity.

Example 13.2 (A market for Internet connectivity)   Suppose there are n potential customers, requiring $x_1, \ldots, x_n$ units of Internet use, where these are independently and uniformly distributed on the interval [0, 1]. Suppose the regulator requires the ISP to charge all customers a flat fee w, without taking account of their actual resource usage. Then, under certain conditions, there may be no profitable production level for Internet services.

Suppose that a customer of type x has a utility for the service $u(x) = x$, and so does not buy service if his surplus of $x - w$ is negative. The network exhibits economies of scale, so that the per unit cost when using total bandwidth b is

$$\rho(b) = a\frac{b}{n/2} + \left(1 - \frac{b}{n/2}\right)$$

which varies linearly from its maximum value 1 when $b = 0$ to its minimum value $a < 1$ when $b = n/2$ (where $n/2$ is the maximum average bandwidth consumed by the customers when all subscribe to the service). If the regulator sets a price w, then only the customers with $x \geq w$ will subscribe, and they will number $n(1 - w)$ on average. The average bandwidth that any one will consume is $(1 + w)/2$, and the average total amount of bandwidth consumed will be $b = n(1 - w)(1 + w)/2$. The average profit per customer of the firm will be

$$w - \tfrac{1}{2}\rho(b)(1 + w) = w - \tfrac{1}{2}[1 - (1 - w^2)(1 - a)](1 + w)$$

For $w = 0$ the profit is $-a < 0$, and for $w = 1$ (the maximum possible charge) the profit is also 0. Numerical calculation shows that when a is greater than 0.7465, the profit of the firm increases with w but is always negative. Hence, no value of w allows stable operation. The reason is adverse selection: given w, only customers with $x \geq w$ subscribe. But w is targeted at the average customer. Adverse selection prevents the average from being favourable. This again illustrates that there are major problems with flat rate pricing.

13.2 Methods of regulation

The following sections describe various methods of monopoly regulation.

13.2.1 Rate of Return Regulation

Under *rate of return regulation* a firm must set its prices, its level of production and its inputs, subject to the constraint that its rate of return on its capital is no more than a 'fair rate of return' set by the regulator. The firm maximizes its profit under this constraint.

The problem with this type of regulation is that the firm has the incentive to inflate the base on which the rate of return is calculated (the so-called *Averch–Johnson effect*). For example, it might substitute more expensive capital for labour, even when this does not minimize its production cost. In other words, production can be inefficient because of an inefficient choice of inputs. However, this might not be bad for the overall efficiency. It can be shown that under rate of return regulation the producer produces more output than he would do if he were unregulated. Since it is the monopolist's reduced level of output (compared with the output under perfect competition) that causes a reduction in social welfare below its maximum, rate of return regulation does improve social welfare.

13.2.2 Subsidy Mechanisms

Price subsidies and taxes can be used to control the point at which the economy of monopoly producer and the consumers lies. The goals are to maximize overall efficiency and redistribute the profits of the monopolist.

The complete information case. The easiest case is that of full information, in which the regulator knows the consumers' demand curve and the cost function of the firm. In this case, a simple policy is to subsidize part of the price set by the firm so that the price seen by the customers are marginal cost prices at the socially optimum production and consumption level of the economy. Then the firm is made to pay a lump-sum tax equal to its profits at this level. Clearly, this strategy maximizes social welfare and reduces the monopolist's profit to zero. More precisely, let p_M and p_{MC} be the monopolist price and the marginal cost price at the levels of output x_M and x^*, that maximize respectively the monopolist profit and the social welfare. Note that $p_M > p_{MC}$. Initially, the monopolist chooses price p_M and has profits

$$p_M x_M - c(x_M)$$

Assume that the monopolist's cost function has decreasing marginal cost. Then the regulator returns to each user an amount $p_M - p_{MC}$ for every unit purchased, and this makes demand rise to the desired point x^*. This increase in demand is welcomed by the monopolist who sees his profits rise even further. To see this, observe that as the derivative of $c(x)$ is decreasing in x,

$$p_M \geq \text{ marginal cost at } x_M \geq \frac{c(x^*) - c(x_M)}{x^* - x_M}$$

Hence

$$p_M x^* - c(x^*) \geq p_M x_M - c(x_M)$$

Now, the regulator exacts from the monopolist a one-time lump-sum tax equal to his profits $p_M x^* - c(x^*)$. Clearly, the monopolist can only continue producing x^*, for zero profit. If he chooses any other production level or price (i.e. p_M) he will suffer a loss.

The total surplus subsidy mechanism. A problem with the above strategy is that to compute the right price subsidy one must know the cost function of the monopolist. This is not required with the following simple mechanism. The regulator only need know the demand curve only, which is often possible. The mechanism generalizes the approach of Section 13.1, using an incentive payment like (13.4), in which

- the monopolist is allowed to set prices and collect the resulting revenue, and
- the regulator pays the monopolist the entire consumer surplus in the form of a subsidy.

Recall that the consumer surplus at consumption level x, given monopolist price $p(x)$, is

$$CS(x) = \int_0^x p(y)\,dy - p(x)x$$

and so can be calculated knowing only the demand curve $p(y)$.

The reason that this mechanism induces social optimality is that the monopolist eventually receives all the social welfare (namely, the sum of the producer's profit and the consumer surplus); thus, his rational choice is to set prices that induce the socially optimum production level.

The problem is that the consumers have no surplus. A remedy would be to auction, as in (13.4), the maximum amount F that a monopolist would be willing to pay as a lump-sum to participate in this market. Since the cost function is not known, the maximum value of F is unknown. If competing firms have different cost functions, the one with the lowest cost would win, and make a profit equal to the difference between its cost and the cost of the competitor with the next lowest cost. The mechanism that follows remedies some of the above problems.

The incremental surplus subsidy mechanism. Unlike those previously described, this mechanism does not work in one step. Although it assumes explicit knowledge of the cost function of the firm, it observes the responses of the firm over time to incentives provided by the regulator, and by adapting to the firm's behaviour eventually settles on the socially optimal operating point, with zero profits for the monopolist. It is an improvement of the average price regulation mechanism that we will briefly mention in Section 13.2.3. In just two rounds, this mechanism achieves output efficiency, zero monopolist profits and *cost minimization*. This latter is key since it provides the incentives to the firm to operate as efficiently as possible, without the presence of actual competition.

Assume that time is divided in periods, $t = 1, 2, \ldots$, and in each period the demand and the cost are the same. At the end of period t, the regulator observes the current and the previous unit price or quantity sold, the expenditure of the firm in the previous period E_{t-1} (taken from the firm's accounting records), and infers the previous accounting profits $\pi_{t-1} = p_{t-1}x_{t-1} - E_{t-1}$. As in the previous section, we suppose the regulator can also calculate the consumer surplus. Knowing this, the regulator

- pays the monopolist a subsidy equal to the *incremental* change in consumer surplus between periods $t - 1$ and t, and
- takes in tax the previous accounting profit π_{t-1}.

To model the fact that the firm might not operate under minimum cost, we suppose that during period t the accounted expenses of the firm are $E_t = c_t + w_t$, where c_t is the actual operating cost and $w_t \geq 0$ is a discrepancy between the actual operating cost and the one declared through the accounting records. Then the actual profits are $\hat{\pi}_t = \pi_t + w_t$. Let $W(x)$ denote the social welfare when the output level is x. Given all the above, the producer makes a profit in period t of

$$\pi_t + \left\{ \mathrm{CS}(x_t) - \mathrm{CS}(x_{t-1}) \right\} - \pi_{t-1}$$
$$= \pi_t + \left\{ [W(x_t) - \hat{\pi}_t] - [W(x_{t-1}) - \hat{\pi}_{t-1}] \right\} - \pi_{t-1}$$
$$= \pi_t + \left\{ [W(x_t) - (\pi_t + w_t)] - [W(x_{t-1}) - (\pi_{t-1} + w_{t-1})] \right\} - \pi_{t-1}$$
$$= W(x_t) - W(x_{t-1}) - w_t + w_{t-1}$$

Summing over periods $t = 2, \ldots, \tau$, we obtain $W(x_\tau) - w_\tau - W(x_1) + w_1$, and see that for all $\tau \geq 2$ the monopolist maximizes his total accumulated profit to time τ by choosing x_τ to maximize the social welfare $W(x_\tau)$, and truly declaring his actual costs, so that $w_\tau = 0$. Notice that once he does this, his profit in period τ is 0, for all $\tau > 2$. Maximizing social welfare at time τ provides the incentive to operate as efficiently as possible, i.e. to choose the smallest possible function $c(x_t)$.

13.2.3 Price Regulation Mechanisms

Price regulation mechanisms are those that directly control the monopolist's prices. The general idea is that the regulator specifies a set of constraints on the firm's prices (called *price caps*), which are defined relative to a reference price vector. The firm is free to set any prices that satisfy these constraints. The aim in that (a) the social surplus increases relative to the reference set of prices, and (b) the firms have incentives to improve production efficiency. Various schemes have been devised. They differ in respect of the information that they require and the dynamics of the resulting prices movements.

A simple scheme, called *regulation with fixed weights*, requires that prices be chosen from the set

$$\left\{ p : \sum_i p_i q_i(p^0) \leq \sum_i p_i^0 q_i(p^0) \right\}$$ (13.5)

where p^0 is the reference price vector and p is the new price vector. Observe that since the customers can always buy the old quantity $q(p^0)$ under the new prices and pay less, the new price vector can only increase consumer surplus. The weakness of the scheme lies in the choice of an appropriate reference price vector p^0 and in the ability to estimate accurately the demand $q(p^0)$.

An alternative is *dynamic price-cap regulation*. The regulator observes the prices and the corresponding demand during period $t - 1$, and controls the prices for period t to lie in the set

$$\left\{ p^t : \sum_i p_i^t q_i^{t-1} \leq \sum_i p_i^{t-1} q_i^{t-1} \right\}$$ (13.6)

This simple variant of (13.5) is called *tariff-basket regulation* and has a number of desirable properties. First, the consumer surplus is nondecreasing, and it can be shown that under reasonable assumptions and constant production costs the prices converge to Ramsey prices. Secondly, the decoupling of prices from cost provides the firm with an incentive to increase its productive efficiency. However, the lack of connection with cost means that the scheme is not robust; if the firm can change its costs then there can be divergence from marginal cost and the firm may obtain greater profits. One way to further increase the incentive to reduce costs is to multiply the right-hand side of (13.6) by a coefficient $(1 - X)$, where $100X\%$ is the intended percentage increase in production efficiency.

In another dynamic price-cap mechanism, due to Vogelsang and Finsinger, the regulator assumes knowledge of the quantity q^{t-1} produced in $t - 1$ and of the resulting cost to the firm, $c(q^{t-1})$. Then he insists that prices be chosen from the set

$$\left\{ p^t : \sum_i p_i^t q_i^{t-1} \leq c(q^{t-1}) \right\}$$ (13.7)

In the case of firms with increasing economies of scale, and which chooses price myopically (that is, to optimize (13.7) at every step), prices under this scheme converge to Ramsey prices and push the profits of the firm to zero. However, the scheme provides an incentive for nonmyopic firms to inflate temporarily their costs of production, since this allows for greater prices in the future. This can lead to an undesirable reduction in social welfare. Economists have found ways to combine various aspects of the above schemes, to improve them and remedy their shortcomings.

A simpler mechanism, which involves less information, is *average revenue regulation*, in which prices are chosen from the set

$$\left\{ p^t : \sum_i p_i^t q_i^{t-1} \leq (1 - X)\bar{p} \sum_i q_i^{t-1} \right\}$$ (13.8)

where $\bar{p} \sum_i q_i^{t-1}$ is the average revenue in period $t-1$, and X is the rate of increasing production efficiency.

Another mechanism is based on the retail price index (RPI) and called the '*RPI minus X*' mechanism. The key idea is that the various services of the firm are grouped into baskets, and the regulator permits the average price of a service basket to change by no more than RPI$-X$, which is the price index of the basket. Service baskets are defined by major customer classes and have different price indices (different values of X). Since substitution can only occur between services of the same basket, this definition of price caps prevents the firm from cross subsidizing one product by another. The choice of X forces the firm to reduce production costs and improve productive efficiency. Its value depends upon the degree of competition available in the given market. The more competitive is the market, the smaller is the value required for X, since competition forces prices to decrease and motivates efficient production.

Finally, we remark that the dynamic regulatory mechanisms cannot avoid issues of game playing between the firm and the regulator. Clearly, anticipation of future regulatory decisions affects a firm's present policy and decisions. In general, the more uncertain is the regulatory framework, the more difficult it is for the firm to plan future investment and infrastructure. For example, if it is not clear whether the regulator's future policy will permit recovery of sunk costs, then the firm is discouraged from making large investments. This has social cost.

A related issue is the frequency with which regulatory policy should change. If it changes infrequently, then the regulator cannot take enough account of new industry facts and rapidly changing cost structures; this results in greater profits for the firms. However, stability of the regulatory framework over longer time intervals, called *regulatory lags*, allows the industry to adapt and to make optimal improvements of its production facilities. Therefore, the regulator must seek a good compromise between providing motivation for investments and cost reduction, and the inefficiency that results due to greater industry profits.

13.3 Regulation and competition

There are many delicate and conflicting issues that a regulator must consider when deciding how to control competition in a market. As remarked in Chapter 6, competition does not always benefit society. A monopoly can be preferable if there is an infrastructure that requires large sunk costs and production has large economies of scale.

A related notion is that competition can lead to *excessive entry*. The presence of a large number of producers, each producing a relatively small output, can rob society of the cost-reducing advantages of production economies of scale. So, although the prices decrease and consumer surplus increases, individual firms produce less efficiently and the economies of scale are reduced. This may result in an overall decrease in social surplus. Of course if competing firms differentiate their products, then the resulting increase in consumer choice can increase consumer surplus. This may outweigh the reduction in economies of scale.

Another possible negative effect of competition is that new entrants into a monopolist's market may target the most profitable parts, engaging in so-called *cream-skimming*. Such entry may be inefficient, i.e. not at the marginal cost, and may cause the monopoly to collapse. This is because, the sustaining of a monopoly usually requires the most profitable part of the market to subsidize the less profitable part, in order to justify the level of production which is required for economies of scale, or because of universal service obligations. Hence, a competitor may charge above marginal cost and still be able to undercut the monopolist in some profitable parts of its market.

Of course, the disadvantage of a monopoly is the allocative inefficiency, since a monopolist will seek to maximize his profits and this will be at the expense of overall social welfare. Increasing the number of competing firms has the positive effect of reducing prices. This can sometimes be achieved without actually increasing the number of competing firms. The theory of *contestable markets* posits that in a natural monopoly, the monopolist will set efficient prices because he must take account of *potential* competition. That is, he must defend himself against competitors who can enter the market at small cost and make profits on a short time horizon (so-called *hit-and-run entry*; so he is forced to discourage these entrants by offering the lowest possible price. Hence, the advantages of monopoly can be enjoyed without sacrificing efficiency, and without any external regulation.

To introduce competition into a market that is controlled by a monopolist it is not enough simply to remove legal obstacles. This is because there are usually inherent asymmetries between the monopolist firm and its potential competitors. The monopolist is in the strategic position of having the 'first move', and so can dictate some terms of the prospective market in a way that deters entry. He is in the position to make *viable threats*, which convince potential entrants that if they enter the market they will incur losses.

Another way a monopolist can defend his position is through so-called 'predatory pricing', in which he subsidizes his provision of certain services from revenues in other parts of the market where he is under no threat, and so pretends to potential competitors that his costs for providing these services are less than they actually are. Potential competitors assess the true costs as being greater and they are deterred from competing. Also to the monopolist's advantage is the production efficiency that results from large investments in technology and the associated large sunk costs. He also gains from 'learning by doing', as he is already in the market and has experience on his side.

If competitors are to sell services then they need access to bottleneck distribution networks. However, a monopolist can sometimes prevent access to such networks. For example, because the copper loop local access networks are typically controlled by telephone companies they can control access to customers. They can protect themselves by bundling the local access service with other services, so that it cannot be purchased on its own, but only as part of some more expensive service. This makes the bottleneck service prohibitively expensive to potential competitors.

The regulator can prescribe the medicine of *unbundling*. He forces the monopolist to offer the bottleneck service as a stand-alone service, at a price close to its actual cost. Now new services can be created within a competitive framework and entry to the market can take place. However, it can be difficult for the regulator to price unbundled elements. If he sets prices that are too low, then there is no incentive for the monopolist to improve the quality of the service or to efficiently maintain facilities for providing the service. Low prices also discourage the development of alternative technologies. For example, unbundling the copper loop in the local access telephone network, and offering copper wire telephony at very low prices could impede the development of alternative wireless technologies by making it hard for wireless to compete with traditional telephony. Thus the regulator must choose prices for the unbundled elements quite carefully.

13.4   Regulation in practice

13.4.1   Regulation in the US

The telecommunications industry has traditionally been a regulated sector of the US economy. Regulation was necessary because the market for telecommunications services was a natural monopoly, and the large investments needed to provide competitive services meant the market could not be contestable. Entry costs have been significantly reduced by modern advances in transmission technologies (namely, optical networks for the backbone and the metropolitan area, wireless and cable with broadband capabilities for the local access, and the availability for leasing at reasonable cost transmission infrastructure in most parts of the country). However, until recently, such technologies were not available and a second competitor would not survive. Hence, regulation began in the early part of last century, and continues today in parts of the telecommunications sector. Regulation is applied both by the Federal Government through the Federal Communications Commission (FCC) and by each State through a Public Utilities Commission or Public Service Commission.

Competition in the market for telecommunications services and equipment went through various stages after the invention of the telephone. AT&T was always the dominant player in this market and by the late 1920s formed the Bell System that had an overwhelming majority of telephony exchanges and submitted to state regulation. The FCC and federal regulation was instituted by the 1934 Telecommunication Act. In the following years the US Department of Justice has brought two important antitrust lawsuits against AT&T. In the first, United States vs. Western Electric (1949), the US Department of Justice claimed that the Bell Operating Companies practised illegal exclusion by buying telephones and communication equipment only from Western Electric, also a part of the Bell System. The government sought a divestiture of Western Electric. The case was settled in 1956 with AT&T agreeing not to enter the computer market, but retaining ownership of Western Electric.

The second major antitrust suit, United States vs. AT&T, began in 1974 and was eventually settled in 1984. The government claimed that (i) AT&Ts relationship with Western Electric was illegal, and (ii) that AT&T monopolized the long distance market. The Department of Justice sought divestiture of both manufacturing and long distance from local service. The result was that seven Regional Bell Operating Companies (RBOCs) were separated from AT&T, each comprised of a collection of local telephone companies that were former parts of AT&T. These were also referred to as Incumbent Local Exchange Companies (ILECs) or 'Baby Bells'. Each RBOC remained a regulated monopoly, with an exclusive franchise in its region. AT&T was allowed to sell only long-distance services and was permitted to enter the computer market.

The rationale of this regulatory move was that new cost-effective technologies in the long-distance part of the network, such as microwave transmission, made new entry and competition that market possible, weakening the reasons for keeping a natural monopoly. In contrast to the long-distance part, the local part of the network was still a natural monopoly due to the lack of a similar substantial technology innovation for this part of the network. Competition in long distance has been a great success. Customers have benefited from seeing their long-distance charges decrease significantly.

In the years following, ILECs have realized significant profits in the local telephony market due to their monopoly status. The Telecommunication Act of 1996 attempted to extend the competition that has been achieved in the long distance part of the market into the monopolized local markets. The goal was to create a market of services in which a new entrant that misses parts of the network infrastructure that are important for the provision of services to his customers, can lease these bottleneck parts at reasonable prices from the local infrastructure owners (ILECs). Resolving such interconnection issues by unbundling the services that are sold by the incumbents and allowing their competitors to lease any service or infrastructure component at its actual cost is an efficient way to promote competition at the higher layers of the communications service hierarchy without having to duplicate expensive infrastructure that is already in place. These infrastructure services, such as the physical copper wire loop that connects customers to the local exchange of the ILEC, are called *Unbundled Network Elements* (UNEs). (Some regulators have sought to unbundle access services at even higher layers than the physical wire level, such as at the bit stream level provided by DSL equipment). Another service is the ability of a customer of an ILEC to select his own long-distance provider for terminating his long distance calls instead of the default long-distance provider of the ILEC. More specifically, the 1996 Act requires that ILECs (i) lease parts of their network (the UNEs) to Competitive Local Exchange Carriers (CLECs) at cost plus some reasonable profit; (ii) provide to competitors at a wholesale discount any service the ILEC provides; and (iii) charge reciprocal rates in termination of calls to their network and to networks of local competitors. The Act also imposes conditions to ensure that the monopoly power in the local markets is not exported to the long-distance market. Thus, the 1996 Act requires that competition be established in local markets before the ILECs are allowed to provide long distance service.

Clearly, the UNEs must be fairly priced if the Act is to achieve the goal of increasing competition. The proposed methodology is LRIC, which is also sometimes called TELRIC (Total Element LRIC). It says that costs should be calculated on a forward looking basis (current costs), i.e. without taking into account historic costs, and assuming that the most efficient technology is used for the provision of the services by the firm (i.e. using a bottom-up model for constructing the cost function). Using historic costs would end-up in extremely low prices (since most parts of the network are already depreciated), which would provide few incentives to the ILECs to maintain and upgrade their infrastructures. On the other hand, the new entrants pay a fair amount as if the network was built with the most efficient technology existing at the moment. Before choosing LRIC, a number of different pricing rules were examined, including ECPR. See the discussion in Section 7.3.5.

The Telecommunications Act of 1996 preserves the provision of universal service through subsidization of the local telephone service. *Universal service* is the provision of basic local telephone service to the widest possible number of customers, including customers in remote and sparsely populated areas where the provision of such service is exceedingly expensive. Without a government subsidy, the prices of services to these customers would be substantially greater than in urban areas, precluding most of these customers from using them. In an important departure from previous practice, the Act imposes the requirement that subsidization is transparent and that the subsidies are raised in a competitively neutral manner (for instance, using auctions). Traditionally, such subsidies were raised by the ILECs providing universal service through the method of high access charges. Origination and termination access charges are paid by long distance companies to local exchange carriers for originating or terminating long distance calls between LEC customers. This artificially increased the cost of the long-distance service. Long distance calls had to pay these extra subsidy charges in addition to the normal termination fees when using the last part of the access network of the ILECs to reach the parties involved. Keeping these two charges together reduced transparency and it was hard for the regulator to determine the actual cost to the ILEC of providing universal service. ILECs had the incentive to declare a higher than actual cost of universal service to keep access charges high and make more profit.

Many experts feel that the Telecommunications Act of 1996 was a regulatory move in the right direction, but that it failed to have the desired impact due to complex market conditions. The process of unbundling the various network elements is harder than it first seems and has many subtleties, including the quality of the elements, maintenance, timely delivery, availability and cost of collocation space and equipment. In a competitive market, in which the ILECs and the CLECs sell competing services over the ILECs infrastructure, it is natural for the ILECs to unbundle their elements as slowly as possible, and to take advantage of their dominant market position to sell the new broadband Internet access services to their existing customer base. One-stop-shopping and the cost of changing telephone numbers contribute to the cost of switching network service providers. The steady income of the ILECs versus the large uncertainties and risks faced by the CLECs are other explanations for the difficulty faced by the CLECs in founding viable businesses. On the other hand, competition in the local part of the network is changing due to (i) substitute services for fixed telephony such as mobile telephony, and (ii) new broadband access technologies for fast Internet that include cable, wireless and satellite, and which do not need any infrastructure from the ILECs in order to work. These technologies allow long-distance carriers to compete with ILECs by installing their own access networks. Such vertically integrated companies include content provisioning, long-distance and access. For their part, ILECs feel that the conditions for increased competition in the local part of the network are met, and so the Telecommunications Act of 1996 should not preclude them from extending their footprint in the long-distance market.

13.4.2 Current Trends

What new challenges does the regulator face? Clearly, his job is far more complex than when telephony was the main service to be regulated. The convergence of transport services, whereby Internet transport protocols are used to transport any type of information, and the convergence of content services of retrieving and displaying digitized information using the uniform Internet application protocols of the World Wide Web, together with the use of personal computers at the edge of the network for multiple tasks (from telephony to watching interactive video), creates a new ubiquitous computing and communication platform. We have already remarked in Section 1.2 that it is very hard to make money out of a commoditized network in which complexity resides at the edges rather than the core. Telecoms companies must depart from the business of simply transporting bits and somehow diversify their services. This explains why the new access technologies may play a vital role in service differentiation, and in helping companies to make profits. Cable companies with large subscriber bases and wireless access providers will play an increasingly important role in the telecoms value chain. Should the regulator unbundle such networks in a similar way as has been done for the copper local loop? There are many similarities, in the sense that such access companies are essentially monopolies in their footprints. However, if cable and wireless technologies are to become truly interactive and broadband, then crucial investments are still to be made. Forcing operators to open their networks to competitors could deter them from making these investments. One can extend this idea from access to more general interconnection settings. It is clear that refusing or strategically pricing interconnection, or artificially deteriorating the quality of interconnection, can be a very powerful tools to incumbents, especially when they also control the access network. It is not clear if, or to what extent, the regulator should become involved.

We must also mention spectrum allocation. Due to the increasing importance of wireless services, the spectrum becomes an increasingly valuable and scarce commodity. How should spectrum be managed and allocated? How should licenses for providing wireless services be structured, and with what terms? Auctions have traditionally been used by regulators to price such unique goods, and we return to these in Chapter 14.

Another important topic, more concerned with national policy than traditional regulation, is the creation of fibre access networks owned by the customers rather than telecoms operators. In the *condominium fibre model*, large customers such as user communities, hospitals, libraries, universities and schools, get together and subcontract the development of a fibre access network to a third party. Each customer owns part of the common dark fibre infrastructure, similarly to the type of ownership that takes place in a condominium housing project. After installing a large number of fibre strands, which terminate in a carrier-neutral collocation facility, these fibres can be leased to communications service providers, which now compete on an equal basis in offering services to the user community. This model is viable because the price of fibre installation is shared amongst many parties, and benefits the customers by allowing full competition at the service level by smaller innovative communication companies or ISPs who cannot afford to build expensive physical network infrastructure. It can be thought as a 'third generation access network', in which access is over fibre and customers can choose their service provider. This contrasts with the 'second generation access network', in which there is just a single fibre access and service provider reaching each customer, and the 'first generation access network', in which the access wire is copper.

It may be better for government to provide economic incentives for the creation of such customer-owned, broadband access infrastructures, than to wait for an incumbent operator to install the infrastructure and have the regulator subsequently implement the unbundling process. It is not necessary that a community own the fibre or build the network itself. A community can encourage the deployment of condominium fibre networks in its jurisdiction by tendering its existing communications business only to those companies that will deploy such networks. Governments can lead by providing additional funding to make sure that all communities can enjoy the benefits of condominium fibre networks.

In the above, we have an example of *demand aggregation*, a strategy that can be used to facilitate the deployment of broadband services and infrastructures. When access is the expensive bottleneck, a single customer may not be able to justify the cost of broadband services if he has to provide the infrastructure completely on his own. However, costs can be reduced by aggregating customer demands and installing common infrastructure. Aggregation at the supply side has similar advantages. Service providers can aggregate infrastructure and so reduce costs to a level at which they can profit from selling broadband services. For optical networks, this sharing of infrastructure cost can occur at various levels: from the ducts, to fibreoptic cables, network transmission, transport services and even content and applications. It must be a political decision as to the level at which to promote aggregation, and with what incentives, taking account of local market factors. By aggregating, one obtains benefit from the aspects of the market that exhibit strong natural monopoly advantages on both the demand and supply side, and focuses competition where it is more socially efficient. The side effect is that regulation may be needed to set appropriate prices and supervise a fair sharing of the infrastructure. Low initial prices may be essential if there is to be rapid penetration of new services, especially broadband. By ensuring that there is high demand from the start, one obtains the large economies of scale required to reduce costs and make the new technologies affordable to consumers.

13.5 Further reading

For a more detailed mathematical introduction to the issues of asymmetric information, see Chapter 25 of Varian (1992). A good book on incentives and contracts is Macho-Stadler and Perez-Castillo (1997). The method of average price regulation is described by Vogelsang and Finsinger (1979). The total surplus subsidy method is due to Loeb and Magat (1979) and the incremental surplus mechanism is due to Sappington and Sibley (1998). A classic on the economics of regulation is the book of Kahn (1998). A theoretical treatment of regulation issues and policies can be found in Wolfstetter (1999).

Many regulators have excellent web sites. We recommend the sites of the UK and US regulators: Oftel (2002) and FCC (2002b). These contain well-presented material on topics such as the unbundling of the local loop, call termination, network interconnection, universal access, number portability, digital television, broadband access, spectrum allocation, satellite, technology standards, security, and media ownership.

An excellent reference for competition issues in the telecommunications sector, and specially the Telecommunications Act of 1996, is the web page of Economides (2002).

A good starting point for study of condominium fibre customer-owned networks is the web site of CA*net 3 (Canada's Research and Education Internet backbone), and specifically the faq on Dark Fibre, see CA*net (2002). For information on the Stockholm municipal fibre network see Stokab (2002), and for the Chicago CivicNet see CivicNet (2002).

14

Auctions

An auction is a sale in which the price of an item is determined by bidding. Flowers, wine, antiques, US treasury bonds and land are sold in auctions. Takeover battles for companies can be viewed as auctions (and indeed, the Roman empire was auctioned by the Praetorian Guards in A.D. 193). Auctions are commonly used to sell natural resources, such as oil drilling rights, or even the rights to use certain geostationary satellite positions. Government contracts are often awarded through procurement auctions. There is the advantage that the sale can be performed openly, so that no one can claim that a government official awarded the contract to the supplier who offers him the greatest bribe. In Section 9.4.4 we saw how instantaneous bandwidth might be sold in a smart market in which the price is set by auction. In recent years, auctions have been used in the communications market to sell parts of the spectrum for mobile telephone licenses. Some of these have raised huge sums for the government, but others have raised less than expected.

An auction can be viewed as a partial information game in which the valuations that each bidder places on the items for sale is hidden from the auctioneer and the other bidders. The game's equilibrium is a function of the auction's rules, which specify the way bidding occurs, the information bidders have about the state of bidding, how the winner is determined and how much he must pay. These rules can affect the revenue obtained by the seller, as well as how much this varies in successive instances of the auction. An auction is economically efficient, in terms of maximizing social welfare, if it allocates items to bidders who value them most. We emphasize that designing an auction for a particular situation is an art. There is no single auctioning mechanism that is provably efficient and can be applied in most situations. For example, in spectrum auctions some combinations of spectrum licenses are more valuable to bidders than others, and so licenses must be sold in packages, using some sort of combinatorial bidding. As we explain in Section 14.2.2, this greatly complicates auction design. One can prove important theoretical results about some simple auction mechanisms, (such as the revenue equivalence theorem of Section 14.1.3). They are not easily applied in many real life situations, but they do provide insights into the problems involved.

The purpose of this chapter is to provide the reader with an introduction to auction theory and some examples of how it can be used in pricing communications services. Auction theory is now a very well-developed area of research, and we can do no more than give an introduction and some interesting results. We have previously discussed how the mechanism of tatonnement can be used to maximize social welfare in resource allocation problems (Section 5.4.1). In tatonnement, price is varied in response to excess demand (positive or

---

Pricing Communication Networks   C. Courcoubetis and R. Weber
© 2003 John Wiley & Sons, Ltd   ISBN 0-470-85130-9 (HB)

negative) until demand exactly matches supply. One crucial property of any tatonnement mechanism is that prices should be able to increase or decrease until that point is reached. Auction mechanisms do not usually allow prices to fluctuate in both directions. Tatonnement can take a large number of steps. Some auctions take place in just one step, with little information exchange between the buyers and seller. In general, auctions are more restricted than tatonnement, and do not necessarily maximize social welfare. However, they have the advantage that they can be faster and simpler to implement. A second requirement for the tatonnement mechanism to work is that customers should make truthful declarations of their resource needs for given posted prices. This will happen if the market has many customers, with no customer being so large that he can affect the price by the size of his own demand. That is, customers are price takers. Auctions, however, can be efficient even when there are a small number of bidders, although the optimal strategy for some may be not to tell the truth.

There are two important and distinct models for the way bidders value items in an auction. In the *private value model*, each bidder knows the value that he places on a given item, but he does not know the valuations of other bidders. As bidding takes place, his valuation does not change, although he may gain information about other bidders' valuations when he hears their bids. In the *common value model*, all bidders estimate their valuation of the item in the same way, but they have different prior information about that value. Suppose, for example, a jar of coins is to be auctioned. Each bidder estimates the value of the coins in the jar, and as bidding occurs he adjusts his estimate on the basis of what others say. For example, if most bidders make higher bids than his own, a bidder might feel that he should increase his estimate of the value of the coins. In this case, the winner generally over-estimates the value (since he has the highest estimate), and so it is likely that he pays more than the jar of coins is worth. This is known as the *winner's curse* (about which we say more in Section 14.1.7). Sometimes a bidder's valuation is a function of both private information and of information revealed during the auction. For example, suppose an oil-lease is to be auctioned. The value of the lease depends both upon the amount of oil that is in the ground and the efficiency with which it can be extracted. Bidders may have different geological information about the likely amount of oil, and have different extraction efficiencies, and so make different estimates of the value of the lease. During bidding, bidders reveal information about their estimates and this may be helpful to other bidders.

There are many other considerations that come into play when designing auctions. The seller may impose a participation fee, or a minimum reserve price. An auction can be *oral* (bidders hear each other's bids and make counter-offers) or *written* (bidders submit closed sealed-bids in writing). In an oral auction, the number of bidders may be known, but in a sealed-bid auction the number is often unknown. Oral auctions proceed in a progressive manner, taking many rounds to complete, while sealed-bid auctions may take only a single round. All these things can influence the way bidders compete; by making them compete more fiercely, the seller's revenue is increased.

In Section 14.1 we describe some types of auction and summarize some important theoretical results. These concern auctions of a single item. However, one may wish to sell more than a single item. In a *multi-object auction*, multiple units of the same or of different items are to be sold. Such auctions can be *homogeneous* or *heterogeneous*, depending on the items to be sold are identical or not; *discriminatory* or *uniform price*, depending on whether identical items are sold at different or equal prices (this distinction only applies to homogeneous auctions); *individual* or *combinatorial*, depending on whether bids are allowed only for individual items or for combinations of items; *sequential* or *simultaneous*, depending on the whether items are auctioned one at a time or all at once. We take up these issues in Section 14.2. Note, however, that we opt for an informal presentation of the multi-object auction, as there are few rigorous results.

In summary, auctions are mechanisms for allocating resources in situations in which there is incomplete information and traditional market mechanisms do not provide incentives for participants truthfully to declare the missing information. Auction design takes account of this lack of information and can improve the equilibrium properties of the underlying games. We conclude the chapter in Section 14.3 by summarizing its ideas in the context of a highspeed link whose bandwidth is put up for sale by auction.

14.1 Single item auctions

14.1.1 Take it or leave it Pricing

In this section, we consider the sale of a single item by auction. For the purposes of comparison, we begin with analysis of a selling mechanism that is not an auction, but which could be used under the same conditions of incomplete information that pertain when auctions are used.

Suppose a seller wishes to sell a single item. He does this simply by making a take-it-or-leave-it offer, at price p. If any customers wants to buy the item at that price, then it is sold; otherwise it is not sold, and the seller obtains zero revenue. If more than one customer wants the item at the stated price, then there must be a procedure for deciding who gets it. However, the seller still receives revenue of p.

Suppose customers are identical and their private valuations are independent and identically distributed as a random variable X, with distribution function $F(x) = P(X \leq x)$. Given knowledge of this distribution, the seller wants to choose p to maximize his expected revenue. Let $x(p)$ denote the probability the item is sold. Then

$$x(p) = 1 - F(p)^n \tag{14.1}$$

Let $f(p) = F'(p)$ be the probability density function of X. By maximizing the expected revenue, of $px(p)$, we find that the optimal price p^* should satisfy

$$p - \frac{1 - F(p)^n}{nF(p)^{n-1}f(p)} = 0 \tag{14.2}$$

For example, if valuations are uniformly distributed on $[0, 1]$, then $F(x) = x$, and we find that the optimal price is $p^* = (n + 1)^{-1/n}$. The resulting expected revenue is $n(n + 1)^{-(n+1)/n}$. For $n = 2$, the optimal price is $p^* = \sqrt{1/3}$. The seller's expected revenue is $(2/3)\sqrt{1/3}$ ($= 0.3849$).

Note that, because there is a positive probability that the item is not sold, this method of selling is not economically efficient. We have seen this before in Chapter 6; if a monopolist seeks only to maximize his own revenue then there is often a social welfare loss. For the example above, the maximum valuation is the maximum of n uniform random variables distributed on $[0, 1]$; it is a standard result that this has expected value $n/(n + 1)$. This is the expected social welfare gain if the item is allocated to the bidder with the highest valuation. For $n = 2$, this is $2/3$ ($= 0.6666$). However, under take it or leave it pricing, the expected social welfare gain can shown to be $1 - p^* - (1 - p^*)^{(n+1)}/(n + 1)$. For $n = 2$, this is only 0.6094.

In all the above, we have assumed that the seller knows the distribution of the bidders' valuations. If he does not have this information, then he cannot determine the optimal 'take it or leave it' price. He also has a problem if his prior beliefs are mistaken. Suppose, for the example with $n = 2$, he believes that valuations are uniformly distributed on [0, 1] and sets the price optimally at $p^* = \sqrt{1/3}$. Say, however, he is mistaken: bidders valuations are actually uniformly distributed on [0, 0.5]. Then, as $p^* > 0.5$, he never sells. It would have been better if he had auctioned the item, thus ultimately selling it to the highest bidder.

Even if the seller does know the distribution of bidders' valuations, he can do better by auctioning. As we see below, one can design auction rules that increase the expected revenue and make auctioning the most profitable selling method. One way to do this is to introduce a minimum price that must be paid by the auction's winner. This *reserve price* has the effect of increasing the average price paid by the winner. In our example, he could set a reserve price of 1/2 and would obtain expected revenue of 0.4167 (see Section 14.1.4).

14.1.2 Types of Auction

We now describe some of the most popular types of auction. In the *ascending price auction* (or *English auction*), the auctioneer asks for increasing bids by raising the price of the item by small increments, until only one bidder remains. Or perhaps bidders place increasing bids by shouting. The item is awarded to the last remaining bidder, at the price of the last bid at which all other bidders had withdrawn. It is clear that in this type of auction the winner is the bidder with the highest valuation, and he pays a price equal to the second highest valuation. Unique items, such as artworks, tend to be sold in English auction, in order to find an unknown price. Another version of this auction is used in Japan; the price is displayed on a screen and raised continuously. Any bidder who wishes to remain active keeps his finger on a button. When he releases the button he quits the auction and cannot bid again.

In a reverse procedure to the English auction above, the *Dutch auction* starts by setting the price at some initial high value. A so-called 'Dutch clock' displays the price and continuously decreases it until some bidder decides to claim the item at the price displayed. Multiple items (such as fish or flowers) tend to be sold in Dutch auctions; this speeds up the time the sale takes. The price is lowered until demand matches supply.

In the next two types of auction, bidders submit sealed-bids and the one with the greatest bid wins. The auctions differ in the price charged to the winner. Under the *first-price sealed-bid auction*, the winner pays his bid. In this auction, the bidder has to decide off-line how much he should bid. This is equivalent to deciding off-line at what price he would claim the item in a Dutch auction, since in that auction no information is revealed until the first bid, at which point the auction also ends. Thus, we see that the Dutch auction and first-price sealed-bid auction are completely equivalent.

In the *second-price sealed-bid auction*, the winner pays the second highest bid. This is also known as a *Vickrey auction*, after its inventor. An important property of the Vickrey auction is that it is optimal for each bidder to bid his true valuation. To understand why this is so, note that a bidder would never wish to bid more than his valuation, since his expected net benefit would then be negative. However, if he reduces his bid below his valuation, he reduces the probability that he wins the auction, but he does not affect the price that he pays if he does win (which is determined by the second highest bidder). Thus, he does best by bidding his true valuation. The winner is the bidder with the greatest valuation and he pays the second greatest valuation. But this is exactly what happens in the English auction, in which a player drops out when the price exceeds by a small margin his valuation, and so the winner pays the valuation of the second-highest bidder. Thus, we see that the English and Vickrey auctions are equivalent.

Other auctions include the *all-pay auction*, in which all bidders pay their bid but the highest bidder wins the object, and the *k-price auction*, in which the winner pays the kth largest bid. Some of these auctions can be easily extended to multiple units. For example, in the two-unit first-price sealed bid auction the participants with the two greatest bids are winners and pay the third largest bid. Multi-unit auctions require bidders to follow much more complex strategies. We return to multi-unit auctions in Section 14.2.

14.1.3 Revenue Equivalence

A simple auction model for which we can give a full analysis is the *Symmetric Independent Private Values* (SIPV) model. It concerns the auction of a single item, in which both seller and bidders are risk neutral. To understand the idea of being risk neutral, imagine that a seller has a utility function that measures how he values the payment he receives. If his utility function is linear he is said to be *risk-neutral*. His average utility (after repeating the auction many times) is the same as his utility for the average payment, and hence the variability of the payment around its mean does not reduce the average utility of the seller. If the utility function is concave then the seller is *risk-averse*; now the average utility is less than the utility of the average payment, and this discrepancy increases with the variability of the payment.

Suppose each bidder knows his own valuation of the item, which he keeps secret, and valuations of the bidders can be modelled as independent and identically distributed random variables. Some important questions are as follows.

1. Which of the four standard auctions of the previous section generates the greatest expected revenue for the seller?

2. If the seller or the bidders are risk-averse, which auction would they prefer?

3. Which auctions make it harder for the bidders to collude?

4. Can we compare auctions with respect to strategic simplicity?

Let us begin with an intuitive, but important, result.

Lemma 1 In any SIPV auction in which (a) the bidders bid optimally, and (b) the item is awarded to the highest bidder, the order of the bids is the same as the order of the valuations.

Proof Suppose that under an optimal bidding strategy a bidder whose valuation is v bids so as to win with probability $p(v)$. Let $e(p)$ be the minimal expected payment that such a bidder can make if he wants to win the item with probability p. Assume v_1 and v_2 are such that $v_1 > v_2$, but $p(v_1) < p(v_2)$. If this is true, then it is simple algebra to show that, with $p_i = p(v_i)$,

$$[p_1 v_2 - e(p_1)] + [p_2 v_1 - e(p_2)] > [p_1 v_1 - e(p_1)] + [p_2 v_2 - e(p_2)]$$

Thus, either $p_1 v_2 - e(p_1) > p_2 v_2 - e(p_2)$, or $p_2 v_1 - e(p_2) > p_1 v_1 - e(p_1)$. In other words, either it is better to win with probability p_1 when the valuation v_2, or it is better to win with probability p_2 when the valuation is v_1, in contradiction to our assumptions. We are forced to conclude that $p(v)$ is nondecreasing in v. By assumption (b) in the lemma statement, this means that the optimal bid must be nondecreasing in v. ∎

We say that two auctions have the same *bidder participation* if any bidder who finds it profitable to participate in one auction also finds it profitable to participate in the other. The following is a remarkable result.

Theorem 4 (revenue equivalence theorem) The expected revenue obtained by the seller is the same for any two SIPV auctions that (a) award the item to the highest bidder, and (b) have the same bidder participation.

We say this is a remarkable result because different auctions can have completely different sets of rules and strategies. We might expect them to produce different revenues for the seller. Note that revenue equivalence is for the expectation of the revenue and not for its variance. Indeed, as we see in Section 14.1.5, auctions can have quite different properties so far as risk is concerned.

Proof of the revenue equivalence theorem Suppose there are n participating bidders. As above, let $e(p)$ denote the minimal expected payment that a bidder can make if he wants to win with probability p. The bidder's expected profit is $\pi(v) = pv - e(p)$, where $p = p(v)$ is chosen optimally and so, since π must be stationary with respect to any change in p, we must have $v - e'(p) = 0$. Hence,

$$\frac{d}{dv}e(p(v)) = e'(p)\frac{dp}{dv} = v\frac{dp}{dv}$$

Integrating this directly and then by parts gives

$$e(p(v)) = e(p(0)) + \int_0^v w\frac{dp(w)}{dw}\,dw = vp(v) - \int_0^v p(w)\,dw \qquad (14.3)$$

where clearly $e(p(0)) = e(0) = 0$, since there is no point in bidding for an item of value 0. Thus, $e(p(v))$, which is the expected amount paid by a bidder who values the item at v, depends only upon the function $p(\cdot)$. We know from Lemma 1 that if bidders bid optimally then bids will be in the same order as the valuations. It follows that if F is the distribution function of the valuations, then $p(w) = F(w)^{n-1}$, independently of the precise auction mechanism. The expected revenue can therefore be computed from (14.3) as $\sum_{i=1}^n E_{v_i} e(p(v_i)) = nE_v e(p(v))$. ∎

Notice that there is also 'expected net benefit equivalence' for the bidders. To see this, observe that the bidders obtain an expected net benefit that is equal to the expected value of the item to the winner of the auction, minus the expected total payment made to the seller. Since the expected value of both these quantities are independent of the auction rules, it follows that the expected net benefit of the bidders is also independent of the auction rules. Since bidders are symmetric they share this surplus equally.

It should be clear that all four auctions described in Section 14.1.2 satisfy the conditions of the revenue equivalence theorem. Let us work through an example in which the valuations, say $v_1, \ldots, v_n$, are random variables, independent and uniformly distributed on $[0, 1]$. Let $v_{(k)}$ denote the kth largest of $v_1, \ldots, v_n$ (the k-order statistic). A standard result is that $E[v_{(k)}] = k/(n + 1)$. Hence in the Vickrey and English auctions the expected revenue is $E[v_{(n-1)}] = (n - 1)/(n + 1)$.

Using this, we can find the optimal bid in the first-price sealed-bid auction. By the theorem the expected revenue in this auction is the same as in the English auction, i.e. $(n - 1)/(n + 1)$. Also, recall that $p(v) = F(v)^{n-1} = v^{n-1}$. Using (14.3), we easily find $e(p(v)) = (n - 1)v^n/n$. This must be $p(v)$ times the optimal bid. So a bidder who values the item at v has an optimal bid of $(n - 1)v/n$. This is a shaded bid, equal to the expected value of the second-highest valuation, given that v is the highest valuation.

14.1.4 Optimal Auctions

An important issue for the seller is to design the auction to maximize his revenue. We give revenue-maximizing auctions the name *optimal auctions*. It turns out that a seller who wants to run an optimal auction can increase his revenue by imposing a *reserve price* or a *participation fee*. This reduces the number of participants, but leads to fiercer competition and higher bids on the average, which may compensate for the probability that no sale takes place. Let us illustrate this with an example.

Example 14.1 (Revenue maximization) Consider a seller who wishes to maximize his revenue from the sale of an object. There are two potential buyers, with unknown valuations, v_1, v_2, that are independent and uniformly distributed on $[0, 1]$. He considers four ways of selling the object:

1. A take it or leave it offer.

2. A standard English auction.

3. An English auction with a participation fee c (which must be paid if a player chooses to submit a bid). Each bidder must choose whether or not to participate before knowing whether the other participates.

4. An English auction with a reserve price, p. The bidding starts with a minimum bid of p.

Case 1 was analysed in Section 14.1.1. The best 'take it or leave it' price is $p = \sqrt{1/3}$ and this gives an expected revenue of $(2/3)\sqrt{1/3}$ $(= 0.3849)$.

Case 2 was analysed above. The expected revenue in the English auction was $1/3$ $(= 0.3333)$.

Case 3. To analyse the auction with participation fee, note that a bidder will not wish to participate if his valuation is less than some amount, say v_0. A bidder whose valuation is exactly v_0 will be indifferent between participating or not. Hence $P(\text{winning} \mid v = v_0)v_0 = c$. Since a bidder with valuation v_0 wins only if the other bidder has a valuation less than v_0, we must have $P(\text{winning} \mid v = v_0) = v_0$, and hence $v_0^2 = c$. Thus, $v_0 = \sqrt{c}$.

To compute the expected revenue of the seller, we note that there are two ways that revenue can accrue to the seller. Either only one bidder participates and the sale price is zero, but the revenue is c. Or both bidders have valuation above v_0, in which case the revenue is $2c$ plus the sale price of $\min\{v_1, v_2\}$. The expected revenue is

$$2v_0(1 - v_0)c + (1 - v_0)^2[2c + v_0 + (1 - v_0)/3]$$

Straightforward calculations show that this is maximized for $c = 1/4$, and takes the value $5/12$ $(= 0.4167)$.

Case 4. In the English auction with a reserve price p, there is no sale with probability p^2. The revenue is p with probability $2p(1 - p)$. If $\min\{v_1, v_2\} > p$, then the sale price is $\min\{v_1, v_2\}$. The expected revenue is

$$2p^2(1 - p) + \left(\tfrac{1}{3} + \tfrac{2}{3}p\right)(1 - p)^2$$

This is maximized by $p = 1/2$ and the expected revenue is again $5/12$, exactly the same as in case 3.

That Cases 3 and 4 in the above example give the same expected revenue is not a coincidence. These are similar auctions, in that a bidder participates if and only if his valuation exceeds $1/2$. Let us consider more generally an auction in which a bidder participates only if his valuation exceeds some v_0. Suppose that with valuation v it is optimal to bid so as to win with probability $p(v)$, and the expected payment is then $e(p(v))$. By a simple generalization of (14.3), we have

$$e(p(v)) = e(p(v_0)) + \int_{v_0}^{v} w \frac{dp(w)}{dw} \, dw = vp(v) - \int_{v_0}^{v} p(w) \, dw$$

Assuming the SIPV model, this shows that a bidder's expected payment depends on the auction mechanism only through the value of v_0 that it implies. The seller's expected revenue is

$$\begin{aligned} nE_v[e(p(v))] &= n \int_{v=v_0}^{\infty} \left[vp(v) - \int_{w=v_0}^{v} p(w) \, dw \right] f(v) \, dv \\ &= n \int_{v=v_0}^{\infty} vp(v) f(v) \, dv - n \int_{w=v_0}^{\infty} \int_{v=w}^{\infty} p(w) f(v) \, dw \, dv \\ &= n \int_{v=v_0}^{\infty} \left\{ vf(v) - [1 - F(v)] \right\} F(v)^{n-1} \, dv \end{aligned}$$

Now differentiating with respect to v_0, to find the stationary point, we see that the above is maximized where

$$v_0 f(v_0) - [1 - F(v_0)] = 0$$

We call v_0 the *optimal reservation price*. Note that it does not depend upon the number of bidders. For example, if valuations are uniformly distributed on $[0, 1]$, then $v_0 = 1/2$. This is consistent with the answers found for Cases 3 and 4 of Example 14.1.

If bidders' valuations are independent, but heterogenous in their distributions, then one can proceed similarly. Let $p_i(v)$ be the probability that bidder i wins when his valuation is v. Let $e_i(p)$ be the minimum expected amount he can pay if he wants to win with probability p. Suppose that bidder i does not participate if his valuation is less than v_{0i}. Just as above, one can show that the seller's expected revenue is

$$\sum_{i=1}^{n} E_{v_i} e_i(p_i(v_i)) = \sum_{i=1}^{n} \int_{v=v_{0i}}^{\infty} \left[v - \frac{1 - F_i(v)}{f_i(v)} \right] f_i(v) p_i(v) \, dv \qquad (14.4)$$

The term in square brackets can be interpreted as 'marginal revenue', in the sense that if a price p is offered to bidder i, he will accept it with probability $x_i(p) = 1 - F_i(p)$, and so the expected revenue obtained by this offer is $px_i(p)$. Differentiating this with respect to x_i, we define

$$MR_i(p) = \frac{d}{dx_i}(px_i(p)) = \frac{d}{dp}(px_i(p)) \Big/ \frac{dx_i}{dp} = p - \frac{1 - F_i(p)}{f_i(p)}$$

Note that the right-hand side of (14.4) is simply $E[MR_{i*}(v_{i*})]$, where i^* is the winner of the auction. This can be maximized simply by ensuring that the object is always awarded to the bidder with the greatest marginal revenue, provided that marginal revenue is positive. We can do this provided bidders reveal their true valuations. Let us assume that

$MR_i(p)$ is increasing in p, for all i. Clearly, v_{0i} should be the least v such that $MR_i(v)$ is nonnegative. Consider the auction rule that always awards the item to the bidder with the greatest marginal revenue, and then asks him to pay the maximum of v_{0i} and the smallest v for which he would still remain the bidder with greatest marginal revenue. This has the character of a second-price auction in which the bidder's bid does not affect his payment, given that he wins. So bidders will bid their true valuations and (14.4) will be maximized.

Example 14.2 (Optimal auctions) An interesting property of optimal auctions with heterogeneous bidders is that the winner is not always the highest bidder.

Consider first the case of homogeneous bidders with valuations uniformly distributed on $[0, 1]$. In this case, $MR_i(v_i) = v_i - (1 - v_i)/1 = 2v_i - 1$. Hence the object is sold to the highest bidder, but only if $2v_i - 1 > 0$, i.e. if his valuation exceeds $1/2$. The winner pays either $1/2$ or the second greatest bid, whichever is greatest. In the case of two bidders, with the seller's expected revenue is $5/8$. This agrees with what we have found previously.

Now consider the case of two heterogeneous bidders, say A and B, whose valuations are uniformly distributed on $[0, 1]$ and $[0, 2]$, respectively. So $MR_A(v_A) = 2v_A - 1$, and $MR_B(v_B) = 2v_B - 2$. Under the bidding rules described above, bidder B wins only if $2v_B - 2 > 2v_A - 1$ and $2v_B - 2 > 0$, i.e. if and only if $v_B - v_A > 1/2$ and $v_B > 1$; so the lower bidder can sometimes win. For example, if $v_A = 0.8$ and $v_B = 1.2$, then A wins and pays 0.7 (which is the smallest v such that $MR_A(v) = 2v - 1 \geq 2v_B - 2 = 0.4$).

14.1.5 Risk Aversion

As we have already mentioned, the participants in an auction can have different attitudes to risk. If a participant's utility function is linear then he is said to be *risk-neutral*. If his utility function is concave then he is *risk-averse*; now a seller's average utility is less than the utility of his average revenue, and this discrepancy increases with the variability of the revenue. Hence a risk-averse seller, depending on his degree of risk-aversion, might choose an auction that substantially reduces the variance of his revenue, even though this might reduce his average revenue.

The revenue equivalence theorem holds under the assumption that bidders are risk-neutral. One can easily see that if bidders are risk-averse, then first-price sealed-bid and Dutch auctions give different results from second-price sealed-bid and English auctions. For example, in a first-price auction, a risk-averse bidder prefers to win more frequently even if his average net benefit is less. Hence, he will make higher bids than if he were risk-neutral. This reduces his expected net benefit and increases the expected revenue of the seller. If the same bidder participates in a second-price auction, then his bids do not affect what he pays when he wins, and so his strategy must be to bid his true valuation. Hence, a first-price auction amongst risk-averse bidders produces a greater expected revenue for the seller than does a second-price auction. However, it is not clear which type of auction the risk-averse bidders would prefer. In general, this type of question is very difficult.

The seller may also be risk-averse. In such a case, he prefers amongst auctions with the same expected revenue those with a smaller variance in the sale price. Let us compare a first and second-price auction with respect to this variance. Suppose bidders are risk-neutral. Let $v_{(n)}$ and $v_{(n-1)}$ be the greatest and second-greatest valuations. In a second-price auction, the winner pays the value of the runner-up's bid, i.e. $v_{(n-1)}$. In a first-price auction he pays his bid, which is the conditional expectation of the valuation of the runner-up, conditioned on his winning the auction, i.e. $E(v_{(n-1)}|v_{(n)})$. Let $Y = (v_{(n-1)}|v_{(n)})$ and apply the standard fact that $(EY)^2 \leq EY^2$. This gives

$$E_{v_{(n)}} \left[E_{v_{(n-1)}} (v_{(n-1)}|v_{(n)})^2 \right] \leq E_{v_{(n)}} \left[E_{v_{(n-1)}} (v_{(n-1)}^2|v_{(n)}) \right] = E v_{(n-1)}^2$$

Subtracting from both sides the square of the expected value of the winner's bid, i.e. $E(v_{(n-1)})^2$, we see that the winner's bid has a smaller variance in the first-price auction, and so a risk-averse seller would prefer a first-price auction. Let us verify this for two bidders whose valuations are uniformly distributed on [0, 1]. In the first-price auction, each bidder bids half his valuation, so the revenue is $(1/2) \max\{v_1, v_2\}$. In the second-price auction each bids his valuation and the revenue is $\min\{v_1, v_2\}$. Both have expectation 1/3, but the variances are 1/72 and 1/18, respectively. Thus, a risk-averse seller prefers the first-price auction.

14.1.6 Collusion

It is important when running an auction to take steps to prevent bidders from colluding. Collusion occurs when two or more bidders make arrangements not to bid as high as their valuations suggest, and so reduce the seller's revenue. Antique auctions are notorious for this. A number of bidders form a 'ring' and agree not to bid against one another and on whom the winner will be. This lowers the winning bid. Later, the winner distributes his gain amongst all the bidders, in proportion to their market power, so that all do better than they would have done by not colluding. In some spectrum auctions in the US, there have been instances of bidders using the final four digits of their multimillion dollar bids to signal to one another the licenses they want to buy. Thus, a critical characteristic of an auction is how susceptible it is to collusion. This depends upon what incentives there are for players to stand by the promises they make to one another when agreeing to collude.

We can see that an ascending English auction is susceptible to collusion. Suppose the bidders meet and determine that bidder 1 has the greatest valuation. They agree that bidder 1 should make a low bid and win the object for a payment close to zero. No other bidder has an incentive to bid against bidder 1, since he cannot win without ultimately outbidding bidder 1; yet if he does so he would incur a loss. Thus, the agreement between the bidders is 'self-enforcing' and the auction is susceptible to collusion.

In contrast, collusion is difficult in a Dutch auction, or in a first-price sealed-bid auction. There is nothing to stop a ring member bidding higher than was agreed. His defecting action becomes obvious, but the auction is over before anyone can react. This is one reason why first-price sealed-bid auctions are often preferred when auctioning large government contracts.

There is also a matter of trusting the seller. He might want to manipulate the auction to raise prices. One way he can do this is by soliciting fake bids. In a first-price sealed-bid auction, such bids do not make any sense, since they could prevent the sale of the object (and the seller could anyway use a reserve price). In a second-price auction, fake bids could benefit the seller. If the seller has approximate knowledge of the highest bidder's valuation, he could solicit a 'phantom' bid with a slightly smaller value, and hence obtain almost all the surplus of the bidder.

14.1.7 The Winner's Curse

Thus far we have discussed the private values model. In the common values model, i.e. where the item that is auctioned has a common unknown value, the winner is the bidder who has the most optimistic estimate of the item's value. If bidders' estimates are unbiased, then the highest estimate will be likely to exceed the item's actual value, and the winner will suffer a loss. To remedy this, a bidder should shade his bid to allow for the fact that if he wins, he has the highest estimate. He should find the item's expected value conditional on his initial estimate being the highest among all initial estimates of the other bidders, i.e. conditional on his being the winner.

To illustrate this, suppose that the item has a random value V. Each bidder receives a signal s_i that is an estimate of the value of the item. These signals are independent and uniformly distributed on $[V - \epsilon, V + \epsilon]$. Since $E[V|s_i] = s_i$, a straightforward approach is for bidder i to bid s_i. But he will suffer the winner's curse. To remedy this, bidder i must assume that he wins the auction because his estimate is the highest and correct estimate for the value of the item. One can show that

$$E[V|s_i = \max\{s_1, \ldots, s_n\}] = s_i - \epsilon \frac{n-1}{n+1} < E[V|s_i] \tag{14.5}$$

This should inform any bid he makes. Note that as n becomes very large, this estimate converges to $s_i - \epsilon$, which is the most conservative estimate he could make. The impact of the winner's curse can be surprising. As in this example, when there are more bidders, they must bid more conservatively, because the effect of the winner's curse is greater. This effect might more than make up for the increase in competition due to there being more bidders, and so the expected sale price might actually decrease!

14.1.8   Other Issues

We have mentioned above the issue of strategic simplicity. A strong argument in favour of the second-price sealed-bid auction is that each bidder's strategy is simple: he just bids his valuation. In contrast, the bidder in a first-price sealed-bid auction must estimate the second-highest valuation amongst his competitors, given that his valuation is greatest.

It is interesting to ask whether it is advantageous for the seller to disclose the number of bidders. It can be proved that the first-price sealed-bid auction results in more aggressive bidding when the number of bidders is unknown, and so the seller may prefer this to be the case.

In the SIVP model we assumed bidders are identical. If this is not so, then things can become very complicated. Suppose there are two bidders, say A and B, with valuations distributed uniformly on $[0, 1]$ and $[1, 2]$, respectively. In a second-price auction both will bid their valuations and B will always win, paying A's valuation. However, in a first-price auction A will bid very near his valuation, but B will shade his bid substantially under his valuation, since he knows A's bids are much lower than his. Now there is a positive probability that A wins. Note that the outcome can be inefficient, in the sense that the object may not be sold to the bidder who values it most. Also, since the item will sometimes sell for more than A's valuation the seller's expected revenue is greater than in the second-price auction.

Suppose the distributions of the bidders' valuations are correlated, rather than being independent. This is sometimes called *affiliation*. The effects of affiliation are complex to analyse precisely, but we can give some intuition. In the presence of affiliation, it turns out that ascending auctions lead to greater expected prices than second-price sealed-bid auctions, and these lead to greater expected prices than first-price sealed-bid auctions. An intuitive way to see this is as follows. A player's profit when he is the winner arises from his private information (his 'information rent'). The less crucial is this information advantage, the less profit the player can make. In the case of the ascending auction, the sale price depends upon all other bidders' information, and because of affiliation, it captures a large part of the winner's information. In the second-price auction, the price depends upon just one other bidder's information, and hence it captures less information about the winner's information. Finally, the price in the first-price auction is determined by the winner's strategy which takes full advantage of his information rent. Hence, the profit of the winner must be higher in the first-price auction than in the second-price auction, and the ascending auction has least profit. The seller's revenues under the three auctions are ordered in the reverse direction.

This discussion motivates the 'Linkage Principle': the more the price can be linked to information that is affiliated with the winner, the less is the value of the information rent of the winner and the less his net profit. Thus, if the seller has access to any information that reduces the value of the winner's private information, it is to his advantage to reveal it to the other bidders.

14.2 Multi-object auctions

14.2.1 Multi-unit Auctions

We now turn to auctions of multiple objects. These auctions can be homogeneous or heterogeneous. In a *homogeneous auction* a number of identical units of a good are to be auctioned, and we speak of a *multi-unit auction*. Multi-unit auctions are of great practical importance, and have been applied to selling units of bandwidth in computer networks and satellite links, MWs of electric power, capacity of natural gas and oil pipelines. In the simplest multi-unit auction, each buyer wants only one unit. The auction mechanisms we have already described can be generalized. For example, in a *simultaneous auction* of k units, all bidders could make closed sealed-bids, and the k objects could be awarded to the k highest bidders. In a first-price auction each bidder would pay his own bid. In a generalization of the Vickrey auction, the k highest bidders would pay the value of the highest losing bid. It can be shown that the revenue-equivalence theorem still holds for these auctions.

Note that, in the first-price auction, the successful bidders pay differently for the same thing; we call this is a *discriminatory auction*. By contrast, the Vickrey auction is called a *uniform auction*, because all successful bidders pay the same. A uniform auction is intuitively fairer, and also more likely to reduce the winner's curse.

Things are more complex when each bidder may want to buy more than one object, or if objects are not the same. The particular auction rules can significantly affect the seller's revenue and the buyer's profits. Consider first the case where objects are identical. We could run a sealed-bid uniform auction by asking each bidder to submit his whole demand function, as a set of values (p, q), where q is the number of units he would purchase at price p. The auctioneer would set the 'stop-out' price as the price at which total demand equals supply. Unfortunately, this type of auction suffers from a *demand reduction problem*: since the stop-out price increases with the total demand, there is an incentive for each bidder to misrepresent his demand function, demanding at every price less than his true demand curve dictates. This reduces the stop-out price. Although each bidder gets fewer units, he gets them at a much lower price, and so with greater total surplus. As each bidder gets fewer units there will be an inefficient allocation of the units amongst the bidders. Discriminatory auctions are less susceptible to such strategic bidding. However, since bidders know that they will pay their bids, they still tend to shade them to obtain some surplus.

Another way to run a uniform auction is with an ascending price clock. The auctioneer gradually raises the price, p, and each bidder states the quantity that he desires at price p. The auction terminates when demand matches supply. Compared to the sealed-bid auction, the dynamic revelation of information that takes place in this auction promotes competition and aggressive bidding. However, if competition is very low, bidders can reach tacit collusion. Neither type of auction is incentive compatible. It can be proved that there is no individually rational, uniform-price auction with multi-unit allocations that is incentive compatible, i.e. such that each bidder's optimal strategy is to bid his true valuation.

An alternative to the simultaneous auction, is a *sequential auction*, in which objects are sold one after the other in separate auctions. When auctions of this type have been used to sell lots of wine and timber, it has been observed that prices tend to decline in the later auctions. One explanation for this is that, since fewer bidders participate in the later sales, the competition declines. However, one could also argue that because bidders know they can win in later auctions, they may be willing to take greater risks in earlier auctions, bidding less in those auctions than in later ones.

14.2.2 Combinatorial Bidding

The objects to be sold in a multi-object auction may be different, and complementary. For example, the value of holding two cable television licenses geographically contiguous can be greater than the sum of their values if held alone. This 'synergistic effect' provides another reason that prices might rise in later sales of a sequential auction: competition can intensify in later sales, because some bidders have already bought licenses that are complementary to those that are sold later.

The possibility of complementarities between objects means that it can be advantageous to allow combinatorial (or 'package') bidding. In combinatorial bidding, bidders may place bids on groups of objects as well as on individual objects. A generalization of the Vickrey auction that can be used with combinatorial bidding is the Vickrey–Clarke–Groves (VCG) mechanism (which we have also met in Section 11.5.2). Each bidder submits bids for any combinations of objects that he wishes. The auctioneer allocates the objects to maximize the aggregate total of their values to the bidders. Each bidder who wins a subset of the objects pays the 'opportunity cost' that this imposes on the rest of the bidders.

More specifically, let L be the set of objects and let P be the set of their possible assignments amongst the bidders. Each bidder submits a bid that specifies a value $v_i(T)$ for each non-empty subset T of L. An assignment $S \in P$ is a partition of L into subsets S_i, with one such subset per bidder i (possibly empty). If social welfare maximization is the objective, then the auctioneer chooses the partition $S^* = \{S_1^*, \ldots, S_n^*\}$ that maximizes $\sum_{i=1}^{n} v_i(S_i^*)$. Each bidder i pays an amount p_i, where

$$p_i = \max_{S \in P} \sum_{j \neq i} v_j(S_j) - \sum_{j \neq i} v_j(S_j^*) \tag{14.6}$$

The first term on the right of (14.6) is the greatest value that could be obtained by the other bidders if i were not bidding. The final term is the value that is obtained by the other bidders when bidder i does participate, and so influences the optimal allocation and takes some value for himself.

This type of auction is incentive compatible, in the sense that each bidder can do no better than submit his true valuations, and it leads to an economically efficient allocation of the objects. It is clearly more difficult to implement than a sequential auction, but it has the advantage that the whole market is available to bidders and they can freely express their preferences for substitutable or complementary goods. However, there are drawbacks.

First, the complex mechanism of a VCG auction can be hard for bidders to understand. It is not intuitive and bidders may well not follow the proper strategy. Secondly, it is very hard to implement. This is because each bidder must submit an extremely large number of bids, and the auctioneer must solve a *NP*-complete optimization problem to determine the optimal partition. No fast (polynomial-time) solution algorithm is available for *NP*-complete problems, so the 'winner determination' problem can be unrealistically difficult to solve. There are several ways that bidding can be restricted so that the optimal partitioning problem becomes a tractable optimization problem (i.e. one solvable in polynomial time). Unfortunately, these restrictions are rather strong, and are not applicable in many cases of practical interest. One possibility is to move the responsibility for solving the winner determination problem from the seller to the bidders. Following a round of bidding, the bidders are challenged to find allocations that maximize the social welfare.

14.2.3 Double Auctions

Another interesting type of multi-unit auction is the *double auction*. In this auction, there are multiple bidders and sellers. The bidders and sellers are treated symmetrically and participate by bidding prices (called 'offers' and 'asks') at which they are prepared to buy and sell. These bids are matched in the market and market-clearing prices are generated by some rule. The double auction is one of the most common trading mechanisms and is used extensively in the stock and commodity exchanges.

In an asynchronous double auction, also called a *Continuous Double Auction* (CDA), the offers to buy and sell may be submitted or retracted at any time. A public order book lists, at each time t, the currently highest buy offer, $b(t)$, and currently lowest sell offer, $s(t)$. As soon as $b(t) \geq s(t)$, a sale takes place, and the values of $b(t)$ and $s(t)$ are updated. Today's stock exchanges usually work with CDAs, and they have also been used for auctions conducted on the Internet.

In a synchronized double auction, all participants submit their bids in lock-step and batches of bids are cleared at the end of each period. Most well-known double auction clearing mechanisms make use of a generalization of Vickrey–Clarke–Groves mechanism. For example, suppose that there are m sell offers, $s_1 \leq s_2 \leq \cdots \leq s_m$ and n buy offers, $b_1 \geq b_2 \geq \cdots \geq b_n$. Then the number of units that can be traded is the number k such that $s_k \leq b_k$, but $s_{k+1} > b_{k+1}$.

The specification of the buy and sell prices are a bit complicated. However, it is interesting to study them, to see once more how widely useful is the VCG mechanism. We suppose that the market maker receives all the offers and asks, and then computes k, as above, and a single price p_b to be paid by each buyer and a single price p_s to be received by each seller. In general, $p_b > p_s$. The buyer price is $p_b = \max\{s_k, b_{k+1}\}$. Thus, p_b is the best unsuccessful offer, as long as this is more than the greatest successful ask; otherwise, it is the greatest successful ask. To see that p_b is indeed an implementation of the VCG mechanism, assume that all participants bid their valuations. Let V be the sum of the valuations placed on the items by those who hold them at the end of the auction. Thus $V = \sum_i s_i + \sum_{i \leq k}(b_i - s_i)$. For any i, let $V(i)$ be defined as V, but excluding the valuation placed by i on any item that he holds at the end of the auction. Suppose i is a successful bidder. If i did not participate and $s_k < b_{k+1}$, then the best unsuccessful bidder becomes successful and obtains value b_{k+1}; so $V(i)$ increases by b_{k+1}. However, if $s_k > b_{k+1}$, then the best unsuccessful bidder has not bid more than s_k and so seller k retains the item for which his valuation is s_k and which he would have sold to buyer i; thus $V(i)$ increases by s_k. Thus, p_b is indeed the reduction due to buyer i's participation in the sum over all other participants of the valuations they place on the items that they hold after the auction concludes. Similarly, each successful seller is to receive the same amount of money from the market maker, namely $p_s = \min\{b_k, s_{k+1}\}$. One can make a similar analysis for p_s, and also check that under these rules it is optimal for each participant to bid his true valuation. Note, however, that this auction has the 'problem' that $p_b > p_s$, so its working necessitates that the market maker make a profit!

Other double auctions are generalizations of the auctions described in Section 14.1.2. The 'Double Dutch auction' uses two clocks. The buyer price clock starts at a very high price and decreases until some buyer stops the clock to indicate his willingness to buy at that price. Now the seller price clock starts from a very low price and begins to increase, until stopped by a seller who indicates his willingness to sell at that price. At this point, one pair of buyer and seller are locked in. The buyer price clock continues to decrease again, until stopped by a buyer, then the seller price clock increases, and so on. The auction is over when the two prices cross. Once this happens, all locked-in participants buy or sell one item at the crossover point. Note that some items may not be sold.

The 'Double English auction' is similar and also uses two clocks. The difference is that the seller clock is initially set high and the buyer clock is initially set low. The maximum quantities that buyers and sellers would be willing to buy or sell at these prices are privately submitted and then revealed to all, say $x(p_1)$ and $y(p_2)$, respectively. If there is an excess demand, $x > y$, then p_1 is gradually increased until $x(p_1') = y(p_2) - 1$. Similarly, if $y > x$, then p_2 is gradually decreased until $y(p_2') = x(p_1) - 1$. This continues, the clocks being alternately modified. The price at which the clocks eventually cross defines the clearing price. There may be a small difference between supply and demand at the clearing price, but this difference is probably negligible and can be resolved arbitrarily.

The 'Dutch English auction' uses one clock, which is initially set at a high price and made to gradually decrease with time. From the buyer's viewpoint the clock is Dutch, while from the seller's viewpoint it is English. As in the Double English auction, the auction ends when the revealed supply and demand match, and the market price is set to the price shown on the clock. Research indicates that the Double Dutch and Dutch English auctions perform extremely well in terms of efficiency under a variety of market conditions.

14.2.4   The Simultaneous Ascending Auction

One type of multi-unit auction that has been extensively analysed is the *Simultaneous Ascending Auction* (SAA). This is a type of auction for selling heterogeneous objects that was developed for the FCC's sale of radio spectrum licenses in the US in 1994. In that auction, 99 licenses were sold for a total of about $7 billion. More recently, in 2000, the UK government sold five third-generation mobile phone licenses for $34 billion. One rationale for choosing an ascending auction over a sealed-bid auction is that, because bidders gradually reveal information as the auction takes place, it should be less susceptible to the winner's curse.

In general, the SAA is considered efficient, revenue maximizing, fair and transparent. However, in cases of low competition it can produce poor revenue. An analysis of this type of auction is very interesting and points up the many issues of complexity, gaming and auction design that are relevant when trying to auction heterogeneous objects to bidders that have different valuations for differing combinations of objects. Issues of complementarity and substitution between objects are important and affect bidding strategies.

Let us briefly describe the rules for a simultaneous ascending auction . Bidding occurs in rounds. It continues as long as there is bidding on at least one of the objects (hence the name simultaneous). In each round, the bidders make sealed-bids for all the objects in which they are interested. The auctioneer reads the bids and posts the results for the round. For each object, he states the identity of the highest bidder and his bid. As the auction progresses, the new highest bid for each object is computed, as the maximum of the previous highest bid and any new bids that occur during the round. In each round, a minimum bid is required for each object, which is equal to the previous highest bid incremented by a predetermined small value. There are rules about whether bidders may withdraw bids, and 'activity' rules, which control bidders' participation by restricting the percentage of objects that a single bidder can bid upon, or possibly win, and which also provide incentives for bidders to be active in early rounds rather than delaying their bidding to later rounds. There are more details, but we omit these as they are not relevant to the issues we emphasize. Note that although a SAA can be modified to allow combinatorial bidding, in its most basic form this is not allowed. We have just a set of individual auctions taking place simultaneously.

14.2.5   Some Issues for Multi-object Auctions

Inefficient allocation

We look now at some issues and problems for multi-object auctions. Ideally, an auction should conclude with the objects being allocated to bidders efficiently, i.e. in a way that maximizes the social welfare. Clearly, this happens in any single-object SIVP auction, since the object is sold to the bidder who values it most. In a single-object SIVP auction, efficiency and revenue maximization are not in conflict. However, in multi-unit auctions they are. The seller has the incentive to misallocate the units to maximize revenue, thus ruining efficiency. As we have seen above, even when revenue maximization is not a consideration, a uniform payment rule can lead to demand reduction. A pay-your-bid rule can result in differential bid shading. In both cases, social welfare is not maximized.

Efficiency is obtained in a multi-object auction if the prices that are determined by the high bid for each of the objects are such that each object is demanded by just one bidder, and the induced allocation of objects to bidders maximizes the sum of the bidders' valuations for the combinations of the objects they receive. It can be proved that this happens if all objects are substitutes for every bidder. However, as we now illustrate, things can be very different if there is even one bidder for whom some objects are complements.

Consider a sale of spectrum licenses in which a pair of licenses for contiguous geographic regions are complementary, i.e. they are more valuable taken together than the sum of their valuations if held alone. Suppose two bidders, called 1 and 2, bid for licenses A and B. For bidder 1 the licenses are complements: he values them at 1 and 2 on their own, but values them at 6 if they are held together. For bidder 2 the licenses are substitutes; he values them individually at 3 and 4, but only at 5 if held together (Table 14.1).

Social welfare is maximized if bidder 1 gets both licenses. However, if bidder 2 is not to purchase either A or B, the high bid for A must be at least 3, and for B at least 4. However, at such prices bidder 1 would not want either license on its own, or both licenses together.

A related problem is the so-called *exposure problem*. A bidder who wants to acquire two objects, which are together valuable to him because of a complementarity effect, is exposed to the possibility of winning just one object at a price higher than he values this object when held alone. In the above example, suppose that prices for both licenses are raised continuously with an increment of ϵ until the prices for A and B are $p_A = 1$, $p_B = 2$. Up

Table 14.1 The socially optimal allocation is to award AB to bidder 1. Note that A and B are substitutes for bidder 2, but are complements for bidder 1. There are no prices at which the socially optimal allocation is obtained

Bidder	v_A	v_B	v_{AB}
1	1	2	6
2	3	4	5

to this point, both bidders remain in the game. Prices now become $p_A = 1 + \epsilon$, $p_B = 2 + \epsilon$. Should bidder 1 participate? If he does, he takes a chance, since if he ends up being the winner of just one of the licenses he incurs a loss. In fact, this is what will happen, since bidder 2 will outbid bidder 1 on at least one license. Hence in this auction the outcome is inefficient, in that prices cannot reflect the fact that the licenses are complements for bidder 1 and that he is willing to pay for that value. A possible solution to this problem is to allow combinatorial bidding, though that has its own problems.

Incentives to delay bidding

If a competitor has a budget constraint, a bidder may wish to delay his bidding until his competitor has committed most of his budget to some objects; then he can safely bid for other objects without committing any of his own budget to objects he will not obtain anyway. Since the sum of a bidder's outstanding bids can never exceed his budget, it is crucial that he allocate his effort in winning situations.

To illustrate this, suppose there are three bidders, each with a budget 20, and valuations for objects as shown in Table 14.2. Bidder 1 wants to maximize his net benefit, i.e. his valuations minus the amount he pays. His strategy depends on bidder 3's valuation for B. If it is known beforehand to be 5, then the optimal strategy for bidder 1 is to bid for both objects, and since he will win them, paying $10 + 5$, and making 30 units profit. If bidder 3's valuation is known to be 15, then bidder 1's budget constraint means he will not be able to win both objects. He should concentrate on winning B, from which he can make 15 units of profit by paying 15, and abstain from bidding for A. The danger is that if during the bidding for A he allocates more than 5 units of budget, then he cannot then win B.

Table 14.2 Bidders 1 and 2 know that bidder 3 values B at 5 or 15, with probabilities 0.9 and 0.1 respectively. This partial information about bidder's 3 valuation makes delayed bidding advantageous for bidders 1 and 2, while they wait to see how bidder 3 bids

Bidder	v_A	v_B	Budget
1	15	30	20
2	10	0	20
3	0	5 w.p. 0.9 15 w.p. 0.1	20

If bidders 3's valuation of B is not known, then the optimal strategy for bidder 1 is to bid on B, but delay bidding on A until he learns the bidder 3's valuation. Then, if enough budget remains, he bids for A. Unfortunately, the optimal strategy of bidder 3, if his valuation for B is 15, is also to delay his bidding until bidder 1 commits a large part of his budget to bidding for A, since then bidder 3 can safely win B. Hence both players choose to delay their bidding, in which case there is no equilibrium strategy. For this reason, it is usual to introduce 'activity rules', which force bidders to bid if they wish to remain in competition for winning objects.

The free rider problem

We have been assuming that bidders are only bidding for single objects, albeit simultaneously. If combinatorial bids are allowed then further problems can arise. Suppose that at the end of the auction, the winning bids are chosen to be non-overlapping and maximizing of the seller's total revenue. This can lead to the so-called 'free rider' problem described in Section 6.4.1. Basically, to displace a combination bid, it is enough for a single bidder to increase his bid on a single object in the combination, say A. By doing so, he ends up winning A, but at a higher price than he would have paid if someone else had played the 'altruistic' role of making a bid to displace the combination bid, say by having raised the bid on some other object, say B. As a result, an equilibrium can occur in which no one displaces the combination bid (because everyone hopes someone else will do it), and the bidder for the combination wins, even though his valuation is less than the sum of the valuations of the other single bidders.

We can see this in the example, with valuations shown in Table 14.3. Here, bidder 1 has valuations $v_A = 4$, $v_B = 0$, and bidder 2 has valuations $v_A = 0$, $v_b = 4$. Both bidders have a budget of 3 units. Bidder 3 has $v_A = v_B = 1 + \epsilon$, $v_{AB} = 2 + \epsilon$, and budget 2.

Bids must be in integers. Suppose that in first round, bidder 1 bids 1 for A, bidder 2 bids 1 for B, and bidder 3 bids 2 for AB (and nothing for A or B). It has been announced that if no further bids are received then AB will be awarded to bidder 3. In this circumstance, bidder 1 prefers to wait until bidder 2 raises his bid for B to 2, after which bidder 3 cannot profit by bidding any further: A and B will be awarded to bidders 1 and 2, respectively. By exactly similar reasoning, bidder 2 prefers to wait for bidder 1 to raise his bid for A to 2. Hence, both may decide not to raise their bids in the next round, and bidder 3 is awarded AB, even though this is not socially optimal. In fact, at the end of the first round, the payoff matrix for bidders 1 and 2 (as row and column players, respectively) in the subgame is as shown in Table 14.4.

Table 14.3 The free rider problem. Suppose bidders 1 and 2 have bid 1 for A and B, respectively. But the combination AB will be awarded to bidder 3, who has bid for this 2, unless bidders 1 or 2 bid more. Bidder 1 prefers to wait for bidder 2 to make a bid of 2 for B. But bidder 2 prefers to wait for bidder 1 to make a bid of 1 for A

Bidder	v_A	v_B	v_{AB}	Budget
1	4	0	0	3
2	0	4	0	3
3	$1 + \epsilon$	$1 + \epsilon$	$2 + \epsilon$	2

Table 14.4 The payoff matrix of the subgame
for bidders 1 and 2

	Raise bid	Don't raise
raise bid	2, 2	2, 3
don't raise	3, 2	0, 0

The equilibrium strategy is a randomizing one, with $P(\text{raise}) = 2/3$, $P(\text{don't raise}) = 1/3$, and so there is a probability $1/9$ of inefficient allocation (when neither bidder 1 or 2 raises his bid).

The definition of objects for sale

We have assumed so far that the objects for sale are given. In many cases, these are defined by the auctioneer by splitting some larger objects into smaller ones. For example, in spectrum auctions, the government decides the granularity of the spectrum bands and the geographical partitioning, and so defines the spectrum licenses to be auctioned. This defining of objects affects the auction's social efficiency and the revenue that is generated. It turns out that the goals of revenue maximization and social efficiency can conflict. The finer is the object definition, the more flexibility bidders have to choose precise sets of objects that maximize their valuations. However, if the object definition is coarser, then a bidder may be forced to buy a larger object just because this is the only way to obtain a part of it that he values very much, and the rest of the object is wasted and cannot be used by somebody else (who could not afford to buy the combined object).

On the other hand, 'bundling' of objects can result in higher revenue for the seller. As an example, consider selling two licenses A and B, or the compound license AB. There are two bidders. Bidder 1 has $v_A = 9$, $v_B = 3$; bidder 2 has $v_A = 1$, $v_B = 10$. For both, $v_{AB} = v_A + v_B$. Auctioning licenses A and B separately (say, using two Vickrey auctions) results in selling prices of $p_A = 1$ and $p_B = 3$, and the total value generated is 19. Auctioning the single license AB will result in bidder 1 winning it, with $p_{AB} = 11$ and the value generated is 12. Hence, the revenue and social welfare sum to 23 in both auctions, but the seller does better by selling the two licenses as a bundle.

14.3 Auctioning a bandwidth pipeline

We conclude this chapter by summarizing its ideas in the context of auctioning a high bandwidth communication link, or pipeline. The pipeline is to be sold for a period of time, such as a year. Its bandwidth is to be divided in discrete units, and these units sold in a multi-unit auction. We suppose that the potential bidders are several companies that wish to use a part of the pipeline's capacity. These companies may use the capacity to transfer their own information bits or, assuming resale is permitted, they may act as retailers that provide service to end-customers. Let us review the major issues involved.

Information model, competition and collusion

The nature of the market is the most crucial factor in determining the auction design. If the bidders have private values or no clear idea of how much the units to be auctioned are worth to them, then an open ascending auction can be considered. If demand and competition are substantial, then this type of auction helps bidders to discover the goods' actual market values. Such discovery limits the winner's curse, promotes competition, ensures the auction is efficient and produces good revenue.

However, if competition is expected to be mild, then an open auction is vulnerable to collusion or tacit collusion. Bidders can collude to divide the goods amongst themselves, proportionally to their market powers. Although this reduces the seller's revenue, it may be acceptable if efficiency is the seller's primary goal.

If the auction's goals are either to produce high revenue or estimate the actual market value of the goods, then a sealed mechanism should be considered. If the bidders have a good idea of the market value of the units to be auctioned, then this is straightforward. Their bids will be near or equal to their valuations, and the lack of feedback will limit the opportunity for tacit collusion. Thus, the seller's revenue should be good, even when it is not possible to set a reserve price (because it is difficult for the auctioneer to estimate successfully such a price when demand is low).

Design becomes trickier if there is mild competition, collusion is probable, but bidders are not confident in their estimates of the actual market value of the units auctioned. If the bidders are retailers and the bandwidth market has just been formed, then they may be uncertain of the customer demand for bandwidth. In a sealed bid auction there is no 'dynamic price discovery' and so bidders cannot bias their estimates. Thus, they are vulnerable to winner's curse and the submission of erroneously high or low bids is probable. In these circumstances, a uniform or VCG payment rule is more appropriate than 'pay-your-bid'. This should reduce the bidder's fear of the winner's curse and the auction's performance should be better than in an ascending format. However, the sealed format means there will be substantial demand reduction, and hence smaller revenues.

Participation

If there are bidders with both low and high valuations, then use of an ascending auction will discourage participation; bidders with low valuations will expect to be outbid by those with high valuations and therefore choose not to participate. This will reduce both the size of the market and the seller's revenue (as fewer players can more easier manipulate prices to stay low). Since the size of market and competition within it are extremely important for a viable bandwidth market, the result will be disappointing. For these reasons, a sealed auction should be preferred. It increases the chance that a low valuation bidder can win (by exploiting the fact that the high valuation bidders will shade their bids). This promotes a market with more players and so greater competition.

We can promote the participation of bidders who wish for smaller amounts of bandwidth by adding a rule to the auction that bidders may not compete for both large and small contracts. Such a rule can be easily defined for a FCC-type auctions, in which the types and sizes of contracts are part of the auction design.

Bidder heterogeneity

There are two types of bidder heterogeneity. Both affect the auction design. The first concerns whether bidders have demands for small or large quantities of items. Suppose that in one auction there are many 'small' bidders, each desiring about the same small quantity of bandwidth, and in the other auction there are just a few bidders, each with large demand. The two auctions can have the same aggregate demand, but completely different outcomes. The reason is that in the first auction competition will be fierce and there will be little demand reduction or bid shading compared to that which will occur in the second auction. Furthermore, if there are just a few strong bidders, and their total demand exceeds the available capacity, then the open format of a FCC-type auction will discourage entry by the smaller bidders.

Bidder heterogeneity can also occurs in differences amongst the bidders' utility functions. The theory of some auction mechanisms depends upon assumptions about the bidders' utility functions that do not hold in practice. In our example, some bidders might have utility functions that are positive only for a finite number of bandwidth quantities (for example, retailers who are trying to fulfil their contracts with certain large industrial customers with fixed bandwidth demands). Such bidders do not have utility functions with decreasing marginal utility, though this assumption underlies the theory of many auction mechanisms (for instance, in a VCG auction one needs to supply a price for each additional unit, and be ready to receive any number of units). One may redefine the rules of the auctions to take account of other types of utility function. However, this will destroy the nice incentive compatibility properties of the original design and complicate the strategy of the bidders, and may result in a loss of social welfare, a loss of seller's revenue and, worst of all, a loss of the seller's creditability in the market (since a bidder may be awarded a piece of bandwidth that is of no value to him).

Efficiency or revenue

It is important to decide whether the primary aim of the auction is to maximize social welfare or the seller's revenue. Even if the former is the primary aim, good revenue for the seller may still be an important. In our example, if competition is expected to be low and seller's revenue is important, then a sealed 'pay-your-bid' auction is preferable to an ascending or sealed uniform price auction.

Transparency

Transparency may be important to the seller's credibility. If the bidders believe that the pipeline owner may wish to favour some particular bidders, because of his business strategy or strategic alliances, then he should disprove this by implementing an open auction. Although there is the danger of collusion in an open auction, the success of the auction, bidder participation and competition depend upon the pipeline owner proving his creditability.

Uniform price

In auctions of power transmission and transfer, satellite link bandwidth, and TV broadcast licenses, the law may explicitly require that all market players be treated similarly, and so that uniform pricing be used. Also, national or international law may prohibit differentiated pricing, as being 'politically incorrect'.

Liquid versus less-liquid designs

Thus far, we have assumed that the pipeline's capacity is auctioned in small equal units. This a 'liquid' design since it allows the market to decide on the number and the size of the winning bids, and winning bidders. A less liquid design could simulate competition, reduce the opportunity of collusion and increase revenue. For example, in the FCC-type auction a number of nonidentical and carefully defined contracts are offered for sale. The contracts are defined to ensure that no matter how the bidders attempt to allocate the contracts between themselves, some bidders will end up as losers. This means that they will find it impossible to collude and so must compete. Note that there are no generic rules for defining such heterogenous contracts; one must take into account the particular market demand. We illustrate this idea in the following simple example.

Example 14.3 (A FCC type auction) Suppose that the auctioneer is aware that there is moderate total demand, for between 100% and 150% of capacity. Specifically, he knows that there are five large companies, each of whom wishes to reserve 15-25% of the pipelines capacity, and five smaller companies, each of whom wishes to reserve 5% of the capacity. If the auctioneer decides to split the capacity into many small units and then auction them with a uniform payment rule, then the outcome is almost certain: the five large companies will each successfully bid for 15% of the capacity at nearly zero price, leaving room for the small companies to each obtain 5% at a slightly higher price. Note that a large company cannot benefit by raising its bid to gain an extra 5%, as that will result in a much higher sale price, which will then also be applied to its first 15%.

The use of a FCC-type simultaneous ascending auction would simplify things and improve revenue. The auctioneer could arrange to auction three contracts of 15% capacity and five contracts of 5% capacity. This would ensure that between three and four of the large companies would each win 15% or more of capacity. The fifth large company would be displaced from the market and the fourth would have to bid against some (or all) the smaller companies. This would intensify competition amongst the bidders since each of the larger bidder is at risk of being displaced from the market. The seller should obtain a good revenue, though the bidders may complain that the auction design attempts to 'fix' the market by restricting the number of winners and the size of the winning bids.

Another advantage of the FCC-type auction is that it is suitable for auctioning items that are described by multiple parameters (which we might call called polyparametric auctions). The seller may want to sell the bandwidth of the pipeline separately for peak and off-peak periods, because there is differing demand in these periods. This is easily done in a FCC-type auction, by defining each contract as being either peak or off-peak. He auctions 16 contracts, rather than 8 contracts, each of which is well-defined. This is not the case with the traditional multi-unit auction in which the ranking of polyparametric bids and winner determination may be complicated.

When the auction is conducted for short timescales and repeated often (in our example if the pipeline's capacity is auctioned on a daily basis), many nice properties of the auction regarding incentive compatibility no longer hold. Auction repetition can be used by bidders to enforce tacit collusion, the same way that they would use the feedback of an open auction. The more frequently they compete against each other in repeated auctions, the greater is the possibility that they will tacitly collude. Limiting feedback or switching to a sealed format seems to be the most appropriate action in such a setting.

14.4 Further reading

An excellent introduction to the theory of auctions can be found in Chapter 7 of Wolfstetter (1999). A more extensive review of the literature on auctions can be found in Klemperer (1999).

The revenue equivalence theorem is due to Vickrey (1961). Proposals for simultaneous ascending auctions were first made by McAfee, and by Milgrom and Wilson. Optimal auctions are explained by Riley and Samuelson (1981) and Bulow and Roberts (1989). The design of the 1994 FCC spectrum auction and its successful working in practice are described by McMillan (1994) and McAfee and McMillan (1996). The UK's auction of third-generation mobile phone licenses is very interestingly described by Binmore and Klemperer (2002). The effects of the winner's curse are explored by Bulow and Klemperer (2002). The double auctions are presented in McCabe, Rassenti and Smith (1992). The material in Sections 14.2.4–14.2.5 on the simultaneous ascending multi-unit auction is from Milgrom (2000). The FCC web pages contain interesting information about spectrum auctions; see FCC (2002a).

Appendix A

Lagrangian Methods for Constrained Optimization

A.1 Regional and functional constraints

Throughout this book we have considered optimization problems that were subject to constraints. These include the problem of allocating a finite amounts of bandwidth to maximize total user benefit, the social welfare maximization problem, and the time of day pricing problem. We make frequent use of the Lagrangian method to solve these problems. This appendix provides a tutorial on the method. Take, for example,

$$\text{NETWORK} : \underset{x \geq 0}{\text{maximize}} \sum_{r=1}^{n_r} w_r \log x_r , \quad \text{subject to } Ax \leq C$$

posed in Chapter 12. This is an example of the generic constrained optimization problem:

$$P : \underset{x \in X}{\text{maximize}} f(x), \quad \text{subject to } g(x) = b$$

Here f is to be maximized subject to constraints that are of two types. The constraint $x \in X$ is a *regional constraint*. For example, it might be $x \geq 0$. The constraint $g(x) = b$ is a *functional constraint*. Sometimes the functional constraint is an inequality constraint, like $g(x) \leq b$. But if it is, we can always add a *slack variable*, z, and re-write it as the equality constraint $g(x) + z = b$, re-defining the regional constraint as $x \in X$ and $z \geq 0$. To illustrate, we shall use the NETWORK problem with just one resource constraint:

$$P_1 : \underset{x \geq 0}{\text{maximize}} \sum_{i=1}^{n} w_i \log x_i, \quad \text{subject to } \sum_{i=1}^{n} x_i = b$$

where b is a positive number.

A.2 The Lagrangian method

The solution of a constrained optimization problem can often be found by using the so-called *Lagrangian method*. We define the *Lagrangian* as

$$L(x, \lambda) = f(x) + \lambda(b - g(x))$$

Pricing Communication Networks C. Courcoubetis and R. Weber
© 2003 John Wiley & Sons, Ltd ISBN 0-470-85130-9 (HB)

For P_1 it is

$$L_1(x, \lambda) = \sum_{i=1}^{n} w_i \log x_i + \lambda \left(b - \sum_{i=1}^{n} x_i \right)$$

In general, the Lagrangian is the sum of the original objective function, and a term that involves the functional constraint and a 'Lagrange multiplier' λ. Suppose we ignore the functional constraint and consider the problem of maximizing the Lagrangian, subject only to the regional constraint. This is often an easier problem than the original one. The value of x that maximizes $L(x, \lambda)$ depends upon the value of λ. Let us denote this optimizing value of x by $x(\lambda)$.

For example, since $L_1(x, \lambda)$ is a concave function of x it has a unique maximum at a point where f is stationary with respect to changes in x, i.e. where

$$\partial L_1 / \partial x_i = w_i / x_i - \lambda = 0 \quad \text{for all } i$$

Thus, $x_i(\lambda) = w_i / \lambda$. Note that $x_i(\lambda) > 0$ for $\lambda > 0$, and so the solution lies in the interior of the feasible set.

Think of λ as knob that we can turn to adjust the value of x. Imagine turning this knob until we find a value of λ, say $\lambda = \lambda^*$, such that the functional constraint is satisfied, i.e. $g(x(\lambda^*)) = b$. Let $x^* = x(\lambda^*)$. Our claim is that x^* solves P. This is the so-called Lagrangian Sufficiency Theorem, which we state and prove shortly. First, note that, in our example, $g(x(\lambda)) = \sum_i w_i / \lambda$. Thus, choosing $\lambda^* = \sum_i w_i / b$, we have $g(x(\lambda^*)) = b$. The next theorem shows that $x = x(\lambda^*) = w_i b / \sum_j w_j$ is optimal for P_1.

Theorem 5 (Lagrangian Sufficiency Theorem) Suppose there exist $x^* \in X$ and λ^*, such that x^* maximizes $L(x, \lambda^*)$ over all $x \in X$, and $g(x^*) = b$. Then x^* solves P.

Proof

$$
\begin{aligned}
\max_{\substack{x \in X \\ g(x)=b}} f(x) &= \max_{\substack{x \in X \\ g(x)=b}} [f(x) + \lambda^*(b - g(x))] \\
&\leq \max_{x \in X} [f(x) + \lambda^*(b - g(x))] \\
&= f(x^*) + \lambda^*(b - g(x^*))] \\
&= f(x^*)
\end{aligned}
$$

Equality in the first line holds because we have simply added 0 on the right-hand side. The inequality in the second line holds because we have enlarged the set over which maximization takes place. In the third line, we use the fact that x^* maximizes $L(x, \lambda^*)$ and in the fourth line we use $g(x^*) = b$. But x^* is feasible for P, in that it satisfies the regional and functional constraints. Hence x^* is optimal. ∎

Multiple Constraints

If g and b are vectors, so that $g(x) = b$ expresses more than one constraint, then we would write

$$L(x, \lambda) = f(x) + \lambda^\top (b - g(x))$$

where the vector λ now has one component for each constraint. For example, the Lagrangian for NETWORK is

$$L(x, \lambda) = \sum_{r=1}^{n_r} w_r \log x_r + \sum_j \lambda_j (C_j - \sum_j A_{jr} x_r - z_j)$$

where z_j is the slack variable for the jth constraint.

A.3   When does the method work?

The Lagrangian method is based on a 'sufficiency theorem'. The means that method can work, but need not work. Our approach is to write down the Lagrangian, maximize it, and then see if we can choose λ and a maximizing x so that the conditions of the Lagrangian Sufficiency Theorem are satisfied. If this works, then we are happy. If it does not work, then too bad. We must try to solve our problem some other way. The method worked for P_1 because we could find an appropriate λ^*. To see that this is so, note that as λ increases from 0 to ∞, $g(x(\lambda))$ decreases from ∞ to 0. Moreover, $g(x(\lambda))$ is continuous in λ. Therefore, given positive b, there must exist a λ for which $g(x(\lambda)) = b$. For this value of λ, which we denote λ^*, and for $x^* = x(\lambda^*)$ the conditions of the Lagrangian Sufficiency Theorem are satisfied.

To see that the Lagrangian method does not always work, consider the problem

$$P_2 : \text{minimize} -x, \quad \text{subject to } x \geq 0 \text{ and } \sqrt{x} = 2$$

This cannot be solved by the Lagrangian method. If we minimize

$$L(x, \lambda) = -x + \lambda(2 - \sqrt{x})$$

over $x \geq 0$, we get a minimum value of $-\infty$, no matter what we take as the value of λ. This is clearly not the right answer. So the Lagrangian method fails to work. However, the method does work for

$$P_2' : \text{minimize} -x, \quad \text{subject to } x \geq 0 \text{ and } x^2 = 16$$

Now

$$L(x, \lambda) = -x + \lambda(16 - x^2)$$

If we take $\lambda^* = -1/8$, then $\partial L/\partial x = -1 + x/4$, and so $x^* = 4$. Note that P_2 and P_2' are really the same problem, except that the functional constraint is expressed differently. Thus, whether or not the Lagrangian method will work can depend upon how we formulate the problem.

We can say something more about when the Lagrangian method will work. Let $P(b)$ be the problem: *minimize $f(x)$, such that $x \in X$ and $g(x) = b$*. Define $\phi(b)$ as min $f(x)$, subject to $x \in X$ and $g(x) = b$. Then the Lagrangian method works for $P(b^*)$ if and only if there is a line that is tangent to $\phi(b)$ at b^* and lies completely below $\phi(b)$. This happens if $\phi(b)$ is a convex function of b, but this is a difficult condition to check. A set of sufficient conditions that are easier to check are provided in the following theorem. These conditions do not hold in P_2 as $g(x) = \sqrt{x}$ is not a convex function of x. In P_2' the sufficient conditions are met.

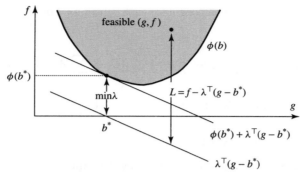

Figure A.1   Lagrange multipliers

Theorem 6   If f and g are convex functions, X is a convex set, and x^* is an optimal solution to P, then there exist Lagrange multipliers $\lambda \in \mathbb{R}^m$ such that $L(x^*, \lambda) \leq L(x, \lambda)$ for all $x \in X$.

Remark Recall that f is a *convex function* if for all $x_1, x_2 \in X$ and $\theta \in [0, 1]$, we have $f(\theta x_1 + (1 - \theta)x_2) \leq \theta f(x_1) + (1 - \theta) f(x_2)$. X is a *convex set* if for all $x_1, x_2 \in X$ and $\theta \in [0, 1]$, we have $\theta x_1 + (1 - \theta)x_2 \in X$. Furthermore, f is a *concave function* if $-f$ is convex.

The proof of the theorem proceeds by showing that $\phi(b)$ is convex. This implies that for each b^* there is a tangent hyperplane to $\phi(b)$ at b^* with the graph of $\phi(b)$ lying entirely above it. This uses the so-called 'supporting hyperplane theorem' for convex sets, which is geometrically obvious, but some work to prove. Note the equation of the hyperplane will be $y = \phi(b^*) + \lambda^\top (b - b^*)$ for some multipliers λ. This λ can be shown to be the required vector of Lagrange multipliers, and Figure A.1 gives some geometric intuition as to why the Lagrange multipliers λ exist and why these λs give the rate of change of the optimum $\phi(b)$ with b.

A.4   Shadow prices

The maximizing x, the appropriate value of λ and maximal value of f all depend on b. What happens if b changes by a small amount? Let the maximizing x be $x^*(b)$. Suppose the Lagrangian method works and let $\lambda^*(b)$ denote the appropriate value of λ. As above, let $\phi(b)$ denote the maximal value of f. We have

$$\phi(b) = f(x^*(b)) + \lambda^* \left[b - g(x^*(b))\right]$$

So simply differentiating with respect to b, we have

$$\frac{\partial}{\partial b}\phi(b) = \sum_{i=1}^{n} \frac{\partial x_i^*}{\partial b}\frac{\partial}{\partial x_i^*}\left\{f(x^*(b)) + \lambda^*[b - g(x^*(b))]\right\} + \frac{\partial}{\partial b}\{f(x^*(b))$$

$$+ \lambda^*[b - g(\bar{x}(b))] + \frac{\partial \lambda^*}{\partial b}\frac{\partial}{\partial \lambda^*}\{f(x^*(b)) + \lambda^*[b - g(x^*(b)]\}$$

$$= 0 + \lambda^* + [b - g(x^*(b)]\frac{\partial \lambda^*}{\partial b}$$

where the first term on the right-hand side is 0, because $L(x, \lambda^*)$ is stationary with respect to x_i at $x = x^*$ and the third term is zero because $b - g(x^*(b)) = 0$. Thus,

$$\frac{\partial}{\partial b}\phi(b) = \lambda^*$$

and λ^* can be interpreted as the rate at which the maximized value of f increases with b, for small increases around b. For this reason, the Lagrange multiplier λ^* is also called a *shadow price*, the idea being that if b increases to $b + \delta$ then we should be prepared to pay $\lambda^*\delta$ for the increase we receive in f.

It can happen that at the optimum none, one, or several constraints are active. For example, with constraints $g_1(x) \le b_1$ and $g_2(x) \le b_2$ it can happen that at the optimum $g_1(x^*) = b_1$ and $g_2(x^*) < b_2$. In this case we will find have $\lambda_2^* = 0$. This makes sense. The second constraint is not limiting the maximization of f and so the shadow price of b_2 is zero.

A.5   The dual problem

By similar reasoning to that we used in the proof of the Lagrangian sufficiency theorem, we have that for any λ

$$\phi(b) = \max_{\substack{x \in X \\ g(x)=b}} f(x)$$

$$= \max_{\substack{x \in X \\ g(x)=b}} [f(x) + \lambda(b - g(x))]$$

$$\le \max_{x \in X} [f(x) + \lambda(b - g(x))].$$

The right-hand side provides an upper bound on $\phi(b)$. We make this upper bound as tight as possible by minimizing over λ, so that we have

$$\phi(b) \le \min_{\lambda} \max_{x \in X} [f(x) + \lambda^*(b - g(x))]$$

The right-hand side above defines an optimization problem, called the *dual problem*. The original problem is called the *primal problem*. If the primal can be solved by the Lagrangian method then the inequality above is an equality, and the solution to the dual problem is just $\lambda^*(b)$. If the primal cannot be solved by the Lagrangian method we will have a strict inequality, the so-called *duality gap*.

The dual problem is interesting because it can sometimes be easier to solve, or because it formulates the problem in an illuminating way. The dual of P_1 is

$$\underset{\lambda}{\text{minimize}} \sum_{i=1}^{n} w_i \log(w_i/\lambda) + \lambda \left(b - \sum_{i=1}^{n} w_i/\lambda \right)$$

where we have inserted $x_i = w_i/\lambda$, after carrying out the inner maximization over x. This is a convex function of λ. Differentiating with respect to λ, one can check that the stationary point is the maximum, and $\sum_i w_i/\lambda = b$. This gives λ, and finally, as before

$$\phi(b) = \sum_{i=1}^{n} w_i \log \left(\frac{w_i}{\sum_j w_j} b \right)$$

The dual plays a particularly important role in the theory of *linear programming*. A linear program, such

$$P: \text{ maximize } c^\top x, \quad \text{subject to } x \geq 0 \text{ and } Ax \leq b$$

is one in which both the objective function and constraints are linear. Here $x, c \in \mathbb{R}^n$, $b \in \mathbb{R}^m$ and A is a $m \times n$ matrix. The dual of P is D:

$$D: \text{ minimize } \lambda^\top b, \quad \text{subject to } \lambda \geq 0 \text{ and } y^\top A \geq c^\top$$

D is another linear program, and the dual of D is P. The decision variables in D, i.e. $\lambda_1, \ldots, \lambda_m$, are the Lagrange multipliers for the m constraints expressed by $Ax \leq b$ in P.

A.6   Further reading

For further details on Lagrangian methods of constrained optimization, see the course notes of Weber (1998).

Appendix B

Convergence of Tatonnement

B.1   The case of producers and consumers

In this appendix we prove that under the tatonnement mechanism price converge. Consider first the problem of maximizing social surplus:

$$\underset{x \in X, y \in Y}{\text{maximize}} \left[u(x) - c(y) \right], \quad \text{subject to } x = y$$

Assuming that u is concave, c is convex, and that both X and Y are convex sets, this can be solved as the sum of two problems:

$$\underset{x \in X}{\text{maximize}} \left[u(x) - \bar{p}^\top x \right] + \underset{y \in Y}{\text{maximize}} \left[\bar{p}^\top y - c(y) \right]$$

for some Lagrange multiplier $\bar{p}$.

Suppose that $\bar{x}$ and $\bar{y}$ are the maximizing x and y. Let x, y be maximizing values at some other value p. Then

$$\bar{p}^\top y - c(y) \leq \bar{p}^\top \bar{y} - c(\bar{y})$$
$$u(x) - \bar{p}^\top x \leq u(\bar{x}) - \bar{p}^\top \bar{x}$$
$$p^\top y - c(y) \geq p^\top \bar{y} - c(\bar{y})$$
$$u(x) - p^\top x \geq u(\bar{x}) - p^\top \bar{x}$$

By some algebra these give

$$(\bar{p} - p)^\top z \geq (\bar{p} - p)^\top \bar{z} = 0$$

where $z = x - y$, $\bar{z} = \bar{x} - \bar{y} = 0$. The inequality is strict unless all of the above four inequalities are equalities.

Let us suppose that at least one is strict, and so we have $(\bar{p} - p)^\top z > 0$. Suppose that prices are adjusted by the rule

$$\dot{p} = z$$

Define the Lyapunov function $V(t) = [p(t) - \bar{p}]^2$. Then

$$\dot{V} = 2 \left[p(t) - \bar{p} \right]^\top z < 0$$

So for all initial price vectors $p = p(0)$, tatonnement converges to the equilibrium socially optimal price vector $\bar{p}$.

Pricing Communication Networks   C. Courcoubetis and R. Weber
© 2003 John Wiley & Sons, Ltd   ISBN 0-470-85130-9 (HB)

B.2 Consumers with network constraints

Suppose we wish to solve the following problem:

$$\text{maximize}_{x_r \geq 0} \sum_r u_r(x_r), \quad \text{subject to } Ax \leq C$$

A typical instance of this problem is when x_r is the flow on route r through the network generating value $u_r(x_r)$. A route is a set of links; $A_{jr} = 1$ if route r uses link j. C_j is the capacity of link j.

There exists a vector Lagrange multiplier $\bar{\mu}$ such that the problem is equivalent to solving

$$\text{maximize}_{x_r \geq 0} \left[\sum_r u_r(x_r) - \bar{\mu}^{\top}(C - Ax) \right]$$

Suppose that the maximum is achieved at $\bar{x}$, so that for all other x,

$$\sum_r u_r(\bar{x}_r) - \bar{\mu}^{\top}(C - A\bar{x}) > \sum_r u_r(x_r) - \bar{\mu}^{\top}(C - Ax)$$

Given a μ, let x maximize $\sum_r u_r(x_r) - \mu^{\top}(C - Ax)$, so

$$\sum_r u_r(x_r) - \mu^{\top}(C - Ax) > \sum_r u_r(\bar{x}_r) - \mu^{\top}(C - A\bar{x})$$

Hence, using the above, and the fact that $\bar{\mu}^{\top}(C - A\bar{x}) = 0$, we have

$$(\mu^{\top} - \bar{\mu}^{\top})(C - Ax) > 0$$

Since $C - Ax$ is the vector excess demand and μ is the price vector, we can repeat the last steps in Section B.1, and show that again in this case tatonnement will converge to the socially optimal vector of prices.

References

Altmann, J., H. Daanen, H. Oliver and A. Sańchez-Beato Suaŕez (2002) How to market-manage a QoS network. *IEEE InfoCom2002*, New York.

Andrews, M., S. Khanna and K. Kumaran (1999) Integrated scheduling of unicast and multicast traffic in an input queued switch. *Proceedings of IEEE INFOCOM '99*, pp. 1144–1151.

Anick, D., D. Mitra and M. M. Sondhi (1982) Stochastic theory of a data-handling system with multiple sources. *The Bell System Technical Journal* **61**(8), 1872–1894.

Atkinson, J. and C. Barnekov (2000) A competitively neutral approach to network interconnection. Technical report, OPP Working Paper No. 34, Federal Communications Commission. www.fcc.gov/Bureaus/OPP/working_papers/oppwp34.pdf.

ATM Forum (2002) Home page. www.atmforum.com/.

AT&T (2000) Evolving SLAs to SLM. www.ipservices.att.com/realstories/attprofiles/iss4art4.cfm.

Band-X (2002) Home page. www.band-x.com/.

Barlow, R. E., D. J. Bartholomew, J. M. Bremner and H. D. Brunk (1972) *Statistical Inference Under Order Restrictions: The Theory and Application of Isotonic Regression.* London: Wiley.

Baumol, W. J. (1983) Some subtle pricing issues in railroad regulation. *International Journal of Transport Economics* **10**(1–2), 341–355.

Baumol, W. J. (1986) *Superfairness: Applications and Theory.* Cambridge, MA: MIT Press.

Berger, A. W. and W. Whitt (1998) Extending the effective bandwidth concept to networks with priority classes. *IEEE Communications Magazine* **36**(8), 78–83.

Binmore, K. (1992) *Fun and Games.* Lexington, MA: D. C. Heath and Co.

Binmore, K. and P. Klemperer (2002) The biggest auction ever: The sale of the British G3 telecom licences. *The Economic Journal* **112**, C74–C96.

Bolot, J.-C. and T. Turletti (1998) Experience with control mechanisms for packet video in the Internet. *SIGCOMM Computer Communication Review* **28**(1), 4–15.

Pricing Communication Networks   C. Courcoubetis and R. Weber
© 2003 John Wiley & Sons, Ltd   ISBN 0-470-85130-9 (HB)

Botvich, D. D. and N. Duffield (1995) Large deviations, the shape of the loss curve, and economies of scale in large multiplexers. *Queueing Systems* **20**, 293–320.

Braden, R., L. Zhang, S. Berson, S. Herzog and S. Jamin (1997) Resource ReSerVation Protocol (RSVP) — Version 1 Functional Specification. `ftp://ftp.isi.edu/in-notes/rfc2205.txt`.

Braess, D. (1968)   Uber ein paradoxon aus der verkehrsplanung [a paradox of traffic assignment problems]. See `homepage.ruhr-uni-bochum.de/Dietrich.Braess/`.

Briscoe, B. (1999) The direction of value flow in multi-service connectionless networks. *Proceedings of Networked Group Communication, '99, Pisa, Italy*, Volume 1736, pp. 244–269. `http://www.btexact.com/projects/mware/`.

Bulow, J. and P. Klemperer (2002) Prices and the winner's curse. *Rand Journal of Economics* **33**(1), 1–21.

Bulow, J. and J. Roberts (1989) The simple economics of optimal auctions. *The Journal of Political Economy* **97**(5), 1060–1090.

Cameron, D. (2001) *Optical Networking: A Wiley Tech Brief.* New York: Wiley.

CA*net (2002) FAQ on dark fibre. `www.canet3.net/gigabit/faqfibre.html`.

Casner, S. and S. Deering (1992)   First IETF Internet audiocast.   *ACM Computer Communication Review* **22**(3), 92–97.

Cisco (2001) Service-level management: Defining and monitoring service levels in the enterprise. `www.cisco.com/warp/public/cc/pd/wr2k/svmnso/prodlit/srlm_wp.htm`.

Cisco (2002a)   Cisco IOS release 12.0 quality of service solutions configuration guide.   `www.cisco.com/univercd/cc/td/doc/product/software/ios120/12cgcr/qos_c/`.

Cisco (2002b) Deploying tight-SLA services on an IP backbone. `www.ripe.net/ripe/meetings/archive/ripe-41/presentations/eof-tight-sla/`.

Cisco (2002c) Internetworking technology overview. `www.cisco.com/univercd/cc/td/doc/cisintwk/ito_doc/index.htm`.

Cisco (2002d) Networking essentials for small and medium-sized businesses. `www.cisco.com/warp/public/779/smbiz/netguide/`.

Cisco (2002e)   Outsourcing managed network services — Is the decision right for your network? `www.cisco.com/warp/public/cc/so/neso/wnso/power/osmnt_wp.htm`.

Cisco (2002f) Product documentation. `www.cisco.com/univercd/home/home.htm`.

Cisco (2002g)   Technology innovations.   `newsroom.cisco.com/dlls/innovators/`.

Cisco (2002h) Virtual private networks. `www.cisco.com/univercd/cc/td/doc/product/software/ios120/12cgcr/qos_c/`.

CivicNet (2002) Request for information Chicago CivicNet. `www.cityofchicago.org/CivicNet/civicnetRFI.pdf`.

Commweb (2002) Commweb tutorial channel. `www.commweb.com/tutorials/`.

Constantiou, I. and C. Courcoubetis (2001) Information asymmetry models in the internet connectivity market. *4th Internet Economics Workshop, Berlin.*

Cook, G. (2002) The Cook Report on Internet. `www.cookreport.com`.

Cormen, T. H., C. E. Leiserson and R. L. Rivest (1995) *Introduction to Algorithms.* Cambridge, MA: MIT Press New York: McGraw-Hill.

Courcoubetis, C., A. Dimakis and G. Stamoulis (2002) Traffic equivalence and substitution in a multiplexer with applications to dynamic available capacity estimation. *IEEE/ACM Transactions on Networking* **10**(21), 217–231.

Courcoubetis, C., F. P. Kelly, V. A. Siris and R. Weber (2000) A study of simple usage-based charging schemes for broadband networks. *Telecommunication Systems* **15**(3–4), 323–343.

Courcoubetis, C., F. P. Kelly and R. R. Weber (2000) Measurement based charging in communication networks. *Operations Research* **48**(4), 535–548.

Courcoubetis, C., G. Kesidis, A. Ridder, J. Walrand and R. R. Weber (1995) Admission control and routing in ATM networks using inferences from measured buffer occupancy. *IEEE Transactions on Communications* **43**, 1778–1784.

Courcoubetis, C., V. A. Siris and G. D. Stamoulis (1999) Application of the many sources asymptotic and effective bandwidths to traffic engineering. *Telecommunication Systems* **12**, 167–191.

Courcoubetis, C. and J. Walrand (1991) Note on the effective bandwidth of ATM traffic at a buffer. Unpublished manuscript.

Courcoubetis, C. and R. R. Weber (1995) Effective bandwidths for stationary sources. *Probability in the Engineering and Informational Sciences* **9**, 285–296.

Courcoubetis, C. and R. R. Weber (1996) Buffer overflow asymptotics for a switch handling many traffic sources. *Journal of Applied Probability* **33**(3), 886–903.

Courcoubetis, C. A. and M. I. Reiman (1999) Pricing in a large single link loss system. *Proceedings of 16th International Teletraffic Congress.* North-Holland. `www.nossdav.org/2000/abstracts/2.html`.

Cremer, J., P. Rey and J. Tirole (2000) Connectivity in the commercial internet. *Journal of Industrial Economics* **48**(4), 433–472. `www.idei.asso.fr/Commun/Articles/JCremer/connectivity.pdf`.

Crowcroft, J. and P. Oechslin (1998) Differentiated end-to-end Internet services using a weighted proportionally fair sharing TCP. *ACM Computer Communications Review* **28**, 53–67.

Cukier, K. (1998) Peering and fearing: ISP interconnection and regulatory issues. www.ksg.harvard.edu/iip/iicompol/Papers/Cukier.html.

de Veciana, G., C. Olivier and J. Walrand (1993) Large deviations for birth death Markov fluids. *Probability in the Engineering and Informational Sciences* **7**, 237–235.

de Veciana, G. and J. Walrand (1995) Effective bandwidths: call admission, traffic policing and filtering for ATM networks. *Queuing Systems* **20**, 37–59.

Deering, S. and D. Cheriton (1990) Multicast routing in datagram inter-networks and extended LANs. *ACM Transactions on Computer Systems* **8**(2), 85–110.

Digital Fountain (2002) Home page. www.digitalfountain.com.

Downes, L., C. Mui and N. Negroponte (2000) *Unleashing the Killer App: Digital Strategies for Market Dominance.* Boston: Harward Business School Press.

Eatwell, J., M. Milgate and P. Newman (1989) *Game Theory.* The New Palgrave, Macmillian Press.

Economides, N. (1997) The tragic inefficiency of the M-ECPR. www.stern.nyu.edu/networks/paper.html.

Economides, N. (2000) Real options and the costs of the local telecommunications network. *The New Investment Theory of Real Options and its Implications for Cost Models in Telecommunications.* New York: Kluwer. www.stern.nyu.edu/networks/real.pdf.

Economides, N. (2002) Economics of networks. raven.stern.nyu.edu/networks/.

Economides, N. and C. Himmelberg (1995) Critical mass and network size with application to the US fax market. Discussion Paper no. EC-95-11, Stern School of Business, N.Y.U., www.stern.nyu.edu/networks/95-11.pdf.

Edell, R., N. Mckeown and P. Varaiya (1995) Billing users and pricing for TCP. *IEEE Journal on Selected Areas in Communications* **13**(7), 1162–1175.

Edell, R. and P. Varaiya (1999) Providing internet access: What we learn from index. *Infocom '99.* www.path.berkeley.edu/~varaiya/papers_ps.dir/networkpaper.pdf.

Elwalid, A. I. and D. Mitra (1993) Effective bandwidth of general Markovian traffic sources and admission control of high speed networks. *IEEE/ACM Transactions on Networking* **1**, 329–343.

EP.NET (2002) Exchange point information. www.ep.net/.

Farinacci, D., A. Lin, T. Speakman and A. Tweedly (1998) PGM reliable transport protocol specification. Internet-Draft. www.globecom.net/ietf/draft/draft-speakman-pgm-spec-02.html.

FCC (2002a) FCC Auctions Home page. wireless.fcc.gov/auctions/.

FCC (2002b) Federal Communications Commission, home page. www.fcc.gov/.

Feigenbaum, J., C. H. Papadimitriou and S. Shenker (2000) Sharing the cost of multicast transmissions. *ACM Symposium on Theory of Computing*, pp. 218–227.

Gibbens, R. (1996) Traffic characterisation and effective bandwidths for broadband network traces.

Gibbens, R. and P. Hunt (1991) Effective bandwidths for the multi-type UAS channel. *Queueing Systems* **1**, 17–28.

Gibbens, R. J. and F. P. Kelly (1999) Resource pricing and evolution of congestion control. *Automatica* **35**, 1969–1985.

Gibbens, R. J., P. B. Key and F. P. Kelly (1995) A decision-theoretic approach to call admission control in ATM networks. *IEEE Journal on Selected Areas in Communications* **13**, 1101–1114.

Gibbens, R. J., R. Mason and R. Steinberg (2000) Internet service classes under competition. *IEEE Journal on Selected Areas in Communications* **18**(12), 2490–2498. www.soton.ac.uk/~ram2/papers/.

Grossglauser, M. and D. N. C. Tse (1999) A framework for robust measurement-based admission control. *IEEE/ACM Transactions on Networking* **7**(3), 293–309.

Gupta, A., D. O. Stahl and A. B. Whinston (1997) A stochastic equilibrium model of Internet pricing. *Journal of Economic Dynamics and Control* **21**, 697–722.

Hajek, B. (1994) A queue with periodic arrivals and constant service rate. In: F. P. Kelly (Ed.), *Probability Statistics and Optimization, a Tribute to Peter Whittle*, pp. 147–157. Wiley.

Henderson, T. and S. N. Bhatti (2000) Protocol-independent multicast pricing. *Proceedings of NossDav — 10th International Workshop on Network and Operating Systems Support for Digital Audio and Video*, Chapel Hill, NC. www.nossdav.org/2000/abstracts/2.html.

Herzog, S., S. Shenker and D. Estrin (1995) Sharing the "cost" of multicast trees: An axiomatic analysis. *Proceedings of ACM SIGCOMM '95*, pp. 315–327.

Hilton, R., M. Maher and F. Selto (2003) *Cost Management: Strategies for Business Decisions*. McGraw Hill/Irwin.

Hui, J. Y. (1988) Resource allocation for broadband networks. *IEEE Journal on Selected Areas in Communications* **6**, 1598–1608.

Hunt, P. J. and F. P. Kelly (1989) On critically loaded loss networks. *Advanced Applied Probability* **21**, 831–841.

Huston, G. (1998) *ISP Survival Guide: Strategies for Running a Competitive ISP*. (1st ed.). Wiley.

Huston, G. (1999a) Interconnection, peering and settlements. *Proceedings of the INET99 Internet Society Conference*.

Huston, G. (1999b) Interconnection, Peering and Settlements — Part II. *The Internet Protocol Journal* **2**(2). `www.cisco.com/warp/public/759/ipj_2-2/ipj_2-2_ps1.html`.

Internet RFC/STD/FYI/BCP Archives (2002) Home page. `www.faqs.org/rfcs/`.

Isenberg, D. S. and D. Weinberger (2001) The paradox of the best network. `netparadox.com/`.

Jacobson, V. (1998) Congestion avoidance and control. *Proceedings of ACM SIGCOMM '88*, pp. 314–329.

Johari, R. and D. Tan (2001) End-to-end congestion control for the Internet: Delays and stability. *IEEE/ACM Transactions on Networking* **9**(6), 818–832.

Kahn, A. (1998) *The Economics of Regulation: Principles and Institutions.* (7th ed.). Cambridge, MA: MIT Press.

Kalai, E. and M. Smorodinsky (1975) Other solutions to Nash's bargaining problem. *Econometrica* **43**(3), 513–518.

Karlin, S. (1959) *Mathematical Methods and Theory in Games, Programming and Economics.* New York: Dover.

Kelly, F. P. (1991a) Effective bandwidths at multi-class queues. *Queueing Systems* **9**, 5–15.

Kelly, F. P. (1991b) Loss networks. *Annals of Applied Probability* **1**, 319–378.

Kelly, F. P. (1991c) Network routing. *Philisophical Transactions of the Royal Society* **A337**, 343–367.

Kelly, F. P. (1994a) On tariffs, policing and admission control for multiservice networks. *Operations Research Letters* **15**, 1–9.

Kelly, F. P. (1994b) Tariffs and effective bandwidths in multiservice networks. In: J. Labetoulle and J. W. Roberts (Eds.), *The Fundamental Role of Teletraffic in the Evolution of Telecommunications Networks, Proceedings of the 14th International Teletraffic Congress — ITC 14*, Volume 1a of *Teletraffic Science and Engineering*, pp. 401–410. Elsevier Science B.V., Amsterdam Antibes Juan-les-Pins, France.

Kelly, F. P. (1996) Notes on effective bandwidths. In: F. Kelly, S. Zachary and I. Ziedins (Eds.), *Stochastic Networks: Theory and Applications Telecommunications Networks*, Volume 4 of *Royal Statistical Society Lecture Notes Series*, Oxford, pp. 141–168. Oxford University Press.

Kelly, F. P. (2000) Models for a self-managed Internet. *Philosophical Transactions of the Royal Society* **A358**, 2335–2348.

Kelly, F. P. (2001) Mathematical modelling of the Internet. In: B. Engquist and W. Schmid (Eds.), *Mathematics Unlimited — 2001 and Beyond*, pp. 685–702. Berlin: Springer-Verlag.

Kelly, F. P. (2002a) Proportional fairness. `www.statslab.cam.ac.uk/~frank/pf/`.

Kelly, F. P. (2002b) A self-managed Internet. www.statslab.cam.ac.uk/~frank/int/.

Kelly, F. P., A. Maulloo and D. K. H. Tan (1998) Rate control in communications networks: shadow prices, proportional fairness and stability. *Journal of Operational Researsh Society* **49**(3), 237–252.

Kelly, K. (1999) *New Rules for the New Economy: Ten Radical Strategies for a Connected World.* New York: Penguin Books.

Key, P., L. Massoulie and J. Shapiro (2002) Stability of distributed congestion control with heterogeneous feedback delays. *IEEE Transactions on Automatic Control* **47**(6), 895–902.

Klemperer, P. (1999) Auction theory: a guide to the literature. *Journal of Economic Surveys* **13**(3), 227–286.

Kurose, J. and K. Ross (2001) *Computer Networking: A Top-Down Approach Featuring the Internet.* Addison-Wesley.

Lafont, J., S. Marcus, P. Rey and J. Tirole (2001) Internet interconnection and the of-net-cost pricing principle. *Proceedings of IDEI Conference on the Economics of the Software and Internet Industries.* www.idei.asso.fr/Commun/Articles/Tirole/LMRTjuin21.pdf.

Likhanov, N. and R. Mazumdar (1999) Cell loss asymptotics for buffers fed with a large number of independent stationary sources. *Journal of Applied Probability* **36**(1), 86–96.

Loeb, M. and W. A. Magat (1979) A decentralized method for utility regulation. *Journal of Law and Economics* **22**, 399–404.

Low, S. and P. P. Varaiya (1993) A new approach to service provisioning in ATM networks. *IEEE/ACM Transactions on Networking* **1**(5), 547–553.

Low, S. H. and D. E. Lapsley (1999) Optimization flow control i: Basic algorithm and convergence. *IEEE/ACM Transactions on Networking* **7**(6), 861–874.

Luce, R. D. and H. Raiffa (1957) *Games and Decisions: Introduction and Critical Survey.* New York: Dover.

Macho-Stadler, I. and D. Perez-Castillo (1997) *An Introduction to the Economics of Information: Incentives and Contracts.* (1st ed.). New York: Oxford University Press.

MacKie-Mason, J. K. and H. R. Varian (1994) Pricing the Internet. In: B. Kahin and J. Keller (Eds.), *Public Access to the Internet.* Englewood Cliffs, NJ: Prentice-Hall.

MacKie-Mason, J. K. and H. R. Varian (1995) Pricing congestible network resources. *IEEE Journal on Selected Areas in Communications* **13**(7), 1141–1149.

MacKie-Mason, J. and B. Whittier (2002) Telecommunications guide to the Internet. china.si.umich.edu/telecom/telecom-info.html.

Market Managed Internet (2002) Home page. www.m3i.org.

Mason, R. (1998)   Internet telephony and the international accounting rate system. *Telecommunications Policy* **22**(11), 931–944. www.soton.ac.uk/~ram2/papers/.

Massoulié, L. and J. Roberts (1999)   Bandwidth sharing: objectives and algorithms. *Proceedings of IEEE INFOCOM '99*, New York, pp. 1395–1403.

McAfee, R. P. and J. McMillan (1996) Analyzing the airwaves auction. *Journal of Economic Perspectives* **10**(1), 159–175.

McCabe, K., S. Rassenti and V. Smith (1992) Designing call auction institutions: Is double dutch the best? *The Economic Journal* **102**, 9–23.

McCanne, S., V. Jacobson and M. Vetterli (1996) Receiver-driven layered multicast. *ACM SIGCOMM*, Volume 26,4, New York, pp. 117–130. ACM Press.

McMillan, J. (1994)   Selling spectrum rights. *Journal of Economic Perspectives* **8**(3), 145–162.

Mendelson, H. and S. Whang (1990) Optimal incentive compatible pricing priority pricing for the $M/M/1$ queue. *Operations Research* **38**, 870–883.

Metro   Dynamics   (2002)   Congestion.   www.metrodynamics.com/bats/congestion.html.

Metz, C. (2001) On the wire: Interconnecting ISP networks. *IEEE Internet Computing* **5**(2), 74–80.

Milgrom, P. (2000) Putting auction theory to work: The simultaneous ascending auction. *Journal of Political Economy* **108**(21), 245–272.

Mitchell, B. M. and I. Vogelsang (1991) *Telecommunications Pricing Theory and Practice*. Cambridge, UK: Cambridge University Press.

Moulin, H. and S. Shenker (2001)   Strategyproof sharing of submodular costs: Budget balance versus efficiency. *Economic Theory* **18**(3), 511–533.

Nash, J. (1950) The bargaining problem. *Econometrica* **18**, 155–162.

Nortel   Networks   (2002a)   Enabling   ATM   service   level   agreements.   www.nortelnetworks.com/products/library/collateral/87021.25-07-00.pdf.

Nortel   Networks   (2002b)   Enabling   frame   relay   service   level   agreements. www.nortelnetworks.com/products/library/collateral/87002.42-09-00.pdf.

Odlyzko, A. (1997)   A modest proposal for preventing Internet congestion.   www.research.att.com/~amo/doc/modest.proposal.ps.

Odlyzko, A. (2001)   Internet pricing and the history of communications. *Computer Networks* **36**(5–6), 493–517.

Odlyzko, A. (2002) Home page. www.dtc.umn.edu/~odlyzko.

Oftel (2002) Home page. www.oftel.gov.uk/.

Osborne, M. J. and A. Rubenstein (1994) *A Course on Game Theory*. Cambridge, MA: MIT Press.

Paschalidis, I. C. and J. Tsitsiklis (2000a) Dependent pricing of network services. *IEEE/ACM Transactions on Networking* **8**(2), 171–184.

Paschalidis, I. C. and J. N. Tsitsiklis (2000b) Congestion-dependent pricing of network services. *IEEE/ACM Transactions on Networking* **8**(2), 171–184.

Ramaswami, R. and K. Sivarajan (1998) *Optical Networks: A Practical Perspective*. Morgan Kaufmann.

RFC Editor (2002) Home page. www.rfc-editor.org.

Riley, J. G. and W. F. Samuelson (1981) Optimal auctions. *The American Economic Review* **71**(3), 381–392.

Rizzo, L. (2000) pgmcc: a TCP-friendly single-rate multicast. *Proceedings of ACM SIGCOMM 2000*, Stockholm, pp. 17–28.

Rizzo, L., L. Vicisano and J. Crowcroft (1998) TCP-like congestion control for layered multicast data transfer. *Proceedings of IEEE INFOCOM '98*, Volume 3, pp. 996–1003.

Rubinstein, A. (1982) Perfect equilibrium in a bargaining model. *Econometrica* **53**(1), 1151–1172.

Sappington, D. and D. Sibley (1998) Regulating without cost information: The incremental surplus subsidy scheme. *International Economic Review* **29**(2), 297–306.

Schulzrinne, H., S. Casner, R. Frederick and V. Jacobson (1996) RTP: A transport protocol for real-time applications. citeseer.nj.nec.com/schulzrinne99rtp. htmlciteseer.nj.

Shapiro, C. and H. Varian (1998) *Information Rules: A Strategic Guide to the Network Economy*. Harvard Business School Press.

Shapiro, J. K., D. F. Towsley and J. F. Kurose (2002) Optimization-based congestion control for multicast communications. *Networking*, pp. 423–442.

Shenker, S. (1993) Service models and pricing policies for an integrated services Internet. In: B. Kahin and J. Keller (Eds.), *Public Access to the Internet*, pp. 315–337. Harvard University Press.

Shenker, S. (1995) Fundamental design issues for the future Internet. *IEEE Journal on Selected Areas in Communications* **13**(7), 1276–1188.

Shenker, S., D. Clark, D. Estrin and S. Herzog (1996) Pricing in computer networks: Reshaping the research agenda. *ACM SIGCOMM Computer Communication Review* **26**(3), 19–43.

Simonian, A. and J. Guilbert (1995) Large deviations approximation for fluid queues fed by a large number of on-off sources. *IEEE Journal on Selected Areas in Communications* **13**, 1017–1027.

Siris, V. (2002)   Large   deviation   techniques   for   traffic   engineering. www.ics.forth.gr/netgroup/msa/.

Slade, M. E. (1994)   What does an oligopoly maximize?   *Journal of Industrial Economics* **42**(1), 45–61.

SoftSwitch (2002) Home page. www.softswitch.org/.

Songhurst, D. J. (Ed.) (1999)   *Charging Communications Networks: From Theory to Practice.* Amsterdam, NL: Elsevier Science.

Stokab (2002) FAQ on dark fibre. www.stokab.se/.

University of Albany Libraries (2002)   Internet tutorials.   library.albany.edu/internet/.

Upton, D. (2002) Modelling the market for bandwidth. Smith Prize Essay, University of Cambridge, www.statslab.cam.ac.uk/~dsu20/Papers/essay.pdf.

Varian, H. (1992) *Microeconomic Analysis.* (3rd ed.). New York: Norton.

Varian, H. (2002) Home page. www.sims.berkeley.edu/~hal/.

Vickrey, W. (1961) Counterspeculation, auctions and competitive sealed tenders. *Journal of Finance* **16**, 8–37.

Visual Networks, Inc and Telechoice (2002) Carrier service level agreements. www.iec.org/online/tutorials/carrier_sla/.

Vogelsang, I. and J. Finsinger (1979) A regulatory adjustment process for optimal pricing by multiproduct monopoly firms. *Bell Journal of Economics* **10**, 157–171.

Walrand, J. (1998) *Communication Networks: A First Course.* McGraw-Hill.

Walrand, J. and P. Varaiya (2000) *High-Performance Communications Networks.* (2nd ed.). Morgan Kaufmann.

Web Proforum (2002) Web Proforum tutorials. www.webproforum.com.

Weber, R. R. (1998) Lecture notes for Optimization. www.statslab.cam.ac.uk/~rrw1/opt/.

Weber, R. R. (2001) Lecture notes for Mathematics of Operational Research. www.statslab.cam.ac.uk/~rrw1/mor/.

Webopedia (2002) Webopedia, INT Media Group. www.pcwebopedia.com/.

Weiss, A. (1986) A new technique for analyzing large traffic systems. *Advanced Applied Probability* **18**, 506–532.

Willig, R. D. (1979) The theory of network access pricing. *Issues in Public Regulation. Michigan State University Public Utilities Papers. Proceedings of the Institute of Public Utilities Tenth Annual Conference.*

Wilson, R. (1990) Efficient and competitive rationing. *Econometrica* **57**(1), 1–40.

Wischik, D. (1999) The output of a switch, or, effective bandwidths for networks. *Queueing Systems* **32**, 383–396.

Wolfstetter, E. (1999) *Topics in Microeconomics: Industrial Organization, Auctions, and Incentives.* Cambridge University Press.

Zeuthen, F. (1930) *Problems of Monopoly and Economic Welfare.* London: Routledge and Sons.

Index

Pricing Communication Networks C. Courcoubetis and R. Weber
© 2003 John Wiley & Sons, Ltd ISBN 0-470-85130-9 (HB)